Plan B to the Moon: A How-to Guide to Return to and Colonize the Lunar Surface

Stephen W. Long
First Copyright © 2019

Version 1.00 20190314, Version 1.01 Update 20190508

ISBN: 978-1-54396-662-6

Author Contact Information:

Stephen W. Long
246 HWY A1A
Satellite Beach, FL 32937
PlanBtotheMoon24@gmail.com

Images:

- The Front Cover photo is of Apollo 17 Astronaut and Mission Commander Eugene A. Cernan adjusting the US flag deployed upon the Moon. NASA ID: 7036672, 1972-12-12.
- The Rear Cover photo is from Apollo 17, close-up view of the U.S. flag deployed on the moon at the Taurus-Littrow landing site. NASA ID: AS1713420466, 1972-12-12.
- The Page 34 photo is from Apollo 11. Astronaut Edwin E. Aldrin Jr., lunar module pilot, walks on the surface of the moon near the leg of the Lunar Module "Eagle" during the Apollo 11 extravehicular activity. NASA ID: AS11405903, 1969-07-20.
- The Page 92 photo is from Apollo 11. Astronaut Edwin E. Aldrin Jr., poses for a photograph beside the deployed United States flag. NASA ID: AS11405875, 1969-07-20.

Copyright Notices:

Version Notes:

- Version 0.66: Formatting for paper and electronic publishing.
- Version 0.70: Last Pre-Publication Electronic version.
- Version 1.00: First Print and E-book Edition.
- Version 1.01: Updated Print and E-book Editions.

Prologue:

20300712:0900. My colleagues, family and friends lovingly urged me to write this book to capture my remembrances of the events that became the history of the colonization of the Moon. It was a high point in my life to lead the team that brought together the right set of skills to change a world; the world being the natural satellite of planet Earth known as the Moon. This year, 2030, marks my 74th birthday. As memories slowly become hazy remembrances, I have no fear looking forward to the longer-term explorations of the deep unknown future that awaits all humans. I have no fear because Plan_B to the Moon is first and foremost a love story. How love of—the need for—exploration has shaped our very DNA. Every human alive on the earth today is a descendent of explorers and wanderers. 70,000 years ago, during the probable worst years of human existence our most ancient ancestors had to leave their old homes to find new homes. We don't know the distant relatives that stayed behind. Those ancient stay-behind relatives would not adapt and change, so they did not survive to pass-on their genes to us. Only the wanderers survived. Change may be the most difficult human condition to accept.

As you read this book, consider the following four inspirational quotes that have been guiding lights to the Plan_B Team throughout their careers and lives:
- *"Everything is theoretically impossible, until it is done."—Robert A. Heinlein*
- *"In order to attain the impossible, one must attempt the absurd."—Miguel de Cervantes*
- *"The best way to predict the future is to create it."—Abraham Lincoln*
- *"Never doubt that a small group of thoughtful, committed citizens can change the world; indeed, it's the only thing that ever has."—Margaret Mead*

Plan_B was impossible, until after the Chapter 4 XPM and SPM systems got built. Program of Record Contractors said reusing launch vehicles was absurd until SpaceX made it routine. And most profoundly, the technological future of the human race never just happens—it arises from the purposeful skill and sweat of envisioning humans. Small groups of committed citizens are the only force that changes the world. Therefore, Plan_B is impossible, its precepts absurd, and its creators are committable...

My team's earliest designs, planning notes and Program Plan briefings eventually became this book, ***Plan B to the Moon—A How-To Guide to Return to and Colonize the Lunar Surface.*** We started in early 2018 and finished the designs, drawings, spreadsheets and accompanying first draft texts by November 2018. We finished polishing the text during the dark, hard months of the 2018/2019 winter. The early 2019 Version 1.00 publication release was uneventful. Later positive reviews led to a wider audience and eventual enthusiastic acceptance by readers. We shared mirthful joy experiencing Plan-B's ever-growing ripples in the fabric of spacetime.

This book crafts a coherent, readable, ***executable plan;*** that defines how to return humans to the surface and colonize the Moon. The Earth celebrated the return of humans to the Moon in 2024. But I'm getting ahead of the story...

Table of Contents:

Version 1.01 Update:

Plan_B 1.00 was published on March 14, 2019. On March, 26, 2019, Vice President Mike Pence ordered NASA to <u>land</u> on the Moon. "The National Space Council will send recommendations to the president that will launch a <u>major course correction for NASA</u> and reignite that spark of urgency that propelled America to the vanguard of space exploration 50 years ago," Pence said. "...If NASA is not currently capable of landing American astronauts on the moon in five years, <u>we need to change the organization — not the mission</u>." The space agency's initial plan was to send astronauts to the moon by 2028, 18 years after NASA began developing the Space Launch System (SLS) to make the journey. But SLS has suffered massive cost overruns and schedule delays that have put in question whether even 2028 was achievable. "Ladies and gentlemen, that's just not good enough," Pence said. "...It took us eight years to get to the moon the first time 50 years ago when we had never done it before."

Forward:

Plan_B to the Moon outlines the detailed step-by-step **_Program Plan and How-to-Guide_** to return to and colonize the Lunar Surface. This Plan_B book was created from the early Return to the Moon Program Plan documents and briefings. Long before this book was written those Program Plan documents were crafted with their intended primary audience in mind–decision makers in the Government. Government programs live and die based on PowerPoint presentation Charts, so to capture the flavor of the time, this Plan_B book replicates many of those earliest PowerPoint Charts. We've also included the intensely detailed literal program spreadsheets at the end of the Plan_B book; which document the schedule, cost, mass, fuel and summary metric data for Plan_B. Don't worry, you don't have to labor over the spreadsheets unless you enjoy such labors.

Considerable effort went into making this book relatable to non-experts (non-scientists). It is best to consider this Plan B book as three books in one:

- Chapters 9 and 10 are the most technical book, presenting an exquisitely detailed and executable Program Plan to return Humans to the Moon and stay there. If you are an experienced Program Director or executive and you were tasked to send humans to the Moon, these Plan_B Chapters provide you with a step-by-step roadmap to accomplish your mission. Most importantly, these Chapters define costs for each step to return to the moon.
- Chapters 3 - 8 provide the by-design more readable (somewhat tutorial) next outward layer book. Chapter 3 describes the Plan_B Launch vehicles to get into Low Earth Orbit. Chapters 4, 5, 6, and 7 describe the pieces of space hardware you need to build spacecraft / habitats.
- Chapter 8 is a literal Catalog of Spacecraft—where the piece-parts hardware of Chapters 3-7 create functional spacecraft and habitats used in Chapters 9 and 10.
- The Prologue, Forward, Chapter 1, Chapter 2, and portions of Chapter 11 (Conclusions) comprise the outermost-layer book. The outermost-layer was written long after the completion of the Program Plan charts and calculations. Consider the outer layer as the SO WHAT, or WHY Plan_B was written. The outer layer sets the stage and provides reasoned discourse as to why it is important for Humans to Return to the Moon. Chapter 1 is science based, but it may come across as narrative fiction for hard-bitten engineers and literal rocket-scientists. For such professional aerospace colleagues and friends, my advice is to just skip ahead to Chapter 2 if Chapter 1 is not to your tastes. Plan_B was written to be accessible to a wider audience, so Chapter 1 sets the stage for this larger audience. Chapter 1 also includes up-front the original Plan_B Executive Summary (Section 1.5). Section 1.6 is autobiographical; in that it documents how and why the Plan_B Program Plan Team created Plan_B in the first place.

The many months developing Plan_B were full of the joy and thrill of invention and discovery. The Program Plan_B Team hopes you the reader enjoy yourself as you experience our Lunar dreams with your eyes wide open.

Stephen Long, Satellite Beach, Florida; Plan_B Program Director on Behalf of the Program Team

Chapter 1:   Introduction

1.1 Introduction to Plan_B to the Moon

Plan_B to the Moon outlines the detailed step-by-step *Program Plan and How-to-Guide* to return to and colonize the Lunar Surface.

In the formal processes in the US Government known as Acquisition, there is the central premise of Programs-of-Record (PoR). If you ever hear / read about a major acquisition—like a new ship, fighter aircraft, bomber, spacecraft, or other large Government program, such a program is a PoR. Returning to orbit the Moon is one National Aeronautics and Space Administration (NASA) Program of Record. Throughout this Plan_B document, **we call the NASA Program of Record Plan_A**. Plan_A spends billions of dollars, under the oversight of NASA executives. Aerospace companies execute their NASA contracts. Thousands (tens of thousands) of aerospace workers are employed. America needs a successful Plan_A. Plan_A has already spent tens of Billions of dollars. Plan_A will spend more billions before the Space Launch System (see Chapter 2) first launches. Many, many people draw their livings from these endeavors. The Plan_B Program Team is certain that NASA's dedicated public servants developed Plan_A in good faith. They developed Plan_A using the resources, tools, people, and organizations they felt were available to them. However, it is manifestly clear that Plan_A reflects a very traditional PoR government mentality. Such programs seldom incorporate *thinking differently about how inexpensively to solve large problems*.

This book advocates making significant changes from NASA's current Plan_A **to a new path called here Plan_B**. Plan_B is all about thinking about problems differently and saving money. The original intent for Plan_B was to develop a detailed Return to the Moon Plan that would:
- Meet a target expenditure of 85% of the costs of Plan_A.
- Enable Americans (as representatives of humanity) to land on the Moon by the end of 2024 (if funds made available starting in FY 2020).
- Meet or exceed the top-level metric requirements of Plan_A—namely first return to Lunar Orbit, land on the surface and begin the longer-term colonization of the Moon.
- Develop the Plan_B documents using non-technical—laymen's language. Returning to the Moon is for all of humanity, so all of humanity should be able to understand the Plan_B goals and execution elements.

1.2 Tourists or Colonists?

To Return to the Moon, we must first make a fundamental policy decision: Is this a tourist trip or a colonization roadmap? For all their awe-inspiring wonder, Apollo era flights were breathtaking tourist trips. President Kennedy said, we would send humans to the moon and return them safely to the earth. *The current NASA Plan_A (Program of Record) appears to achieve neither tourism nor colonization*. If NASA Plan_A survives, its primary objective is to build and crew a Deep Space Orbital Lunar Gateway. One can infer that from said Gateway that humans would eventually (in the much farther future–beyond 2028) proceed to the surface of the Moon and then proceed to Mars. But no Plan_A funding appears in the available-to-the-public budget documents to carry out those ultimate goals. Incredulously, the upper rocket stage of the Plan_A Space Launch System is on hold. We have not found evidence of funds-on-contract to build this upper stage rocket. Lunar payloads will have trouble reaching orbit without an upper stage. To be fair, it appears NASA is waiting for the SLS first stage to work (launch) before they release even more funds to the prime contractor to build the second stage. When the Plan_A Space Launch System launches, NASA's Plan_A defines ten Lunar Exploration Missions (EM-1 to EM-10). There is an 11th unmanned jaunt to the Jupiter moon Europa. None of these Exploration Missions as of the first publication of Plan_B land on the Moon or anywhere else. Based on documented cost expenditures to date, the later Exploration Missions are not achievable (affordable) within programmed budgets. The NASA Inspector Generator's office continues to issue warnings about cost overruns and delays in the Plan_A Space Launch System.

After much reflection and in contrast to Plan_A, the Plan_B Program Team's decision is that ***Plan_B will define the required steps for Lunar colonization–not tourism***. The key differences between tourism and colonization have to do with up-front patience:

- Tourists can use disposable spacecraft to travel to the Moon. Tourists take photos, eat pre-made packaged meals, and pick up souvenirs (rocks) to bring home. Looking at the Moon out of the window–from Lunar Orbit–is a Tourist trip.
- Colonists need to arrive (land) at the Colony with enough tools, will power and stamina to build homes, grow food, and eventually make and raise babies.

There are precedents and practical lessons looking back through human history about how one colonizes a New World. Consider the settlements of Jamestown in Virginia and later the Pilgrims in New England:

- Some of these travelers know/plan to make a one-way trip. Others know/plan to visit for a few months and return to the point of origin.
- Colonists pack their ships with everything they think they will need–with care given to bring tools and seeds. Because with the right tools you can create what you need from raw materials found in the new land and grow food from the seeds.
- You plan to land with enough food to last you until you can grow food. Part of the legend lost over time about the Pilgrims and Thanksgiving is that they landed at the wrong time of year. Landing in November (winter in New England) did not give the Colonists enough time to plant successful crops. They were anxious; would they have enough food to survive? With the help of the native peoples, the colonists raised or gathered enough food to survive until they could successfully grow seasonal crops.
- Ships crossing long distances are typically not capable of landing upon or even touching the new world literal shore. Historically, smaller boats (the long boat is the historical term) ferry colonists from the ship at anchor to the shore. To colonize the Moon, the Plan_B long-distance-ship will be the Earth-Lunar-Transit-Vehicle (ELTV). The ELTV moves supplies and crews from a Low Earth Orbit (LEO) Station (LEOS) to the Low Lunar Orbit Gateway (LLOG).

From the LLOG much smaller Lunar-Landers (LLs) and Lunar-Long-Boats (LLBs) will travel between the LLOG and the Lunar Surface (LS).

- Colonists traditionally have to live on the transport ship until they can build new houses on the shore. Plan_B Lunar pioneers will spend much of their early Lunar environment time living in the LLOG and the first Lunar-Surface-Habitat-Vehicles (LSHVs). As conditions permit, colonists will build permanent homes on the shore.

Human Colonists on the Moon will have Human needs—Humans have needs besides work, sleep and food. In space history, there has been at least one work-strike by crew members on an orbiting space station. Humans (Spam) up in the Can tired of the humans on the Earth's surface telling them what to do. Spam in a Can is an old astronaut saying. Humans down on the Earth work in ~8-hour shifts. At the end of their daily shifts they get to go home to join their families. They get to relax, rest, sleep, have sex, raise children, recreate, pursue hobbies—living a human life. The weary souls in that space Can tired of being hounded 24-7. The ground shift voices on the radio hounded the crews to work, work, work. The humans that colonize the Moon (or Mars) may feel the same way, they can't be work robots. As the Mars Simulators (on Hawaii and other places) have already proven in today's modern era, the first colonists face endless toil, loneliness and danger. Most any trained professional can camp-out for weeks or even months (think Army troops on maneuvers), but they have to go home eventually to rest and recover. Even the most intrepid Moon and Mars early explorers will need earth-side time to recover. Therefore, Plan_B Version 1.0 only begins the steps needed for the long-term return to the moon. Plan_B Version 2.0 (and later plans) document the expected actions and needs for later waves of humans sent to the Moon. Our most highly skilled explorers and scientists will crew Plan_B Version 1.0 Lunar missions. Version 2.0 missions will add more explorers and colonists. Version 3.0 will bring even larger mixes of explorers / engineers / scientists / construction / agriculture / artists / health care / administrators… a community that can live together. Living together is an often-difficult human endeavor. Even the most loving couples occasionally get under each other's skins. The trick is to not avoid personality conflicts—the trick is to learn how to disagree without being disagreeable. Technical challenges are typically solvable. Human personality conflicts are very hard to resolve.

The Plan_B Program Team encourages readers to look up and view the PBS television series *Colonial House*, first aired in 2004. That TV series recreated daily life in the American New England Colony in 1628. Episodes followed the often-desperate rigors of life for colonists in the early 17th century. Key takeaways from watching this program are that modern humans are physically and psychologically ill-prepared to live lives of endless, relentless toil and isolation. Make no mistake—the first humans to live permanently on the Moon will face intense hardships. Some humans are more psychologically prepared for such stress-filled lives than others. Some extraordinary humans—the true Adventure Gene souls that don't easily fit into our modern pampered life—thrive in such environments.

1.3 Solving Hard Problems–Plan_B Art-of-the-Possible-Solutions and Precepts

Plan_B posits that hard problems fit into three solution categories:

- Science Problems–Can you find solutions without violating known physical laws? If not, you must fund and start additional research to determine how the systems behave and how eventual engineering development could yield useful results.
- Engineering Problems–Can you apply technology and engineering solutions within our scientific knowledge base? Will highly motivated teams of Managers, Engineers and manufacturing experts be able to develop affordable technological solutions to problems?
- Checkbook Problems–Will decision makers have the will to change their preconceptions and reallocate funds required to accomplish the goal? It is a common Program of Record fallacy that all problems are fixable with more money and more time. It seldom occurs to such PoRs to question their program of record mentality. Why not use a mechanical pencil instead of inventing new space pens?

Returning to the Moon is no longer a traditional Science Problem. We know this task is possible because we have already visited, walked upon, and briefly explored the Moon. Returning to the Moon therefore complexly combines Engineering Problems and Checkbook problems. The Plan_B approach provides *Art-of-the-Possible* engineering solutions with defendable checkbook-math budget projections. The PD and Team have across decades of successful careers proven that Art-of-the-Possible approaches and solutions will work. **But only if decision makers choose** and allow said Team to carry out heretofore-thought-impossible missions.

Enabling humans to live human lives in harsh and dangerous environments is a very hard job. Low Earth Orbit (LEO), Low Lunar Orbit (LLO) and the Lunar Surface (LS) are stressful dangerous places. One technique to help ameliorate some of the worst elements of early colony stress is to give people *room to live and breathe*. Plan_B advocates that the early explorers and the colonists will need habitats/living structures that are much larger (more cubic volume) than the historical spacecraft cans of history. Three extraordinary men (heroes) can live in a 20 cubic foot can for a week (the Apollo command and Lunar modules). But they can't live as spam in a can for 3 months. Visiting the historical Skylab space station at NASM and looking at photos of the more modern International Space Station (ISS) provides sobering insights. To us it appears that left-brain engineers–not right brain interior designers must have designed those facilities. Humans Living on the Moon will have left AND right brain human needs and addressing such needs underlie much of Program Plan_B thinking. Plan_B offers space vehicle and habitat designs that emphasize significant habitable volume for comparative low launch masses.

Plan_B offers Art-of-the-Possible solutions to these very hard problems. The fundamental *Art-of-the-Possible-Solutions* behind Plan_B mission plans and resultant spacecraft designs are:

- Apportion the mass required to return to the Moon into smaller, bite-sized payloads that fit within the maximum payload mass for available (already flying) commercial launch vehicles.
- Send compact building blocks into orbit and construct much larger vehicles and habitats out of the building blocks. The terms used for such Plan_B building blocks are *X-Polyhedron-Modules* (XPMs) and their assembled spacecraft offspring, *Space-Poly-Modules*, (SPMs).

The Plan_B two big ideas expand into actionable planning elements, defined here as **Plan_B Precepts**. Precepts further define fundamental technological/engineering approaches required for further development of detailed engineering concepts and plans. There is no (well maybe just a smidgen of) magic to these Precepts. Precepts are important decisions that enable the engineering team (and future execution activities) to further continue with Plan_B designs and cost estimates. Plan_B organizes its precepts into the Six Pillars of Plan_B Moon Colonization (the Six Pillars).

Chart 1 depicts the Six Pillars of Plan_B Moon Colonization Hexagon:
- Pillar 1: Mass into Orbit Affordably (Chapter 3)
- Pillar 2: Use X-Poly-Modules to Lower Spacecraft Costs (Chapter 4)
- Pillar 3: Leverage Existing Human Rated Spacecraft (Chapter 5)
- Pillar 4: Water* as Fuel–Common Conventional CH4/O2 (Methane/Oxygen) Rocket Engines and LARIT Engines (Chapter 6)
- Pillar 5: On-Orbit Refueling and Infrastructure (Chapter 7)
- Pillar 6: Catalog of Plan_B Spacecraft (Detailed Spacecraft designs, Built from Common XPM parts) (Chapter 8)

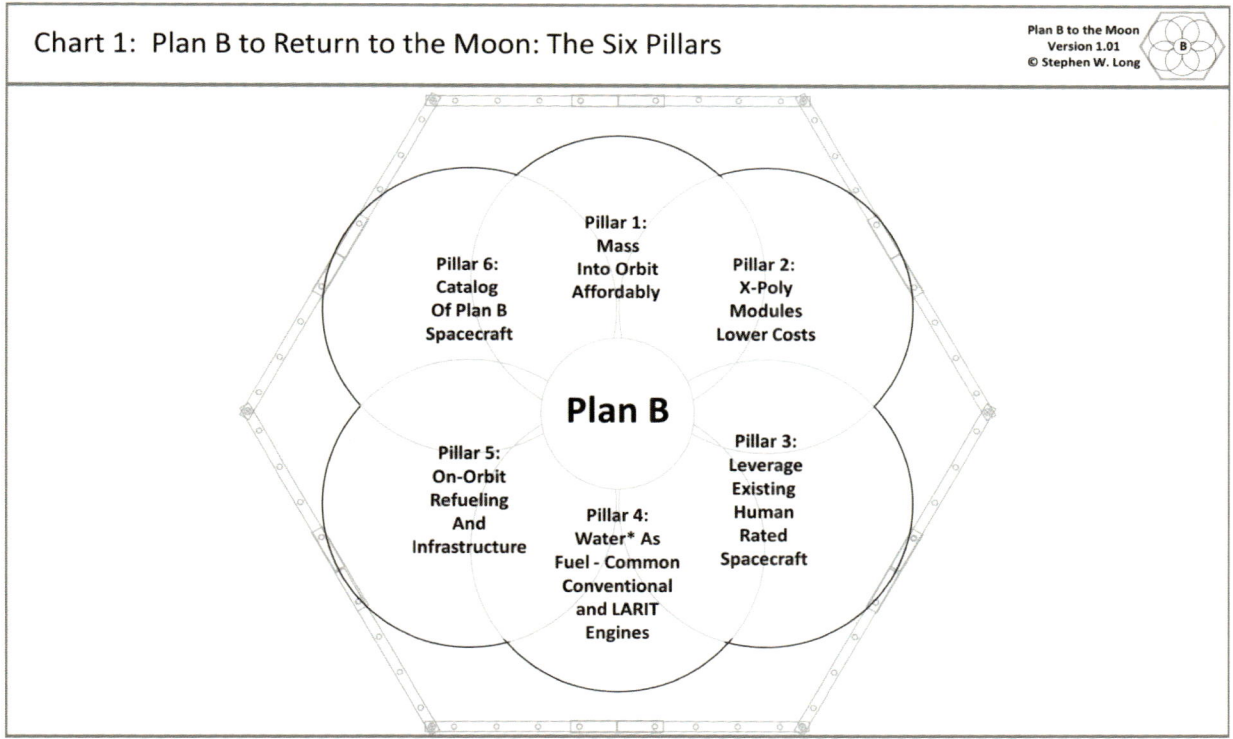

Plan_B Chapter 9 further builds upon the six pillars and defines the step-by-step activities to return to and enable the colonization of the Moon. Plan_B calls these discrete steps **Mission Circles**. As of Plan_B Version 1.00, it takes 804 Mission Circles to land on the Moon and by

Plan B to the Moon: A How-to Guide to Return to and Colonize the Lunar Surface

Mission Circle 899 Lunar Colonization begins. We note here that Plan_B Version 2.0 (written in 2019) picks up at Mission Circle 899 and further defines the first Lunar Colony (Lunar Base 1.0).

Plan_B uses multiple, detailed process-flow-diagrams to depict all of the major step-by-step tasks within a Plan_B Phase. The flow diagrams visually portray the step-by-step execution of the mission. These sequenced steps define the *How-To* elements of the Plan_B title (*A How-to Guide to Return to and Colonize the Lunar Surface*).

Not all Mission Circles are equal—some circles define half billion-dollar spacecraft launches; other Mission Circles report that a Phase is Complete. This phase-is-complete practice borrows from the long heritage of aircraft multi-crew dialogs (known as Challenge and Response). The Challenge crew member always states out loud at the end that a given, "Checklist is Complete."

A significant percentage of Mission Circles describe the Rendezvous phases and exquisite 3D dances of spacecraft as they maneuver around each other. Plan_B added many more Mission Circles over time (Plan_B Version 0.4 had less than 100 steps to land on the Moon). Early on the Team learned we had to *dance with the SPMs*. Holding models in our hands, we danced in three dimensions. We taught ourselves to understand and describe how the dozens of building blocks plug together on-orbit to create complex spacecraft and habitats.

Every Mission Circle has a purpose. When Mission Circles 899 concludes, Plan_B will have defined the literal How-To step-by-step plans to return to and colonize the Moon.

Chart 2 summarizes the Plan_B logical Return-to-the-Moon phases:
- Phase 1A and 1B builds the LEO Station-01 in Low Earth Orbit. LEO Station-01 is the SPM Assembly and Staging location for all later Plan_B activities to return to the Moon.
- Phase 2A and 2B builds the Space-Fuel-Terminal (see Chapter 7.2 for details).
- Phase 3A and 3B builds the first Earth-Lunar-Transfer-Vehicle payloads and delivers them to Low Lunar Orbit. Plan_B Phase 3B (return to LLO) achieves Mission Parity with the NASA Plan_A Exploration Mission 3 (EM-3).
- Phases 4A, 4B, 5A, 5B, 6A and 6B deliver more ELTV payloads to LLO.
- Phase 7A and 7B assemble the final vehicles/components for Humans to return to the Moon's surface.
- Phase 8A Humans Land on the Lunar Surface multiple times, seeking and then choosing a location for the First Permanent Lunar Base (LB). In Phase 8B, the Astronaut Crews build the first LB outpost. If they volunteer to do so, four to eight of those crew members transition from being Astronaut Crews to becoming Lunar Colonists.

Where Chapter 9 is all about the Mission Circle graphics, Chapter 10 is all about the data that define the inner workings of the Mission Circles. MISSION MATRIX data can be found at the end of Chapter 11. The MISSION MATRIX defines all of the numerical details for each Mission Circle. Each Mission Circle has a corresponding Element (row—line of data) in the MISSION MATRIX.

What may appear to be a simple single number in one MISSION MATRIX cell is a linked calculation (numerical answer), derived after extensive and deep analysis.

Chart 2: Plan B to Return to the Moon: Mission Phases	Plan B to the Moon Version 1.01 © Stephen W. Long
• Phase 1A: LEO Station-01 Assembly Start	• Phase 09: Expand Lunar Base-01V2 / Produce Food
• Phase 1B: LEO Station-01 Assembly Complete	• Phase 10: Expand Lunar Base-01V3 / Lunar Mining
• Phase 2A: SFT-01 Assembly Start	• Phase 11: Expand LEO Station-01V2 / Artificial Gravity Module
• Phase 2B: SFT-01 Assembly Complete	• Phase 12: Expand LEO Station-01V3 / Build Deep Space Food Modules
• Phase 3A: ELTV-01 Assembly Start (LLO Gateway-01)	
• Phase 3B: ELTV-01 Arrival - Humans Return to LLO	• Phase 13: Build Prototype LARIT Engine
• Phase 4A: ELTV-02 Lunar Components Delivery	• Phase 14: Build Deep Space Transportation Vehicle
• Phase 4B: ELTV-02 Arrival LLO	• Phase 15: Send Humans to Mars
• Phase 5A: ELTV-03 Lunar Components Delivery	• Phase 16: Comet or Asteroid Rodeo
• Phase 5B: ELTV-03 Lunar Components Delivery	• Phase 17: Europa Exploration
• Phase 6A: ELTV-04 Lunar Components Delivery	
• Phase 6B: ELTV-04 Lunar Components Delivery	
• Phase 7A: LLO Lunar Components Assembly	
• Phase 7B: LLO Lunar Components Assembly	
• Phase 8A: Humans Land on Lunar Surface / Lunar Base-01 Location Selection	
• Phase 8B: Lunar Base-01 Establish	

It is very important to note that MISSION MATRIX calculations rely upon initial specified values, called Plan_B Defined Variables (see Chapter 10.2). These Defined Variables are not just random numbers (most of the time). For instance, Plan_B specifies the mass of a spacecraft based on published data or based on the design-to mass value using XPM/SPM components.

1.4 Why Return to the Moon?

During one of the preliminary review sessions of Plan_B (around Version 0.5), a trusted friend asked a very hard and profound question: **Why Return to the Moon?** After a short, stunned silence, the Team admitted this is actually a reasonable question to ask. We researched and wrote Section 1.4 of Plan_B to answer the question, Why Return to the Moon. We note that answering the emotional question Why has both technical and emotional component answers.

1.4.1 First Movers

Most high technology businesses understand that **first movers** dominate a new market. In the business world, a first mover is a company that gains advantage and typically insurmountable market position by being the first to establish itself in a market. Examples of first movers include Amazon.com (books, now all on-line products), Travelocity (airline tickets), and eBay (online auctions). Competition has emerged in these domains. But no one can deny that early arrival and commitment to becoming the predominant owner of a market is one path to assure success. The Nation that colonizes the Moon—especially the polar regions where ice is waiting—will become the first-movers of the next-century future of the Earth and the solar system.

1.4.2 Access to Rare-Earth Resources

Earth has a small and finite supply of Rhodium, Platinum, Ruthenium, Iridium, Osmium, Palladium, and Rhenium. These rare-earth minerals are vital elements in modern electronics and magnetics. The term Rare-Earth minerals is not an exaggeration, they are dangerously rare.

Once humanity breaks free from the tyranny of Earth's gravitational field, such rare-Earth resources are available for harvesting from space. Ice from the Moon enables significantly lower cost space fuel production. Lowering fuel mass costs is the key to exploring and harvesting rare minerals and other riches from even deeper space—the asteroid and Kuiper belts. Multiple sources estimate towing a single nickel-iron class asteroid from the asteroid belt to LEO would provide Earth with decades of rare-earth minerals. The Plan_B Team would much rather see the US spend treasure on rodeo lassoing an asteroid than fighting a war in some far-flung corner of planet Earth. Do we want our sons and daughters to die fighting wars over neodymium / praseodymium / dysprosium magnet ore deposits?

1.4.3 Access to Helium-3

The Apollo landings identified that the Moon has significantly higher concentrations of Helium-3 than found on earth. Helium-3 is the likely very-rare-earth solution to solving fusion based power production for humanity. ***Our survival may depend on mining the Moon to enable a future of Earth energy independence based on Helium-3***. We won't know if Helium-3 is a viable fusion fuel source until we have commodity quantities of Helium 3 available to researchers to in-turn build fusion-power-production miracles.

By 2028 humanity accepts that carbon-caused warming imperils the second half of the 21st century. Even if sceptics continue to believe Earth's ever-increasing CO_2-caused-warming will NOT cloud Earth's future, there is little harm in seeking technological solutions to create carbon-free energy and to extract CO_2 from the atmosphere. What is our plan just-in-case the climate experts were right all along? What if the second half of the 21st century faces damage from what humans are doing right now? It will be existentially embarrassing to learn we squandered 50 years of clean energy opportunities because we did not seek out Helium-3. Perhaps the reduced-carbon-future of Earth has been staring at us from the Moon since 1969.

1.4.4 Technological Advancement

We learned from the Apollo program that seeking solutions to very hard problems will develop and or discover new technologies unknown of today. As the Program Director is fond of saying, the Apple iPhone did not emerge from a Government Request for Proposals.

Plan B to the Moon: A How-to Guide to Return to and Colonize the Lunar Surface

Non-Government experience civilians may be unaware of a very nuanced secret about how large aerospace contractors answer Government Request for Proposals. Large contractors offer solutions that only deliver what the Government asks for, not what could be. Large contractors happily live inside of their Program Boxes and they discourage outside-of-the-box thinking. Large PoRs have scant patience for new ideas. They have lower tolerances for enthusiastic imagination. From direct personal experience the Plan_B Team knows PoRs discourage inputs that disrupt the status quo.

In the spirit of new ideas and enthusiastic imagination, Plan_B development and space missions will advance new technologies that will:

- Develop and operationalize on-orbit capabilities to convert water and carbon dioxide into usable and portable fuel (Liquid Methane LCH4 and Liquid Oxygen LOX). With the proper scale, such space derived technologies will become pilot plants to prove the feasibility of large-scale Earth surface carbon dioxide extraction from Earth's atmosphere. This is not the only candidate CO_2 extraction technology. But it will be at least one good candidate that can give humanity a specific technological solution / path to extract atmospheric warming gases AND convert those gases into recyclable fuel. Extracting LCH4 and LOX will become an important technology to create a new, near-infinite supply of portable fuel sources so we can stop burning dinosaur goo petroleum. Instead of dinosaur goo, we need to learn to burn LCH4 to power aircraft and industrial processes. We should fight no more wars over foreign oil!

- Build the largest (measured surface area) structure ever assembled on-orbit. The Plan_B Solar Array of the Space-Fuel-Terminal will produce over 504 kW of instantaneous electric power. Plan_B delivers 819 kW total power in space. This magnitude of electric power production will enable future developments of even larger space structures. Large power arrays enable a farther-future of on-orbit (out of atmosphere) manufacturing of advanced materials and technologies.

The 1970s book, "The High Frontier," by Gerard O'Neill (see Chart 3) included concepts to beam microwave electric power back to the earth. While the pictures were pretty, the books never explained how to build or how much it would cost to build giant space colonies or power arrays. Plan_B explains HOW to build and HOW MUCH it will cost to build giant power arrays. Our pictures aren't so pretty, but at least the power production (launch mass) and checkbook math works.

The Program Team asserts that the Plan_B XPM / SPM concepts and eventual-built-systems are examples of technological game-changers. We invented SPMs to solve the high-cubic-volume-for-low-mass requirements of Lunar landing and habitation. They will also influence emergency shelters and portable defense applications on Earth for decades to come. The Team points out no Government RFP asked for Plan_B to develop XPMs. Instead, XPMs emerged as solutions to very hard problems.

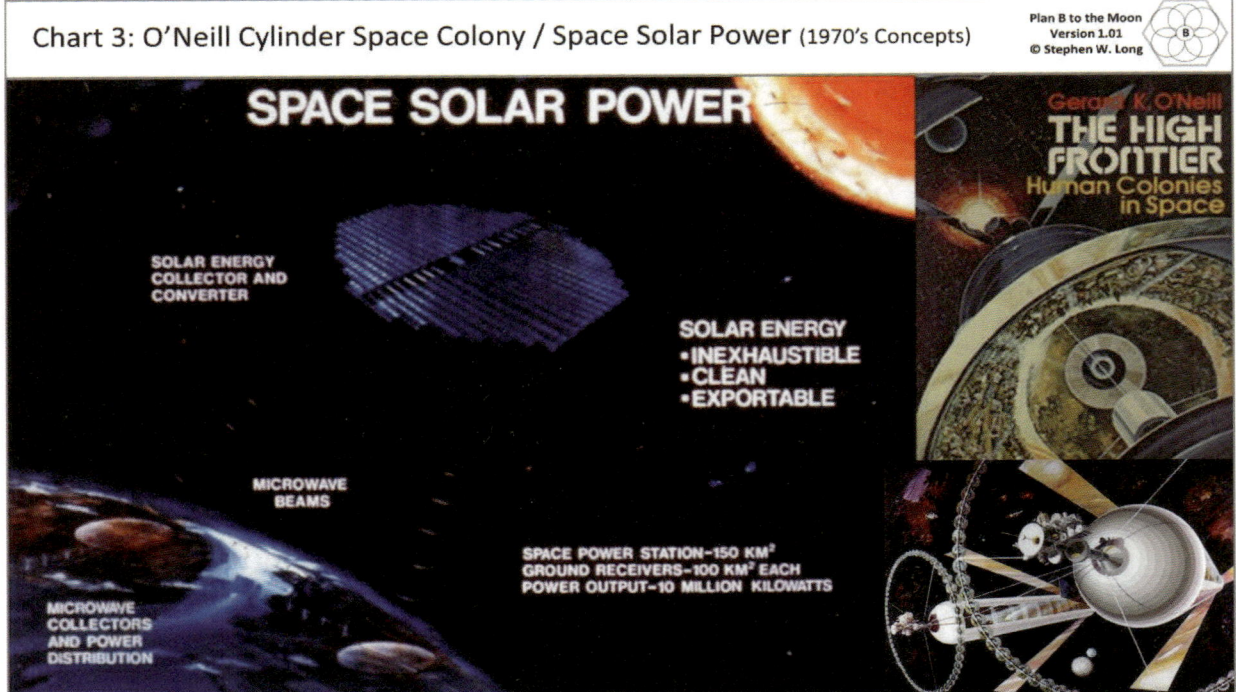

Chart 3: O'Neill Cylinder Space Colony / Space Solar Power (1970's Concepts)

1.4.5 Geopolitics

America is not the only space faring nation. Russia, and especially China understand that current and future geopolitical resource power struggles on Earth will move into space. Energy and mineral treasures are waiting on the Moon, the asteroid and Kuiper belts. Recall that the Apollo program was a direct response to the Soviet launch of Sputnik. Is America willing to wait to return to the Moon until AFTER China has established a mining colony on the South Pole of the Moon? Is it in our national best interest to *wait and then too late* decide that America needs to be there as well? If you think such worrisome views are only remnants of Cold War thinking, please research, consider and explain why China is creating new islands in the South Asian Sea. China is building military base islands in international waters (see Chart 4). By their military presence on those islands, China seeks to claim ownership (has claimed ownership) of the land underneath and the seas around the islands.

America stated to the world on the Apollo 11 lander that, "We Came in Peace for All Mankind." Will China colonize the south pole of the moon and invite all other nations to come share the wealth discovered there? How will American's react if China discovers a unique pocket of scarce Helium-3 on the Moon? What if the Helium-3 claiming Nation dominates fusion energy production on Earth for the next century? Will China share that rare treasure with all other Earth Nations? There is an old saying in America—Possession is 9/10ths of the law. **The first nation to find and mine Helium-3 on the Moon will dictate future earth energy production. The first nation to mine ice on the Moon will dictate the future of space exploration for centuries to come.**

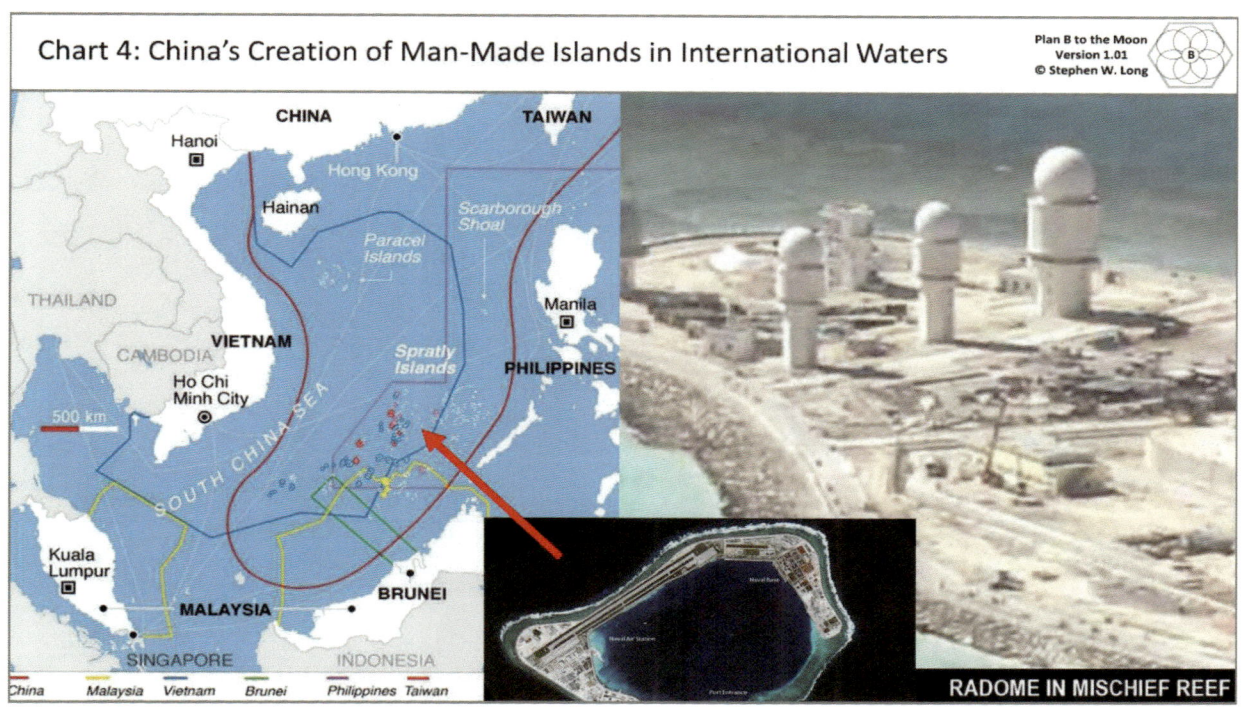

Chart 5 shows spectral images of the far side of the Moon–the Aitken basin. This basin is a most likely place to find ice and maybe other rare elements such as Helium 3 on the Moon. The Chinese Chang'e 4 unmanned spacecraft landed in the Aitken basin in January 2019. The Plan_B

Team notes that while many news organizations reported China landed Chang'e 4 on the far side of the Moon, we did not find a single story about WHY China chose that spot on the Moon. The Plan_B Team knows the answer. China is looking for Lunar Ice.

Said from a more hopeful point of view, from an interview of Neil deGrasse Tyson by Simon Worrall, published in National Geographic (20181019). Neil says, *"... space basically has unlimited resources—energy, minerals, rare-Earth metals. And if you look at the cause of maybe a third or a half of all wars, they were fought over limited access to resources. If space has unlimited resources, there is a chance that the continued exploration of space will be the greatest force in the generation of peace the world has ever seen. That's the hope I carry with me into the future."*

1.4.6   Humanity's Insurance Policy

"The dinosaurs became extinct because they didn't have a space program. And if we become extinct because we don't have a space program, it'll serve us right!" Larry Niven, Author

Consider the haunting images from WWII depicted in Chart 6. Humans will make any necessary sacrifice if we believe we face an existential threat. America and its allies fought World War II because we understood that the fascist Germany/Japan Axis were existential threats to freedom, life and liberty in the world. Several times in WWII, the smallest changes in outcomes would have caused the world we know today to have ceased to exist. It was a very dangerous time for humanity.

Please follow this metaphorical white rabbit for a moment. If we humans have sufficient warning, we would move mountains to ameliorate an existential threat. If we knew an asteroid, comet or other celestial event would destroy our existence on the earth, we would find some way to save ourselves. If we faced a terrestrial disaster such as a massive super-volcano eruption, we would spare no expense; no sacrifice would be too high. However, there is a very serious problem waiting for a threat to appear before you act—you may run out of time. As the old saying goes, nine women can't make a baby in one month. Consider that in the late 1930s Nazi Germany was ahead of the Allies in nuclear research. Would a six-month delay in the strategic commando raids and Allied bombing of German nuclear research centers have enabled Nazi Germany to build an atomic weapon first? If they had such a weapon, would they have used it to conquer and enslave the world? Yes, they would have. Would a one-year funding delay for the Manhattan Project (US atomic weapons program) have caused the Allies to lose the war? It very well may have. When your literal existence might be at stake, the most prudent, most adult course of action is to prepare proactively to respond to existential threats.

Chart 6: Why Return to the Moon?

Plan B to the Moon
Version 1.01
© Stephen W. Long

Existential Fear is a powerful motivator for H. Sapiens. If we collectively believe we must do something drastic to survive, we will do drastic things (think World War II)

Many well-meaning people don't believe the US Government should spend a dime on the space program. Whenever one hears such views, we suggest asking the following questions:

- Question One: Do you buy life insurance? Shockingly, some people don't buy life insurance. But most people do. If they say yes, they buy life insurance to protect their loved one's futures, ask what percentage of their annual income do they spend on life insurance? Internet sources say the average nonsmoker aged 35-45 will pay about 0.7% of their annual income to buy life insurance. Sources report that the US Gross Domestic Product in 2016 was $8.57 trillion dollars. If the USA *needed* life insurance, 0.7% of its annual gross domestic product (its annual income) would equate to an annual expenditure of $60 **Billion** dollars.

- Question Two: How would we prepare for a day when our existence as a species is at risk? We assert it is reasonable for America to spend annually tens of billions of dollars to prepare for and ameliorate potential existential threats. What is our plan to face and deal with the most terrible day in human history? Every ship at sea has lifeboats for its passengers and crew. Humanity does not today have the technology to build a lifeboat for every human on the planet. But we do have the technology to build a big enough lifeboat for our species to continue to exist. The sooner we put an offspring-viable human population onto another celestial body the sooner we will have purchased an insurance policy against the worst day (period) in human history. We should send human colonists to the Moon, Mars, Europa, Ganymede, and eventually other star systems.

If getting hit by a space rock is the future worst day in human history, consider we humans have already previously survived the second-worst-day (period) in human history. Over the course of Earth's 4.7-Billion-year history, there have been five catastrophic mass extinctions of life on Earth (see Charts 7-11). Most people have heard of the dinosaur mass extinction event of 65 million years ago. Few people know of the Great Dying event of 252 million years ago where 90-96% of all species on Earth went extinct. Almost every plant, creature, and living entity died.

Casual observers might claim these Great Dying events are too unlikely to occur ever again to modern humans. Just because this is uncomfortable to think about does not mean it can't happen, for there is the existence proof that near-extinction recently happened to humans. Modern examinations of human DNA show there was a Homo Sapiens near-extinction event 70,000 years ago (see Chart 11). That is only 3500 generations ago. The catastrophic Toba Super Volcano Eruption 70,000 years ago coincides with the near complete die off of our great-great (3500 times) Homo Sapiens grandparents. Modern DNA evidence says all humans alive today came from a winnowed population of as few as 1000 breeding couples. The geophysical evidence dates of the Toba Eruption coincides with the dates of the DNA near-extinction evidence. For there was a terrible calamity—the catastrophic explosion of an island in modern Indonesia—the Toba volcano. The Toba volcano no longer exists as a recognizable mountain.

Today's maps show an ocean-filled-crater lake, about half the size of Lake Erie. The Toba volcano ruinously exploded. It covered the Earth in ash and sulfuric acid clouds, blocking the sun, and it caused a ten-year-long winter. All the other species of hominids (Neanderthals, Denisovans and others yet discovered) appear to have gone extinct during this period. Imagine how horrendous that interminable decade 70,000 years ago would have been for our ancestors. Imagine a life desperately trying to survive in a bitterly cold world of no sun, no green vegetation, and death around the world for lack of food. We as a species only survived this dark-cold-hell-on-earth because we picked up and left where we were and traveled. Our ancestors walked across the planet, to other places to survive.

Why raise these extinction level event facts in this book about returning to the Moon? Because the decision to colonize the Moon—and continue traveling further to colonize Mars and beyond, may be the most important decisions we can make as a modern species. Forget rational geopolitics (if there is such a thing)—how would today's world of modern humans across the globe react to an existential threat? Would our responses be to sit around the campfire, hold hands and sing kumbaya together? Would humanity stay positive and focused on finding solutions, or would the world just descend into chaos? In America, we have recent and devastating experiences of nearly losing a major city to natural disaster. Government, at all levels, was ill prepared to cope with the reality of human basic needs during such a serious crisis. We have choices in how we face existential threats. We can either prepare (buy insurance) or let outside forces decide our destiny. Or we can **just go extinct**.

In a more positive line of thought, a Greek proverb says it best. ***A society grows great when old men plant trees whose shade they know they shall never sit in.*** The Plan_B Program Team understands they will never likely sit under trees grown under a garden dome on the Moon or on Mars. But we are at peace with that. For us, it will be enough to know future children will play under those trees.

Plan B to the Moon: A How-to Guide to Return to and Colonize the Lunar Surface

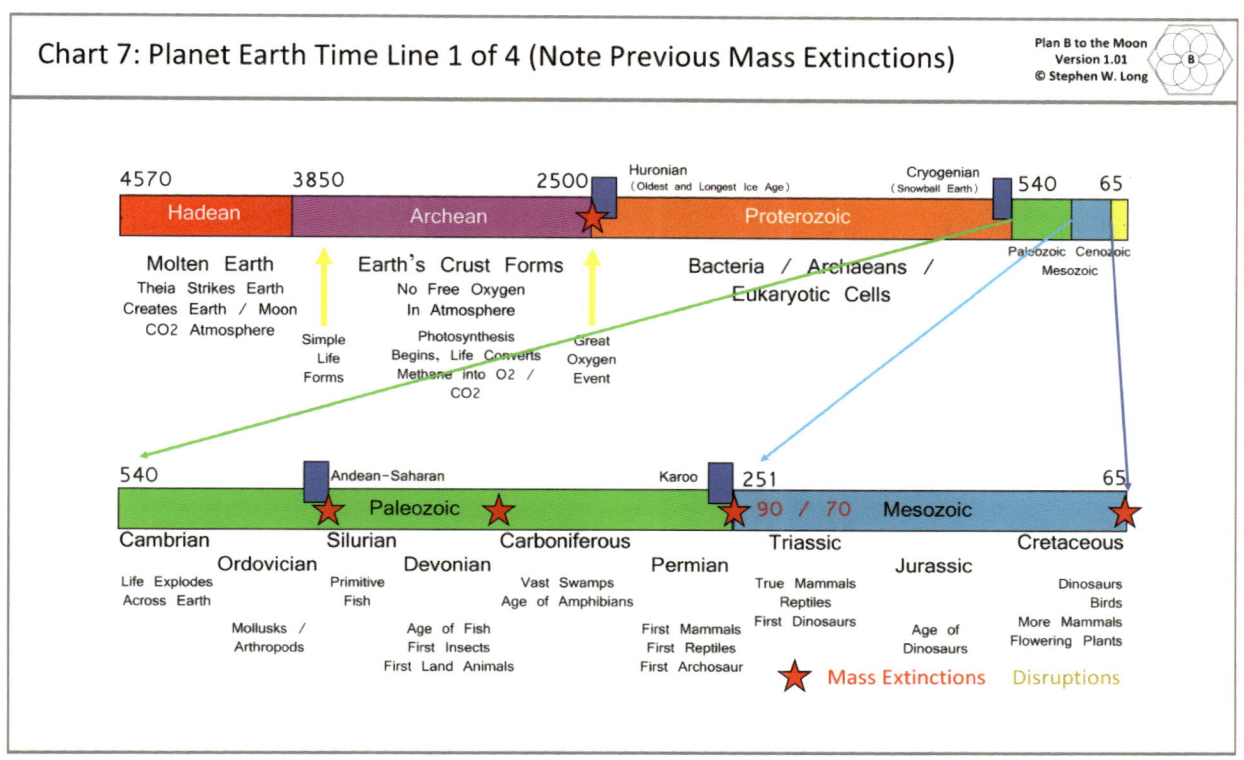

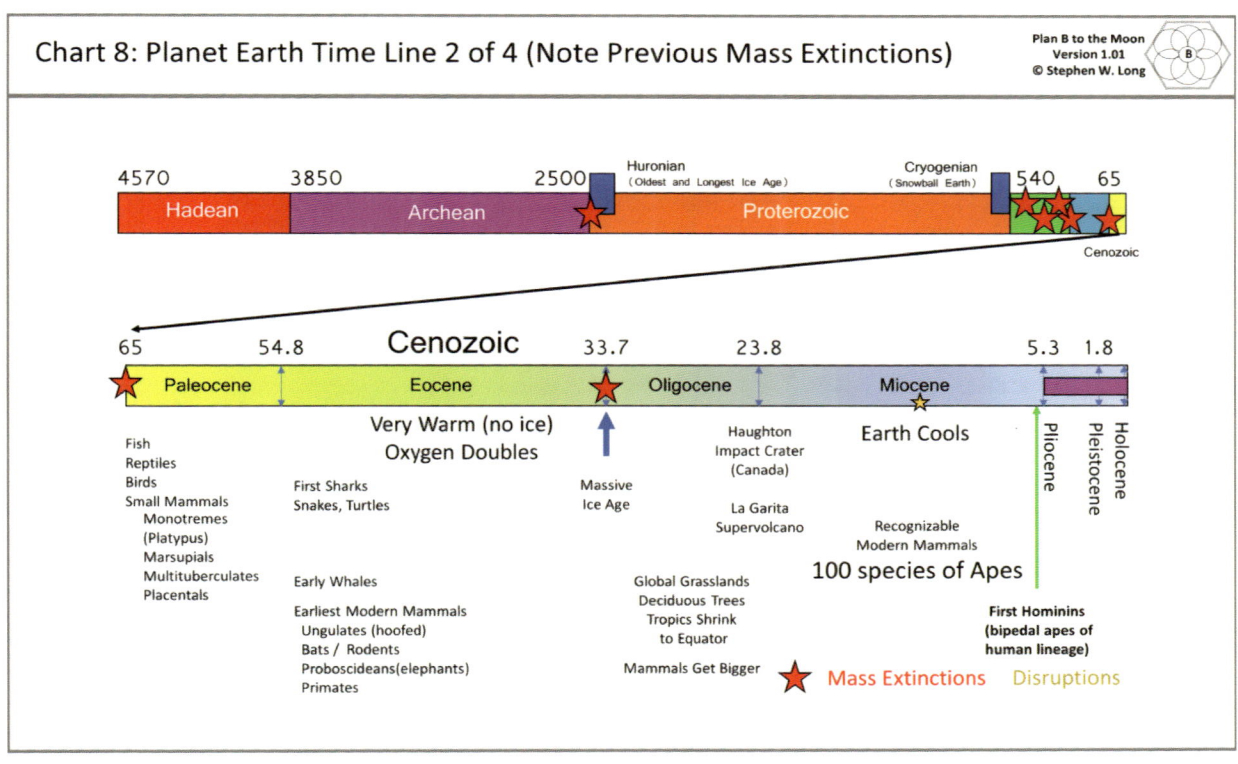

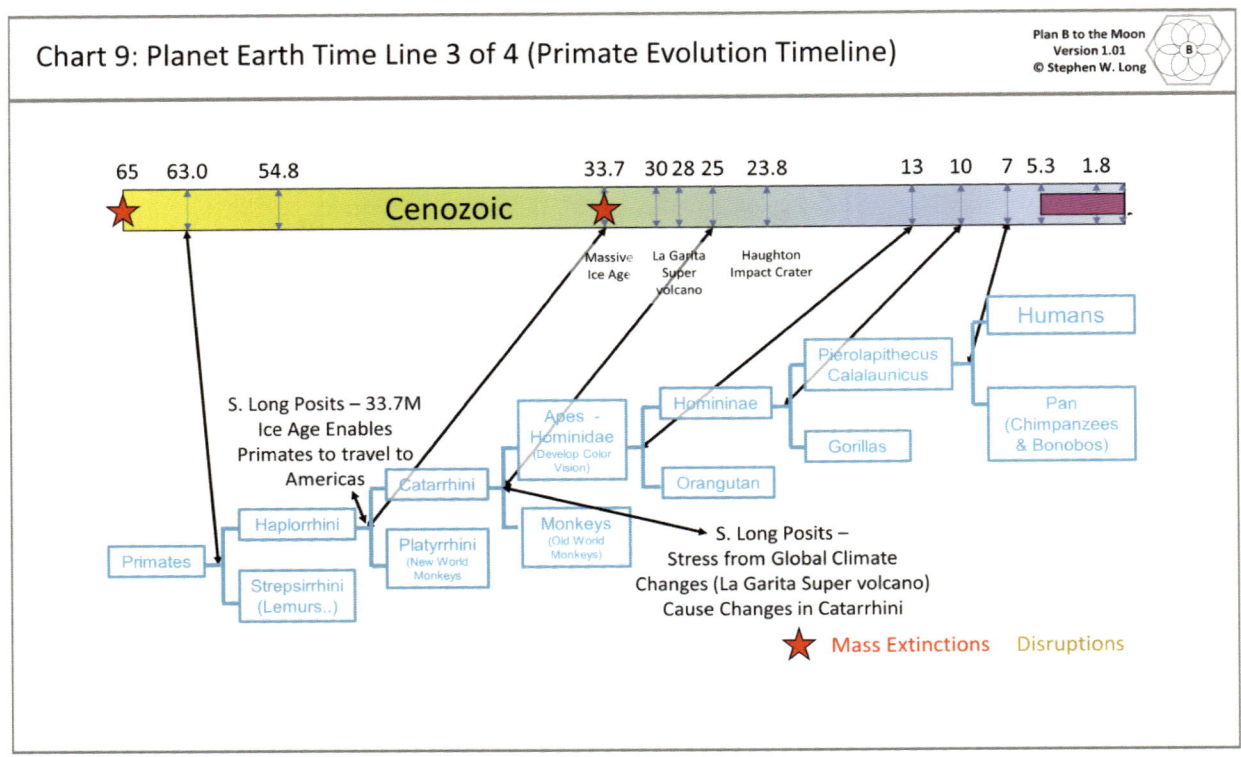

Chart 9: Planet Earth Time Line 3 of 4 (Primate Evolution Timeline)

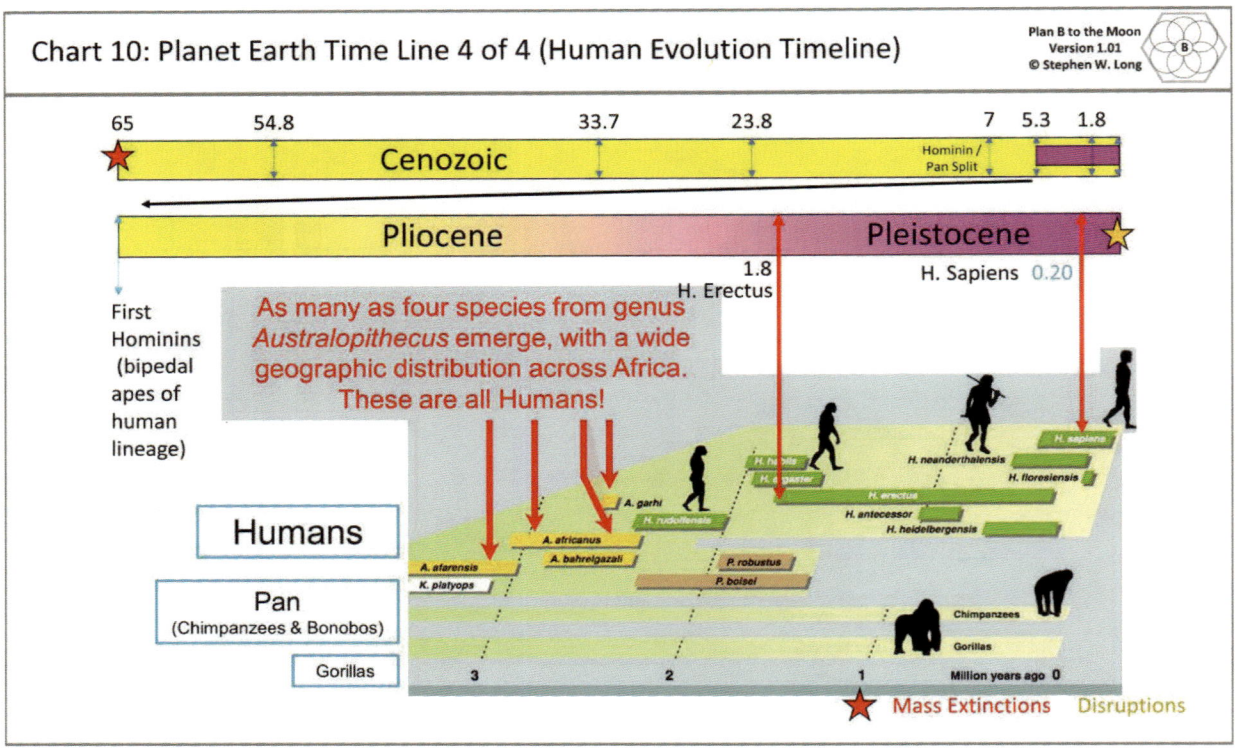

Chart 10: Planet Earth Time Line 4 of 4 (Human Evolution Timeline)

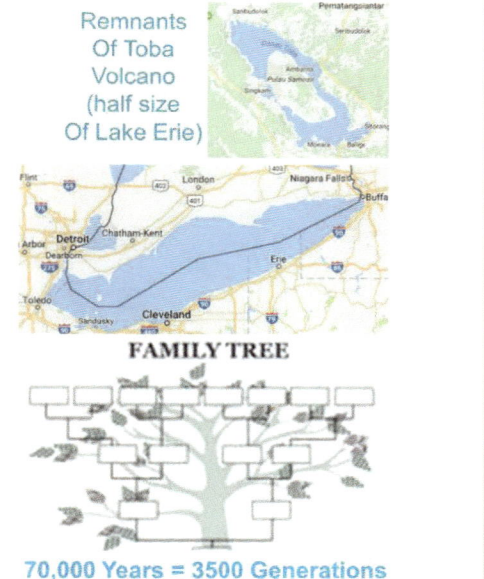

Chart 11: Human Genetic Bottleneck (We Almost Went Extinct)

Plan B to the Moon
Version 1.01
© Stephen W. Long

- DNA evidence points to a near-extinction event for H. Sapiens 70,000 years ago
- ~Coincides with "Toba Super volcano Eruption" 70,000 years ago in today's Indonesia
  - Eruption spewed 100 times the mass of the largest volcanic eruption in recent history (1815 eruption of Mount Tambora in Indonesia) - 1815 eruption caused 1816 "Year Without Summer" (Note the Mount Pinatubo eruption (1991) lowered Northern Hemisphere by 0.6 C in 1992 and 1993)
  - Toba eruption is postulated to have caused a 10 year winter, massive vegetation (food source) collapse across earth
- DNA evidence shows all modern humans come from a population of ~1000 Couples (3000-10,000 total humans - children / older adults)
- Today - 3500 Generations after

Remnants Of Toba Volcano (half size Of Lake Erie)

FAMILY TREE

70,000 Years = 3500 Generations

1.4.7   Because it is Hard

On September 12, 1962, President John F. Kennedy gave a historic and profound speech at Rice University (Chart 12). We know this speech today as the, "We Choose to Go to the Moon" speech and it set the stage for the American Apollo program. By the close of 2030, the President will give a future-history defining speech that borrows from Kennedy, recasting the 1962 speech into a grand challenge for the rest of the 21st century:

We set sail on this new heavenly sea because there is new knowledge to be gained. For interstellar space science, like science and all technology, has no conscience of its own. But why, some say, the Stars? Why choose this as our goal? And they may well ask why climb the highest mountain? Why, 61 years ago, did we Choose to Go to the Moon? We choose to go to the Stars! We choose to go to the Stars in this 21st Century and do the other things, not because they are easy, but because they are hard. Because that goal will serve to organize and measure the best of our energies and skills. Because that challenge is one that we are willing to accept, and one we are unwilling to postpone...

Chart 12: President Kennedy's 1962 "We Choose to Go to the Moon" Speech  Plan B to the Moon
Version 1.01
© Stephen W. Long

"We set sail on this new sea because there is new knowledge to be gained, and new rights to be won, and they must be won and used for the progress of all people. For space science, like nuclear science and all technology, has no conscience of its own. But why, some say, the Moon? Why choose this as our goal? And they may well ask, why climb the highest mountain? Why, 35 years ago, fly the Atlantic?

We choose to go to the Moon! ...We choose to go to the Moon in this decade and do the other things, not because they are easy, <u>but because they are hard</u>; because that goal will serve to organize and measure the best of our energies and skills, because that challenge is one that we are willing to accept, one we are unwilling to postpone, and one we intend to win ..."

Do you the Plan_B reader have within you the profound, burning desire to explore? Were you inspired in your youth by the opening sequence of Star Trek episodes, "To Seek out New Worlds and New Civilizations?" If your answers are lukewarm yes or emphatic no, there may be little anyone can say to you to convince you of the profound need for and premise of space exploration. But consider these facts, we carry in our DNA the need for exploration. Every person alive today is the descendent of explorers and wanderers. Exploration is central to who we are as humans. Only our ancestors that left their old homes to find new homes survived. The people that stayed behind did not survive catastrophic change and their stay-behind DNA no longer has any influence over the future of our species.

If the spirit of exploration does not move you, perhaps you should consider space exploration as the best financial deal in modern human history. Space and space technology will make a great deal of money—once we get out there. It is impossible to calculate the massive financial return on investment that came out of the Apollo program. But we can prove and trace most every example of modern technology we enjoy today came from space science spending. That spending enabled scientists, engineers and companies in the 1960s to shrink transistors down to the size of what we call computer chips today. The Space Race paid for that technology advancement, and from those investments, every human on earth benefits today.

In late 1997 a group of engineers from around the world met in Geneva, Switzerland to agree upon new standards for the future of television and digital cinema. Precise time is a critical enabler to allow different locations to synchronize their TV signals. Synchronization makes possible the seamless integration of signals from different time base sources. Without synchronization, the pictures roll or glitch every time camera shots change. Achieving perfect synchronization is seamless today. This was difficult a few decades ago. During those Geneva meetings, engineers from around the world advanced concepts to use the available precision time signals of the Global Positioning System (GPS). Using GPS time signals solved the synchronization problem. It was the solution agreed upon then, and it has been the global solution used ever since then. America allows the world to use GPS without cost. In that early Geneva meeting, a young engineer from a European country complained that, "he didn't believe the US military should control the GPS system." After a moment's pause, the retired-veteran head of the US Delegation gave a measured response. "If you don't like GPS, your country is welcome to build your own."

If you are a reader that does not support America funding space exploration that is your right to do so. But consider the following—you should walk home—because your car relies on the small embedded computer that enables your car run. Once you get home, throw away your cell phones, your flat screen TVs, your video games, your iPads, your laptops, your desktop computers. Unscrew all of your LED bulbs, plug in incandescent bulbs (if you can find them). Make sure you Will and Last Testament is up to date because you will probably die 20 years sooner than everyone else that enjoys the miracles of modern medicine—which increases life span. Throw out more than half of your food. Try to grow your own food in your back-yard, if you have the room. Good luck if you live in a high-rise building. Modern science-based agriculture is the only thing that feeds the massive population that now fills our world. And the list of space-derived (funded) modern technology goes on. There is an entire office at NASA dedicated to documenting and explaining the derived benefits of space investments. This office helps the American people understand what they get for their spent tax dollars.
Heroes will nobly risk their lives to explore and invent the future. With some passion we advise, if you don't want to explore, please get out of the way. "Shut up and color," as a wonderful art teacher once told the class.

Returning to the Moon—and striving for the Stars—is a love story about the human spirit's existential need and desire to stand on another world. **Step by gritty step, we will eventually stand on that far shore because it will be the hardest thing there is to do. Some few of us thrive on the challenges of hard to do—easy is for the people that get left behind**.

1.5 Executive Summary

Note: If the Reader plans to read the entire Plan_B book we recommend you skip this Executive Summary until you finish the book.

This book defines a Program Plan to Return Humans to the Moon. This book was created from a simple premise. You are an experienced Program Team (and Program Manager) assigned the task to return humans to the Moon, land, and start a Lunar Colony. How would you organize and plan such a complex series of tasks? This book is that Program Plan. To be crystal clear, NASA did not ask for nor did NASA fund Plan_B. Private citizens wrote Plan_B, with no external funding, and at no time does Plan_B state or imply NASA has approved Plan_B. But maybe NASA will adopt Plan_B.

NASA's current program plan is Plan_A (see Chapter 2). This book is Plan_B–an alternative Plan. By its existence, Plan_B proposes a different path forward than NASA's Plan_A. The Plan_B Program Team is certain that NASA's dedicated public servants developed Plan_A in good faith. They developed Plan_A using the tools, people, and resources they honestly felt were available (would be available) to them. Plan_A reflects a traditional large program-of-record mentality that seldom incorporates thinking differently about problems. Plan_B is all about thinking about problems differently.

Plan_B offers Art-of-the-Possible-Solutions to very hard problems. The fundamental **Art-of-the-Possible-Solutions** behind Plan_B mission plans and resultant spacecraft designs will:
- Apportion the mass required to return to the Moon into smaller, bite-sized payloads that fit within the maximum payload mass for available (already flying) commercial launch vehicles.
- Send compact building blocks into orbit and on-orbit assemble much larger vehicles and habitats out of the building blocks. The terms used for such building blocks in Plan_B are *X-Polyhedron-Modules*, or XPMs and their assembled spacecraft offspring, *Space-Poly-Modules* or SPMs.

Plan_B organizes components and spacecraft into actionable planning elements, defined here as **Plan_B Precepts** and these Precepts are technological / engineering foundational decisions for Plan_B engineering designs.

The Six Pillars of Moon Colonization are the organizing framework for **Plan_B Precepts**:
- Pillar 1: Mass into Orbit Affordably (Chapter 3)
- Pillar 2: Use X-Poly-Modules to Lower Spacecraft Costs (Chapter 4)
- Pillar 3: Leverage Existing Human Rated Spacecraft (Chapter 5)
- Pillar 4: Water* as Fuel–Common Conventional CH4/O2 (Methane/Oxygen) Rocket Engines and LARIT Engines (Chapter 6)
- Pillar 5: On-Orbit Refueling and Infrastructure (Chapter 7)
- Pillar 6: Catalog of Plan_B Spacecraft (Detailed Spacecraft designs, Built from Common XPM parts) (Chapter 8)

In some long-term future, organizations with ample checkbooks may easily order new spacecraft from a commercial catalog, just like buying a new car. Since that spacecraft catalog does not exist (yet), the Plan_B team had to create its own Catalog of Spacecraft (see Chapter 8). We used our catalog to develop detailed mission plans and used the plans to calculate program costs. Before we could create the Catalog of Spacecraft, we had to create the catalog of spacecraft components–the X-Polyhedron-Modules (XPMs) and aggregate systems (Space-Poly-Modules). See Plan_B Chapters 4, 5, 6, and 7 for details about XPMs and SPMs.

In Chapter 9, Plan_B defines the step-by-step activities to return to and enable the eventual colonization of the Moon. Plan_B calls these discrete steps **Mission Circles**. As of Plan_B Version 1.00, it takes 804 Mission Circles to land on the Moon and by Mission Circle 899 Lunar Colonization begins. Plan_B Version 2.0 picks up at Mission Circle 899 and starts creation of the first Lunar Colony (Lunar Base 1.0). Plan_B connects the Mission Circles into logical process-flow-diagrams to depict the flows and the major step-by-step tasks within a Plan_B Phase (see Chapter 9). The flow diagrams visually portray the step-by-step execution of the mission. Consider these flow diagrams to be the embodiment of the *How-To* element of the Plan_B title. Chapter 10 provides hard data in spreadsheet matrix format, that defines all of the numerical details for each Mission Circle. Each Mission Circle has a corresponding Element (line of data) in the Summary MISSION MATRIX. What may appear to be a simple single number in one MISSION MATRIX cell is a linked calculation derived after extensive and deep analysis.

Per our tasking, the Program Team reports that Plan_B Accomplished its Cost, Schedule, and Performance objectives, with bottom-line summary details provided in Chart 13. It's not magic. It's a hard checkbook fact that ~37 Falcon Heavy and Falcon 9 Plan_B launches will cost billions less than the few (less than 10) planned Plan_A SLS launches. The many-launches-approach–the Plan_B approach–saves enough program money to send enough mass to LLO to triumphally land on the Moon. And even with all the additional FH Rocket launches to send up the mass required for multiple moon surface landings, Plan_B still costs less than Plan_A.

In true elevator speech fashion, the key Plan_B summary points are: Plan_B is less expensive than Plan_A; Plan_B returns to the Moon in 2024; and Plan_B meets or exceeds (Performance Metrics) the Plan_A planned on-orbit capabilities. See Chart 13 for Summary Cost, Schedule and Performance Metrics.

This book, **"Plan_B to the Moon: A How-to Guide to Return to and Colonize the Lunar Surface,"** has looked back at the 14-month development of Program Plan_B. This Program Plan and later book has sought answers to the ruthlessly hard question: What will we as a species do if NASA's Plan_A stumbles or fails outright? We did not write Plan_B to focus on negative outcomes. Instead, Plan_B was written and proposed as a thought-provoking insurance policy. If NASA's Plan_A PoR stumbles, we as a species will have at least considered another way to return to the Moon and stay there permanently this time.

Chart 13: Plan B Executive Summary: Cost, Schedule and Performance	Plan B to the Moon Version 1.01 © Stephen W. Long

Plan B Summary Costs:					**Plan B Summary Schedule:**	
	Plan B	Plan B $4k	Plan B $7k	No SFT	Authority to Proceed (FY2020):	2019-10-01
FH Payload Mass	4,000,000				First Mission Payload Launch:	2022-12-27
F9 Payload Mass	350,000	☑ Cost			Phase 01 - LEO Station-01 Complete:	2023-07-31
Human Veh No Fuel Mass	299,100				Phase 02 - SFT-01 Complete:	2023-11-28
Vehicle No Fuel Mass	1,883,158				Phase 03A - ELTV-01 Humans LLO:	2024-02-18
Human Veh Costs	$2,991	$2,991	$2,991		Phase 03B - LLO Gateway-01 IOC:	2024-03-03
Vehicle No Fuel Costs	$9,416	$7,533	$13,301		Phase 04 - ELTV-02 Complete:	2024-06-07
FH Launch Costs	$3,962	$3,962	$3,962		Phase 05 - ELTV-03 Complete:	2024-09-11
F9 Launch Costs	$434	$434	$434		Phase 06 - ELTV-04 Complete:	2024-10-20
Plan B Other Costs	$5,212	$5,212	$5,212	-$2,423	Phase 07 - LLO Gateway-01 FOC:	2024-11-01
Plan B Total Costs	$22,015	$20,132	$25,900	$19,592	Phase 08A - Humans Land on Moon:	2024-11-02
Plan A Through 2024 $M	$25,900	$25,900	$25,900	$25,900	Phase 08B - Lunar Base-01 IOC:	2025-01-18
Plan B Savings $M	$3,885	$5,768	$0	$6,308	☑ Schedule	
Plan B to A Ratio	85.00%	77.73%	100.00%	75.65%		

Plan B Summary Metrics:			
Launches (FHR and F9)	37		One (1) LEO Station-01V1, FOC
Cumulative Habitable Vol LEO/LLO/LS	43,780	cu ft	One (1) Space Fuel Terminal-01, FOC
Cumulative Mass Sent to LEO	4,350,000	lbs	One (1) LLO Gateway-01V1, FOC
Cumulative Mass No Fuel to LLO / LS	716,700	lbs	One (1) Lunar Surface Base-01V1, IOC
Water* (from Earth, Shielding/Stock)	297,000	lbs	Two (2) Lunar Landers (Human)
LCH4/LOX (from Earth)	1,161,200	lbs	Two (2) Lunar Long Boats (Cargo)
LCH4/LOX from SFT Production	593,434	lbs	Two (2) Lunar Habitat Vehicles
LCH4/LOX from Earth and SFT	1,754,634	lbs	☑ Performance
Power on Orbit (SFT Array is 504kW)	798	kW	

Plan_B to the Moon is a love story about Returning to the Moon—and eventually striving for the Stars in this 21st century. Plan_B has attempted to put into words the human spirit's existential need and desire to stand on another world. We humans do our best work when we tackle the hardest problems. **Step by gritty step, we will eventually stand on that far shore because it will be the hardest thing there is to do.**

1.6 How and Why Plan B Was Developed

Note: Earliest draft Readers of the Plan_B book asked the Program Director to add commentary about how and why Plan_B came to be tasked and written. Explaining how and why is by definition autobiographical and emotional in nature. Therefore, understand this Section 1.6 is not about the science / engineering behind Plan_B, but about human emotions and seeking answers to the hardest questions. If you only want the engineering and science facts, skip ahead to Chapter 2.

There are tens of thousands of fictional stories based on science that eloquently dream of living in space. Future heroes travel to the stars in city-sized spacecraft. Vast colonies of eager humans scratch out new lives on new worlds. However, Plan_B makes the admittedly rude observation that none of these dreams of the future actually ever tell you HOW to do it. None of such fictional stories explain how to build a functioning spacecraft or how to build a lunar habitat suitable for extended human exploration. Few ever teasingly provide any engineering details (like when the Star Trek Enterprise 1701 blueprints arrived one holiday season). None of these prior stories-based-on-science ever explain what it costs to create these spacecraft marvels. They never provide an engineering road-map describing the practical, needed steps to turn such space dreams into space reality.

The beginnings of Plan_B were highly improbable. The Program Director (PD) received a phone call that changed his life and eventually changed the course of Lunar history. That phone call started the sequence of events that created the Program Plan that became known as Plan_B. If Plan_B had a birthday, it would be February 6, 2018. That was the day the Program Director sent an E-mail to multiple friends in the aerospace community. That E-mail expressed strong emotions arising from witnessing the historic launch of the SpaceX Falcon Heavy Rocket:

"The SpaceX Falcon Heavy (FH) Rocket Launch and Landings inspired me to research NASA's published Plans to Return to the Moon. When through my research I learned that NASA's Program of Record does not land on the Moon, I became incensed. Those strong emotions led me to Think Differently About the Problem. *I witnessed the FH Rocket Launch from my front yard in Satellite Beach, FL and raced inside my home to see the FH Rocket Dual Landings Live on TV. I Knew in that Instant that the World Had Changed and a New Future was Possible."*

That life-altering phone call proposed a new, impossible mission, and asked the Program Director to come out of semi-retirement. Why this team? Why us? The Program Director suspects he got that phone call because he had a deserved and admittedly focused (a polite word for driven) professional reputation for accomplishing impossible missions. The PD excelled at getting things done in impossible time frames. The 2018 phone call proved the maxim that no-good-deed goes unpunished. That Plan_B phone call was a historical echo from way back in 2000 when a thought-impossible mission began with a secure phone call. That secure call asked the Program Director to accept the impossible mission of creating a new intelligence paradigm. The PD accepted the task to build a National capability that did not exist. The small founding team built from scratch a new collection system—essentially an entirely new domain within the intelligence community. And even more crazy than building the new system that first National mission needed to produce intelligence in less than 90 days. Against all odds that historic first mission occurred in 83 days. No one from that time was unmarked by the experience. Working 20-hour days / seven days a week ruins health, staffs burn out, relationships fracture, and people just leave because they can no longer handle the stress. But the small team that had the fortitude to persevere, to continue on, they got the mission accomplished. By their efforts they changed the world. Security measures prevented public thanks for the team for what they did. However, observant viewers will find an obscure railing plaque overlooking an exhibit at the National Air and Space Museum (NASM), hinting at that now long ago first mission. That plaque was enough reward—to know what we did mattered in the righteous cause of defending America.

To get to the point, it was surprising, and admittedly ego boosting to get a second such impossible mission phone call. Once is rare, twice is unprecedented. This time around, the Program Director was semi-retired, rebuilding his life after a heart attack, having survived the Widow-Maker by the thinnest of time margins. Two minutes delay during any part of that drama nearly made his wife a widow. But he survived. During the short but interminable wait for the ambulance to arrive, the Program Director has privately shared having one overpowering and secret prayer, "Please, I'm not done yet." If one believes in such things, the universe was waiting for Plan_B to be written.

The old friend at the other end of the phone line asked the Program Director to form a quick reaction analysis team. The Team needed to come up with an alternative plan to the NASA Program of Record (Plan_A). As the small team dug into the NASA PoR details, we soon found multiple early warning signs. A recent NASA Inspector General Reports warns the Space Launch System (SLS) was not likely to succeed without receiving significantly more and more (over budget) funds. It was not rocket science (pun intended) that someone labeled our efforts, *Plan_B to the Moon*. No government money was available to compensate the small team the Program Director gathered to come up with an alternative plan to the NASA Program of Record. We volunteered our time, months of time, because it would take the government too long to allocate funding and award a formal study contract. There was also the reality that seeking government funds for program planning called *Plan_B to the Moon* would tip off and potentially alarm the Plan_A Prime Contractors. Too often, competitive small research projects get routinely squashed by big programs (of record) before the small competing programs can even be born. So, to be crystal clear, the US Government did not pay for Program Plan_B Version 1.0. There was no-cost to the government. There was a high price—days and nights and weekends—months of time taken away from family and friends. But as the lyrics from *Man of La Mancha* proudly proclaim, "you have to be willing to march into hell for a heavenly cause." Too much of our Nation's future (our species' future) was at stake to not have an alternative course of action to Plan_A.

The internal-voice motivation driving our *march into hell for a heavenly cause* was to seek answers to ruthlessly hard questions. What will we as a species do if Plan_A stumbles, fails outright and gets canceled? Will a single, catastrophic systems or launch failure cancel the return to the Moon program? Maybe. Recall that powerful Congressmen wanted to cancel the Apollo program after the Apollo 1 fire and harvest those funds for new social welfare programs. While critically (politically) important PoRs often become too big—too important to fail—even jaded politicians will kill a program when said program egregiously busts it budgets. If not killed outright, these too-big-to-fail programs become resource-consuming monsters. Speaking from dark personal experiences, one typical response instead of canceling a troubled program of record is for the bureaucracy to steal money from smaller successful programs. Those funds get re-allocated as bill-payers for the failing big program. Only in the Government are large, failing programs rewarded with ever-increasing amounts of more money. During his 15+ years of government civilian senior service, the Plan_B Program Director supervised many successful technology development small programs. Multiple times good programs got canceled, for no other reason than the bureaucracy needed to harvest funds. Small programs get sacrificed to the feed the beast (large failing programs). After this happens more than once, a feisty Program Manager / Director becomes crafty. A PD learns to spend (put on an untouchable contract) all of your allocated funds early in the government fiscal year. Once the funds are on contract, the big programs can't steal your money later during that year. Is this the best way to run a program? No, but sometimes this is the only way to survive the collateral damage and tyranny of too-big-to-fail PoRs. The Program Director recounts one of his more memorable encounters of his years of government service. He received threats from a newly appointed red-faced screaming boss, shouting he would have the PD fired for daring to tie up all the research program's money. Screaming boss wanted to reallocate said funding for his failing (over budget) program of record. The ultimate universal Karma expressed itself when the screaming boss's program catastrophically failed, forcing him to leave the Government. Payback is sometimes the name for female canines.

Instead of focusing on negative outcomes, the Team wrote and proposed Plan_B as an insurance policy. If (when) the NASA PoR stumbles again we as a species will have an alternative approach to return to the Moon and stay there. It is astonishing to report that NASA's published Plan_A has already spent tens of billions of dollars. Plan_A needs even more tens of billions of dollars to finish. As of early March 2019, Plan_A does not Land** humans on the Lunar Surface. Instead, Plan_A missions will only admire the moon from Lunar Orbit. In every setting with family, friends and professional colleagues they are incredulous when we discuss that Plan_A does not land on the moon. They always ask, if we do not land, why even bother?

*** Plan_B Version 1.01 update: Plan_B was published on March 14, 2019. On March, 26, 2019, Vice President Mike Pence ordered NASA to land on the Moon. "The National Space Council will send recommendations to the president that will launch a major course correction for NASA and reignite that spark of urgency that propelled America to the vanguard of space exploration 50 years ago," Pence said. "...If NASA is not currently capable of landing American astronauts on the moon in five years, we need to change the organization — not the mission." The space agency's initial plan was to send astronauts to the moon by 2028, 18 years after NASA began developing the Space Launch System (SLS) to make the journey. But SLS has suffered massive cost overruns and schedule delays that have put in question whether even 2028 was achievable. "Ladies and gentlemen, that's just not good enough," Pence said. "...It took us eight years to get to the moon the first time 50 years ago when we had never done it before."*

The original tasking accepted by the Plan_B team was to develop a detailed Return to the Moon Program Plan that would be less expensive, land by 2024, meet/exceed requirements, non-technical language. The small Plan_B team worked as an extended family from February through December 2018. Stephen, David, Barry, Guy, Mike, Michele and Josée gave up our spring, summer and fall. As did our respective families. Four of us were scarred veterans of the year 2000 miracle first program. One key life lesson learned during that effort was that you must be willing and able to make excruciatingly difficult technical/financial decisions even in the face of insufficient or fuzzy data. As one finds oneself up to your rear in alligators, you still have to keep your eye on draining the swamp. Walt Disney famously said: keep moving forward. One learns through hard experience to keep the swamp drains clean—to prepare for the certain occurrence of and interference by naysayers. The team's new favorite saying is, "Where Do the Naysayers Go When the Impossible Happens?" Plan_B showed us a better way forward. It became the common language and framework for a new, revised roadmap to return to the Moon.

Plan B to the Moon: A How-to Guide to Return to and Colonize the Lunar Surface

Perhaps this book, based on a real Program Plan, written by a Program Manager is the genesis of a new genus of space dreams. This is not a dream book; it is a guidebook to the touchable near future. This book is the story of humankind's return to the Moon–an unabashed love story about the human spirit's desire to stand on another world. This book shows how we modern humans accomplish this grand mission to walk on another heavenly shore–Step by Gritty Step.

Apollo 11, Buzz Aldrin on the Moon, NASA ID: AS11405903

Chapter 2:   NASA's Plan_A

2.1 Programs of Record

The NASA Programs of Record to return humans to Lunar Orbit include the Orion Spacecraft (Orion) and the Space Launch System (SLS), pictures depicted in Chart 14.

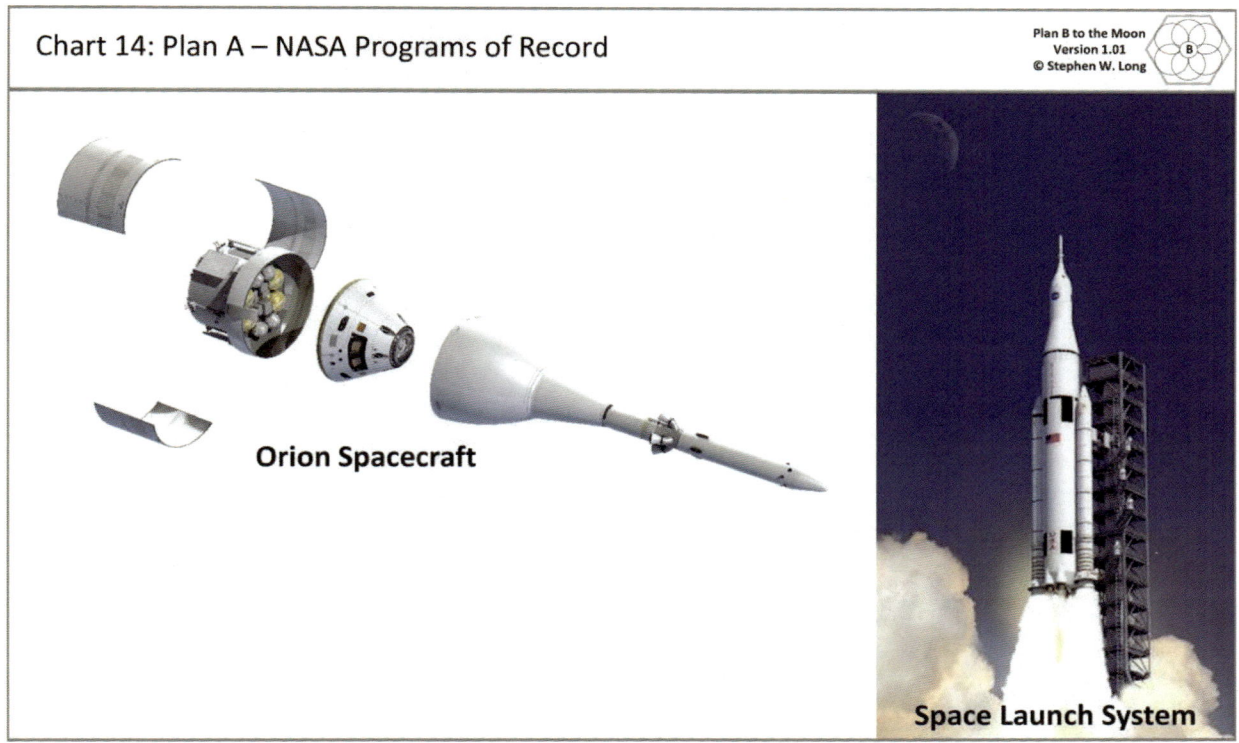

Chart 15 depicts NASA's summary mission goals for Plan_A. Plan_A calls for ten (10) Exploration Missions, EM-1 through EM-10 plus maybe a trip to Europa (EC) and maybe a trip beyond the Moon (EM-11). What is missing from these Exploration Missions? Landing on the Moon is not in the EM plans.

Chart 16 depicts the money math. From 2006 to 2017 NASA has spent $26 Billion to only prepare to return to the Moon. Since budget numbers beyond 2024 are not available to the public, one has to do some educated guesses as to the next ten-year NASA funding plan.

Chart 17 includes a snippet from the official NASA budget documents (Fiscal years). Chart 16 calculations use this budget data.

Chart 15: NASA Programs of Record Missions

Mission	SLS Block	Crewed	Launch	Duration	Mission Summary	Destination
Exploration Mission 1 (EM-1)	1 Crew	No	Dec-19	1 month	Send Orion capsule on trip around the Moon, deploy 6 other small CubeSats.	Lunar orbit
Europa Clipper (EC)	1B Cargo	No	2022		Unmanned Flagship-class mission to explore Europa	Jovian orbit
Exploration Mission 2 (EM-2)	1B Crew	Yes	Jun-22	8-21 days	Orion capsule (crew 4) delivers to lunar orbit the first module of Lunar Orbital Platform-Gateway the 40kW Power/Solar Electric Propulsion (SEP) Bus	Lunar orbit
Exploration Mission 3 (EM-3)	1B Crew	Yes	2024	16-26 days	Orion capsule (crew 4) delivers to lunar orbit the habitation module to the Lunar Orbital Platform - Gateway	Lunar orbit
Exploration Mission 4 (EM-4)	1B Crew	Yes	2025	26-42 days	Orion capsule (crew 4) logistics delivery to the Lunar Orbital Platform - Gateway.	Lunar orbit
Exploration Mission 5 (EM-5)	1B Crew	Yes	2026	26-42 days	Orion capsule (crew 4) delivers to lunar orbit the airlock module to the Lunar Orbital Platform - Gateway.	Lunar orbit
Exploration Mission 6 (EM-6)	1B Cargo	No	2027		Deep Space Transport (DST) to the Lunar Orbital Platform - Gateway.	Lunar orbit
Exploration Mission 7 (EM-7)	1B Crew	Yes	2027	191-221 days	Orion capsule (crew 4) - DST checkout mission.	Lunar orbit
Exploration Mission 8 (EM-8)	1B Cargo	No	2028		DST Cargo Logistics and refuelling mission.	Lunar orbit
Exploration Mission 9 (EM-9)	2 Crew	Yes	2029	1 year	Orion capsule (crew 4) - DST long-duration shakedown mission.	Lunar orbit
Exploration Mission 10 (EM-10)	2 Cargo	No	2030		Cargo DST Logistics and Refuelling Mission	Lunar orbit
Exploration Mission 11 (EM-11)	2 Crew	Yes	2033	2 years	Interplanetary flight	Martian orbit

What is Missing From This Mission List?

Chart 16: NASA PoR Expenditures to Date and Projections to 2033

Plan A Programs of Record		Year	Costs $M (A)	Costs $M (B/E)	Funding Type / Destination
	Prior Orion (2006-2010)	2010	$6,116		
SLS		2011	$1,536		Research and Development
	MPCV		$1,196		Multi Purpose Crew Vehicle (MPCV)
SLS		2012	$1,498		Research and Development
	MPCV Orion		$1,200		
SLS		2013	$1,415		Research and Development
	MPCV Orion		$1,138		
SLS		2014	$1,600		Research and Development
	Orion Program		$1,197		
SLS		2015	$1,678		Research and Development
	Orion Program		$1,194		
SLS		2016	$2,000		Research and Development
	Orion Program		$1,270		
SLS		2017	$2,150		Research and Development
	Orion Program		$1,550		
		2018		$3,700	
		2019		$3,700	Plan A
		2020		$3,700	
		2021		$3,700	2018-2024=
		2022		$3,700	
		2023		$3,700	$25.9B
		2024		$3,700	
		2025		$3,700	
		2026		$3,700	
		2027		$3,700	
		2028		$3,700	
		2029		$3,700	
	EM-10	2030		$3,700	
		2031		$3,700	
		2032		$3,700	
		2033		$3,700	
$7,813 Cost Per Mission / Launch			$26,738	$59,200	
				$85,938	

- PoR Costs 2006-2017 = $26B
- PoR Projected Costs 2018-2033= $59B
- 2018-2024= $25.9B
- Total Program(s): $86B
- Calculated Cost Per Mission / Launch = $7.8B (11 Launches)
  - If you subtract $26B of "sunk" costs, Plan A still comes to $5.45B PER MISSION ($16.35B by EM-3)
- Moon Landings, Mars Landings are NOT in the Current Plan A PoR
- Plan B Precept: Target Ceiling: $25.9B

The Programs of Record (SLS and Orion) projected costs are $59 Billion (through 2033), for a total program cost of $86 Billion or $7.8 Billion PER EXPLORATION MISSION. If you subtract the earlier $26 Billion of sunk costs, it still comes to $5.45 Billion per Exploration Mission.

Plan_A will spend $16.35 Billion by EM-3. Plan_A will spend $25.9 Billion by 2024. This 2024 budget projection value is the chosen design-to-cost-ceiling for Plan_B. If Plan_A delivered Moon landings for this large sum of US Treasury dollars ($86 B), Plan_B would not likely exist today. Disappointment is too gentle a word to describe the Plan_B Team's profound shock when this size of expenditure—without landing on the Moon—was first realized. From that deep well of emotion, the concepts that became Program Plan_B emerged and from that Program Plan we later wrote this book.

The Plan_B team is not the only group of citizens concerned about the Plan_A NASA PoRs—other groups have expressed serious doubts. Chart 18 conveys excerpts from news stories about NASA Inspector General Reports that convey significant concerns about the Space Launch System (SLS). **The NASA IG Report states NASA has spent $11.98 Billion to date, with a $1.28 Billion cost overrun, and claims the first SLS launch should have been in 2017.** The Report states the SLS launch missed its promised 2018 launch, will miss 2019, and the Report is skeptical of claims promising a mid-2020 launch. Sources say it will be 2022 before the SLS would likely be ready. As of early 2019, contract funding for the SLS upper stage has yet to released. If you have questions about NASA budgets, please contact NASA.

Plan_B Version 1.01 update: Plan_B was published on March 14, 2019. On March, 26, 2019, Vice President Mike Pence ordered NASA to land on the Moon. "The National Space Council will send recommendations to the president that will launch a major course correction for NASA and reignite that spark of urgency that propelled America to the vanguard of space exploration 50 years ago," Pence said. "...If NASA is not currently capable of landing American astronauts on the moon in five years, we need to change the organization — not the mission." The space agency's initial plan was to send astronauts to the moon by 2028, 18 years after NASA began developing the Space Launch System (SLS) to make the journey. But SLS has suffered massive cost overruns and schedule delays that have put in question whether even 2028 was achievable. "Ladies and gentlemen, that's just not good enough," Pence said. "...It took us eight years to get to the moon the first time 50 years ago when we had never done it before."

Chart 17: NASA 2018 Budget Data, Program of Record Continues Beyond 2018

Plan B to the Moon
Version 1.01
© Stephen W. Long

- 2018 Budget requests $19.1 billion for NASA (0.8% decrease from 2017)
  - [PoR: $0.624B] Aeronautics Research
    - Over-land commercial supersonic flights and safer, more efficient air travel
  - [PoR: $1.9B] Planetary Science Program
    - Robotic exploration of the Solar System
  - [PoR: $3.7B] Orion Spacecraft, Space Launch System, and Ground System
    - Continued develop. of programs to send American astronauts to deep-space
  - [PoR: $1.8B] Earth Science Portfolio
  - [$] Supports and expands public-private partnerships as the foundation of future U.S. civilian space efforts

Note: There are MANY documents / document summaries that describe the NASA Budget. Every document reports slightly different budget numbers!

Budget Authority ($ in millions)	Operating Plan 2016	Enacted 2017	PBR 2018	Notional 2019	Notional 2020	Notional 2021	Notional 2022
Exploration	**3,996.2**	**4,324.0**	**3,934.1**	**4,259.7**	**4,513.3**	**4,437.9**	**4,449.9**
Exploration Systems Development	3,640.8	3,929.0	3,584.1	3,739.7	3,898.2	3,771.5	3,762.3
Orion Program	1,270.0	1,350.0	1,186.0	1,170.2	1,123.4	1,124.5	1,124.5
Crew Vehicle Development	1,251.5	--	1,175.5	1,159.7	1,112.9	1,114.0	1,114.0
Orion Program Integration and Support	18.5	--	10.5	10.5	10.5	10.5	10.5
Space Launch System	1,971.9	2,150.0	1,937.8	2,083.6	2,265.6	2,177.6	2,177.4
Launch Vehicle Development	1,921.9	--	1,881.7	2,032.7	2,189.9	2,101.1	2,101.1
SLS Program Integration and Support	50.0	--	56.1	50.9	75.6	76.5	76.3

Chart 18: Plan A is in Trouble, October 2018 Inspector General Report

Plan B to the Moon
Version 1.01
© Stephen W. Long

There's a new report on SLS rocket management, and it's pretty brutal

"Boeing's poor performance is the main reason for the significant cost increases."

- **Inspector General Report:**
  - **Boeing Cost Overrun**
  - **11.9B spent to date**
  - **$1.2B overrun**
- **Schedule Slips**
  - **First Launch Promised 2017**
  - **Promised in 2018, delayed**
  - **New Promise, mid 2020**
  - **IG Report says 2020 unlikely**
- **Commentary: Where will the additional funds come from?**

Boeing has been building the core stage of NASA's Space Launch System rocket for the better part of this decade, and the process has not always gone smoothly, with significant overrun and multiyear delays. A new report from NASA's inspector general makes clear just how badly the development process has gone, laying the blame mostly at the feet of Boeing.

"We found Boeing's poor performance is the main reason for the significant cost increases and schedule delays to developing the SLS core stage," the report, signed by NASA Inspector General Paul Martin, states. "Specifically, the project's cost and schedule issues stem primarily from management, technical, and infrastructure issues directly related to Boeing's performance."

As of August 2018, the report says, NASA has spent a total of $11.9 billion on the SLS. Even so, the rocket's critical core stage will be delivered more than three years later than initially planned—at double the anticipated cost. Overall, there are a number of top-line findings in this report, which cast a mostly if not completely negative light on Boeing and, to a lesser extent, NASA and its most expensive spaceflight project.

Schedule slips

The report found that NASA will need to spend an additional $1.2 billion, on top of its existing $6.2 billion contract for the core stages of the first two SLS rockets, to reach a maiden launch date of June 2020. NASA originally planned to launch the SLS rocket on its maiden flight in November 2017.

However, given all of the development problems that the SLS rocket has seen, the report does not believe a mid-2020 date is likely either. "In light of the project's development delays, we have concluded NASA will be unable to meet its EM-1 launch window currently scheduled between December 2019 and June 2020," the report states.

There are other troubling hints about schedule in this new report, too. One concerns facilities at Stennis Space Center in Southern Mississippi, where NASA will conduct a "green run" test of the core stage of the SLS rocket. This is a critical test that will involve a full-scale firing of the rocket's core stage—four main engines along with liquid hydrogen and oxygen fuel tanks—for a simulated launch and ascent into space.

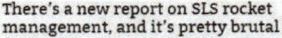

The report found that Boeing's development of "command and control" hardware and software needed to conduct this test is already 18 months behind a schedule established in 2016. This means the Stennis facility won't be ready to accommodate a green run test until at least May 2019, with further delays possible.

This is critical, because often the most serious engineering problems are uncovered during the phase when key rocket components are integrated and tested. The delay in green-run testing means that any problems that crop up during that phase of development will only push the maiden launch of the SLS further into the future.

Reasons why

There is a rather remarkable section of the report that discusses the reasons for these delays. Boeing evidently told the inspector general that initial issues for SLS development were caused by "insufficient funding." Notably, Boeing said the SLS contract was underfunded for 2015, and therefore it could not maintain its delivery schedule for the first two core stages.

The inspector general appeared to be having none of this, however. "By the end of FY 2015 the company had received $706 million, only $53 million less than requested for its work to build two core stages," the report states. "In addition, due to a congressional 'plus-up' the following year, Boeing received approximately $200 million more than what NASA estimated was needed to meet the original 2017 launch schedule. Further, in May 2016 NASA added almost $1 billion in additional contract value—bringing the total contract value to $5.1 billion—with only minimal changes in the scope of work."

NASA officials told the inspector general that they did not believe the schedule slippage could be explained by a lack of adequate funding.

In response to a query from Ars, Boeing issued the following statement: "An unprecedented rocket program has inherent challenges; developing the first unit of a system that will safely carry humans into space, even more so. But the program described in the OIG's report does not represent the Space Launch System program today."

Implications of this

Another thrust of the report is that NASA has improperly awarded tens of millions of dollars to Boeing for performance fees the company has not earned. "We question nearly $64 million in award fees provided to Boeing since 2012 for the 'very good' and 'excellent' performance ratings it received while the SLS Program was experiencing substantial cost increases, technical issues, and schedule delays," the report states.

In his response to the new report, NASA's chief of human spaceflight, Bill Gerstenmaier, essentially shrugs off the criticism by saying building big rockets is hard work.

"The SLS is the largest launch system in the history of space flight," Gerstenmaier's response states. "The design, development, manufacturing, test, and operations of the system are highly complex and represent a national investment in a longterm commitment to deep space exploration."

This may be true. But it seems an increasingly difficult sell after SpaceX developed the not-quite-as-large-or-complex Falcon Heavy rocket for $500 million. It is not clear what will happen next. In the past, Congress has largely ignored criticism of the SLS rocket, even from official sources. After all, the vehicle has 1,100 contractors in 43 states, covering a lot of legislative districts.

However, there are a few critics close to the White House who have been whispering concerns and criticisms about the big, expensive rocket to Vice President Mike Pence, who leads the National Space Council. To be clear: the vice president has been publicly supportive of the SLS rocket to date. But this report will at the very least add fuel to the fire of the criticisms he is hearing.

2.2 Paradigm Entrapment: Plan_A Versus Plan_B

The Plan_B Program Team is certain that NASA's dedicated public servants developed Plan_A in good faith. They developed Plan_A using the resources, tools, people, and organizations they honestly felt were available (would be available) to them. So how is it possible for such smart people to get so stuck in such a near-impossible-situation? This is a human condition, given the term Paradigm Entrapment. Accident investigators like to use the term root-cause-analysis. After an event, all subsequent actions / occurrences of outcomes are traceable back to a root (first) decision or action.

Charts 19 (Apollo Era) and 20 (Space Launch System) show that you need ~200,000 pounds of mass to send ~100,000 pounds of usable mass to land on the Moon. This implies you first have to get 300,000 pounds of mass into Low Earth Orbit.

The Plan_A PoR lunar paradigm appears to be:
- **You must build a single rocket** (SLS) using 200,000 pounds of mass to get 100,000 pounds to the Moon. You must launch all 300,000 pounds of mass into LEO at one time.

- If you have all of that thrust available in your single large rocket you should use (need) all of that thrust to deliver mass to the Moon. Ergo, the Orion crew vehicle and Lunar Orbiting Gateway modules need the SLS to reach the moon.

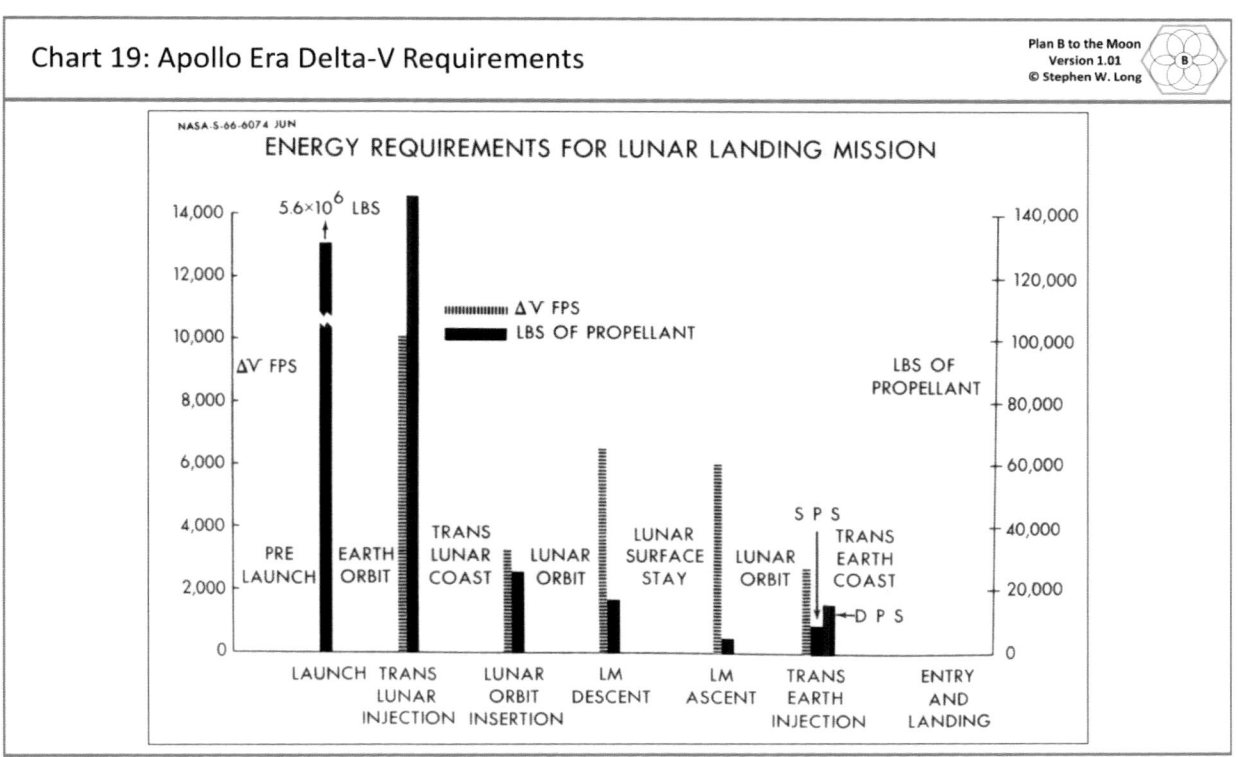

Chart 19: Apollo Era Delta-V Requirements

Chart 20: Mass Needed to Travel from LEO to Low Lunar Orbit

Launch System	Mass to LEO (lbs)	Mass to Moon (lbs)	Calculated Required Mass to Get to Moon
Falcon Heavy Rocket (2018)	140,600		
US Space Launch System Block 1 (2020)	150,000	57,000	
US Space Launch System Block 1B Cargo (2023)	290,000	88,000	202,000
US Space Launch System Block 1B Crew (2022)	290,000	81,000	209,000
US Space Launch System Block 2 Cargo (2024)	290,000	99,000	191,000
US Saturn V Apollo 17 (1972)	310,000	107,100	202,900

We appear to need about 200,000 lbs. of Mass (rocket engines and fuel) to Move 100,000 lbs. of Mass (spacecraft) from LEO to the Moon and Return

All of Plan_A is likely built upon this premise you MUST develop and use a giant rocket. **Once you have made this giant-rocket decision, every subsequent program decision or program element assumes use of the giant rocket**. This is a classic case of paradigm entrapment— "we must use a giant rocket"–paradigm.

In direct contrast to Plan_A, Plan_B mission plans and resultant spacecraft designs apportion the mass required to return to the Moon into smaller, bite-sized payloads. Plan_B payloads fit within the max launch payload mass for smaller, AVAILABLE, commercial launch vehicles. For Plan_B, Version 1.0, the payloads fit within the published constraints of the Falcon Heavy (FH) Rocket and the Falcon 9 (F9) Rocket. It will be wonderful when other commercial rockets are available. But for Plan_B Version 1.0, the mass delivery specifications for those different rockets had not been flight proven as of 2018. No matter the name painted on the side of the commercial launch vehicles, it will take multiple launches to deliver enough mass into low earth orbit. From LEO we then embark to land upon and colonize the Moon. The Plan_B magic that isn't magic is instead a cold-hard-checkbook-fact. 37 Falcon Heavy and Falcon 9 launches will cost billions less than the few (less than 10) planned SLS launches. The Plan_B many-launches-approach–saves enough program money to send enough mass to LLO and then land on the Moon. And even with all of the additional FH launches to send up the mass required for multiple moon surface landings, Plan_B still costs less than Plan_A.

Chapter 3:   Mass into Orbit Affordably

The February 6, 2018 Falcon Heavy (FH) Rocket launch changed the world and inspired the Plan_B Program Plan (and many years later this book). It was thrilling to experience the FH Rocket launch live from the Space Coast of Florida—all space rocket launches are awe-inspiring. It was the FH Rocket sticking the landings that proved that a new future was possible (Chart 21).

Chart 21: Falcon Heavy Rocket 2/6/2018 Launch and Historic Landings

Plan B to the Moon
Version 1.01
© Stephen W. Long

Forward from book "Plan B to the Moon:"
The SpaceX Falcon Heavy (FH) Rocket Launch and Landings inspired me to research NASA's published Plans to Return to the Moon. When through my research I learned that NASA's Program of Record does not land on the Moon, I became incensed. Those strong emotions led me to Think Differently About the Problem. I witnessed the FH Rocket Launch from my front yard in Satellite Beach, FL and raced inside my home to see the FH Rocket Dual Landings Live on TV. I Knew in that Instant that the World Had Changed and a New Future was Possible."

Given the FH Rocket future-changing-breakthrough, would it be possible to use the FH Rocket capabilities to return to the Moon AFFORDABLY? Would it be possible to design lunar missions to fit within the mass launch budgets of commercial rockets? Chart 22 introduces Pillar 1: Mass into Orbit Affordably.

Chart 23 summarizes the major current and historic Launch systems, ranked by launch Mass. The FH Rocket is powerful–it can boost 140,000 pounds into Low Earth Orbit. But it is not as powerful as the Apollo Saturn V which could launch 300,000 pounds of mass into LEO to send 100,000 pounds to the moon.

But what if we carefully apportion the lunar mission masses into smaller packages of 140,000 pounds (or less), each capable of launch by the FH Rocket? And is this approach less expensive than the Plan_A Space Launch System?

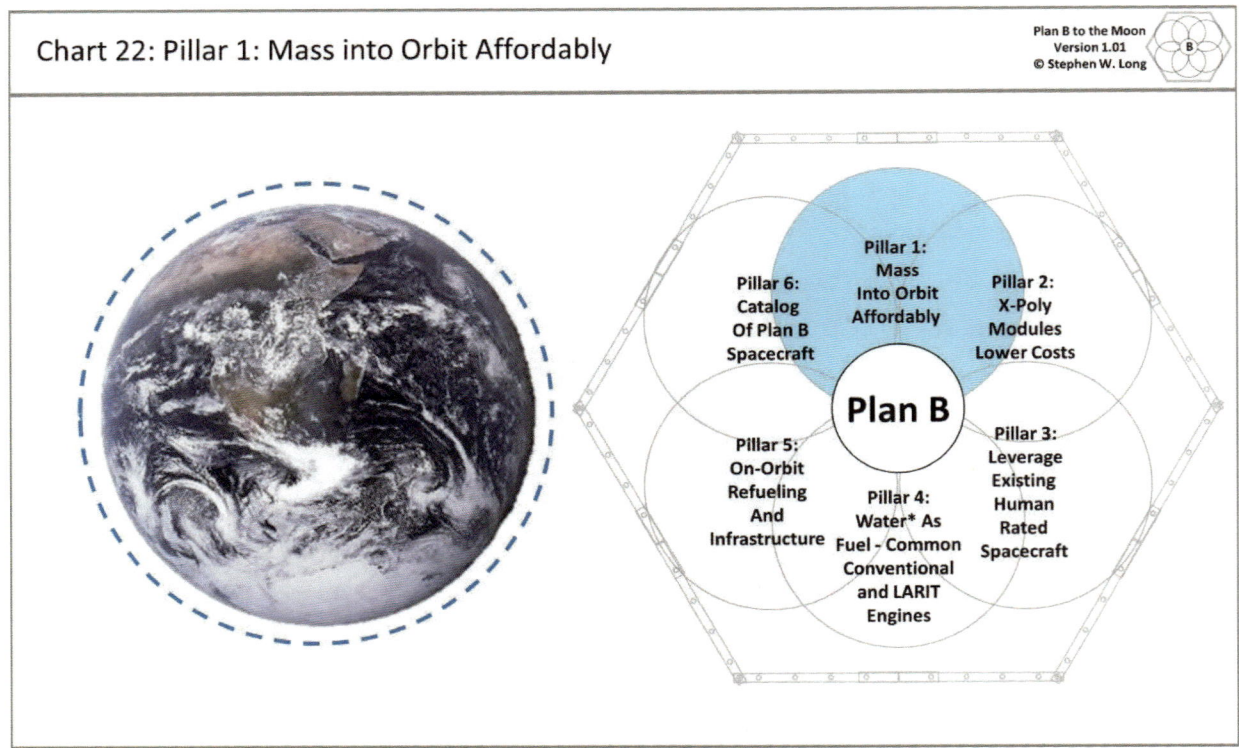

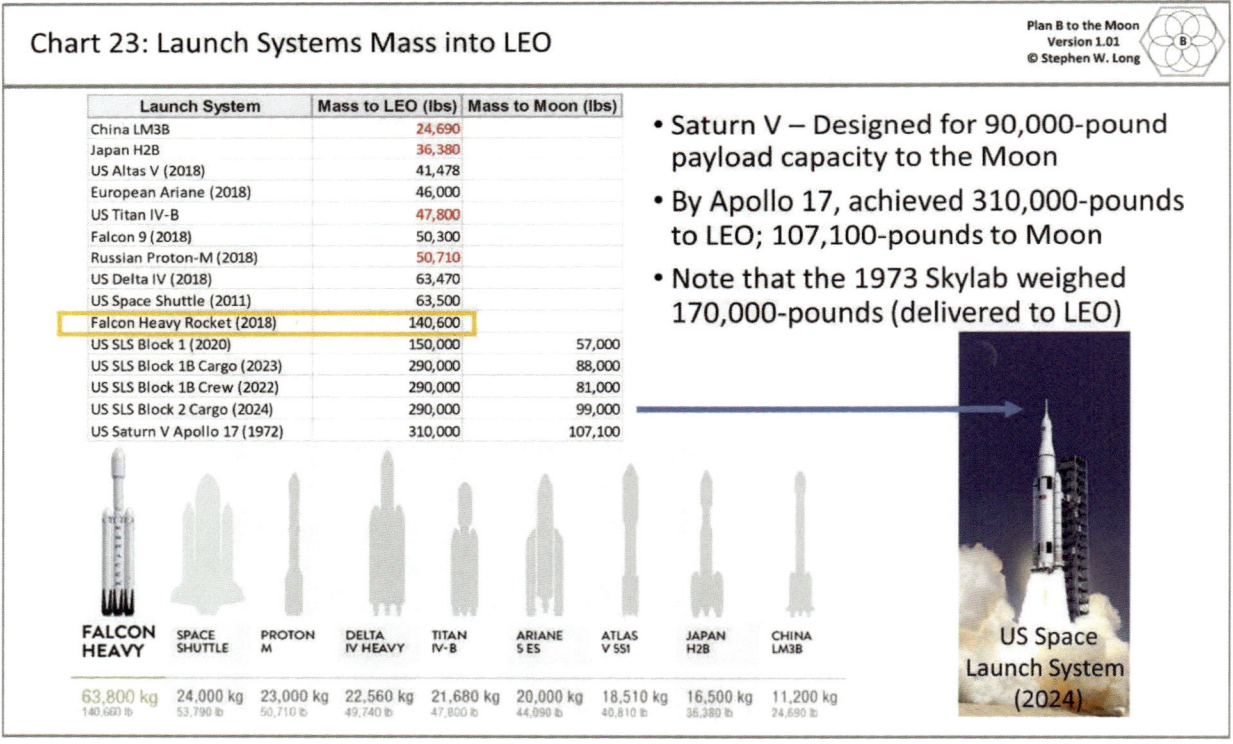

Chart 24 notes the advertised price for a Falcon Heavy Rocket is $90 Million per launch to deliver 140,000 pounds into LEO. A real-world contract from the USAF hired FHR for $130 M. The FHR beat United Launch Alliance's (ULA's) Delta IV costs in a competition under the Evolved Expendable Launch Vehicle program. It is a **Plan_B Precept** to use future FHR launches at a nominal cost of $130 M/launch to deliver 130,000 pounds of usable payload into LEO. This neatly calculates as $1000 per pound. In the past NASA used a cost figure of $10,000 per pound for spacecraft. NASA's Plan_A appears to calculate as $3300 per pound using the SLS.

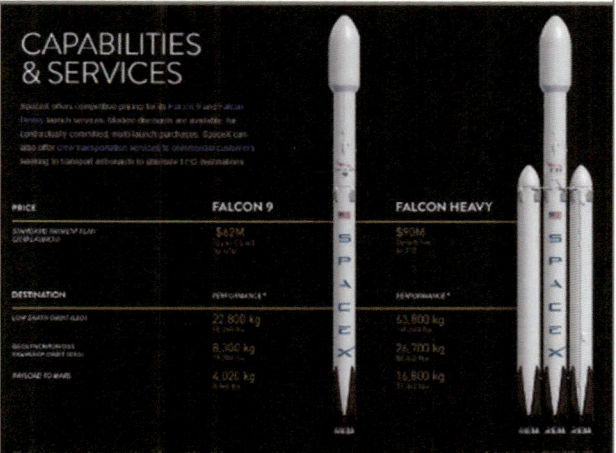

Chart 24: SpaceX's Falcon Heavy Rocket – Breakthrough Low Cost

Plan B to the Moon
Version 1.01
© Stephen W. Long

- "Advertised" price for a Falcon Heavy Rocket is $90M per launch to deliver 140,000lbs to LEO. A real-world contract from the USAF hires a FHR for $130M.

- Plan B Precept: Use future FHR launches at $130M/launch to deliver 130,000 lbs. into LEO. Neatly calculates as $1k per pound. NASA historically uses cost figure of $10k per pound. Plan A SLS appears to calculate as $3.3k per pound.

- Multiple FHR 130,000 lb. launches will be required to deliver mass to return to the Moon. But at $1k per pound, the Plan B approach is still less expensive (and faster) than NASA's Plan A (which relies on ~11+ giant SLS rocket launches)

FHR Enables Plan B Pillar 1 – Mass into Low Earth Orbit at $1000 a pound – Plan B Successfully conforms Lunar-bound- payloads to Fit 130,000 pound FHR Launches

Yes, it will take multiple FHR 130,000 lb. launches to deliver enough mass to return humans to the Moon. The number of launches does not matter, it's the $1000 per pound that matters. The Plan_B approach to Return to the Moon is less expensive (and faster) than NASA's Plan_A. Plan_A relies on ~11+ giant SLS rocket launches which cost ~$5 Billion dollars per launch (See Chapter 2).

Chart 25 provides a handy visualization of FH Rocket payload launch capabilities. Twelve thousand (12,000) gallons of water weighs 100,080 pounds. For a properly designed spacecraft, Plan_B plans to launch water as usable ballast on many FH Rocket launches. This will maximize the dollar value of each launch (no wasted launch payload mass).

It is a **Plan_B Precept** to maximize the value of each FHR delivery and launch by sending Water* (* Water plus CO2) to LEO as usable payload ballast. This approach would squeeze every drop of value out of every launch and use the Water* for mission needs.

Chart 25: Visualizing FH Payload Capabilities

Plan B to the Moon
Version 1.01
© Stephen W. Long

- How to Understand / Visualize 130,000lb Falcon Heavy Payload:
  - 100,080 lbs. of water =12,000 gal
    - Reference: 12,000 gal water tank
    - Dimensions: 142" (11'10") diameter
    - 197" height (16'5")
    - 3.12x10^6 in3
  - For FHR launches, 29,920 lbs. remaining for shroud/payload
- Plan B Precept: Maximize each FHR delivery/launch by sending Water* fuel to LEO on ~every launch – where such adjunct Water* payloads are precisely adjusted to maximize total payload weight (up to 130k lbs). Water* is stored on-orbit for future use...

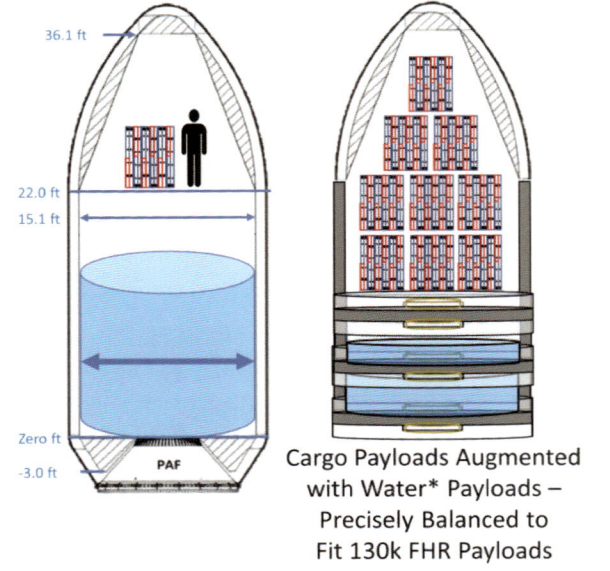

Cargo Payloads Augmented with Water* Payloads – Precisely Balanced to Fit 130k FHR Payloads

Chapter 4: X-Poly-Modules—Building Spacecraft at Lower Costs

4.1 Introduction

As a species we will have trouble returning to the Moon or traveling further into deep space if we spend $7B+ every time we launch a rocket. The Plan_A average cost per mission for 10 Exploration Missions is $7B per launch (see Chapter 2). Chapter 3 has shown we can lower the costs of heavy payload launches by delivering usable mass into LEO for $1000 per pound. This Chapter 4 shows how to lower the construction costs of spacecraft, for LEO, LEO to LLO transport vehicles, on-orbit and Lunar Surface habitats. It is a **Plan_B Precept** that combining lower launch costs and lower spacecraft and habitat costs will enable Plan_B to afford to land on the Moon. **Plan_B lands on the moon within the budgeted funds already allocated for Plan_A to only orbit the Moon**.

A nominated exemplar of a thinking differently to lower spacecraft vehicle costs is to consider and emulate the concepts behind ubiquitous storage containers (CONEX). Ubiquitous CONEX boxes are the backbone for most modern world-wide shipping (Chart 26). The CONEX box emerged during the Korean War and it transported and stored supplies during the Korean and Vietnam wars. Malcom McLean reinvented the CONEX box (see Wikipedia) to form the standard Intermodal shipping container (often called an ISO box, after ISO 668) used by container shipping companies today. The now-obvious breakthrough concept for common containers is for everyone to agree to the standards for the outer/inner diameter and essential structural components of the storage containers. These well-defined ISO Standards enable multiple world-wide manufacturing companies to compete and build the containers.

Chart 26: ISO 668 CONEX Containers

Plan B to the Moon
Version 1.01
© Stephen W. Long

Any standard container-ship vessel can move the storage containers to and from any port in the world. Beyond just deep-sea-crossing vessels, there now exists a planet-wide end-to-end inter-modal system of transport of shipping containers. Cargo from any factory or warehouse in the world can fill the containers and the boxes get loaded onto a truck or train and transported to a port. The boxes are stacked on ships at the port like children's building blocks to cross oceans. At the destination port the containers move from ship to truck to rail to truck. Sometimes the boxes get delivered to the end-user doorstep. Our interconnected global commerce system relies upon this premise of common shipping and storage containers. Who would have forecast 60 years ago that a simple, standardized metal box would transform global commerce?

Let's consider some simple math for standard 20 foot and 40-foot CONEX containers. The average cost of a new container is $1000 to $4000 depending on size and features, or a figure of merit of $100 per linear foot. A 20-foot Steel CONEX (called a dry container) weighs 5072 pounds, with interior dimensions of 19.4 by 7.5 by 7.7 (1,220 cubic feet). Simple math says $2440 / 1220 cubic feet = $2 per cubic foot of usable payload space. Simple math says $2440 / 6,100 pounds = $0.40 per pound. As a math cross check, rounding up to $1.00 per pound, a 20-foot steel CONEX would cost ~$5000 to build.

A 20-foot refrigerated container uses stainless steel, aluminum and insulation between inner and outer walls in their construction. A refrigerated CONEX weighs 6,800 pounds, including the weight and volume of refrigeration equipment. It can carry a payload of 60,400 pounds, provides 1000 cubic feet for payload, and costs about $10,000. The calculated weight is 6.8 pounds per cubic foot of payload. The calculated cost is $1.50 per pound, or $10 per cubic foot. We rounded up these CONEX calculations as quick calculation figures of merit.

Plan_B asks the reader to consider a future, hypothetical standard space container–the Space-Polyhedron-Module (SPM)–that provides 3,888 cubic feet of usable on-orbit habitat space. If we used refrigerated CONEX box Earth surface construction standards, such a container would cost $38,880; weigh 26,500 pounds, or $1.50 per pound. Using FHR at $1000 per pound, it would cost $26.5 Million to put this refrigerated CONEX into LEO.

Earthbound steel storage/shipping containers (CONEX ISO Boxes) are not suitable for space travel. But the raw materials used to build refrigerated CONEX boxes are the same stuff as found in spacecraft–aluminum, stainless steel, and high technology insulation. Note that a CONEX metal box built to support a payload mass of 60,400 pounds under one G load is strong stuff. The CONEX Box is an exemplar of how a single invention can change and revolutionize the world. In this framework it is inspiring to see stacks of ISO Boxes, one on top of the other, filling container ships with a cornucopia of brightly colored standard cargo boxes (Chart 26).

Plan_B asks, would it be possible to design and build a suitable analog–a Space equivalent (on-orbit or on-Lunar Surface) to the ISO CONEX Box? Would such a space ISO box enable significant long-term cost savings, and manufacturing efficiencies? Would such boxes enable high quality/quantity builds of repeatable, well-engineered and tested components? Yes: Plan_B proposes the concept of Space-Polyhedron-Modules (SPMs) for the return to the Moon and deep space. SPMs are much lower cost, extensible, safe, and will become ubiquitous space structures and vessels, orbital and lunar surface modules. Space-Poly-Modules are themselves built from collections of smaller components–building blocks. The building blocks are named X-Poly-Modules–where X is the variable component sub-part changed or assembled based on mission need. Said another way, traditional historic spacecraft are hand-built, they are exquisite machines with resulting exquisite price points. If spacecraft (or at least their building blocks) were ~mass produced, the unit costs will /must come down. It is a **Plan_B Precept** to design, build and send common building blocks into orbit and assemble larger spacecraft and habitats out of the common building blocks.

One of the most astonishing things Plan_B Team members learned in our aerospace careers was the premise of calculating program costs based on the weight of an aircraft (spacecraft). It boggles the mind, but one can approximate the final delivered cost of a complicated craft just by weighing it. Chart 27 shows a plot of the costs of fighter aircraft per pound (Source: Time Magazine, 2011). The newest aircraft, the F-35, costs ~$5000 a pound. So, a 30,000-pound (dry weight) F-35C fighter aircraft maximum cost should be less than $150 Million per copy. The publicly reported manufacturing cost for the F-35C is $107.7 Million, which does NOT include billions of dollars of development costs. If we include the billions of dollars of development costs, one can easily believe the F-35 fighter jets cost $150 Million per unit to build.

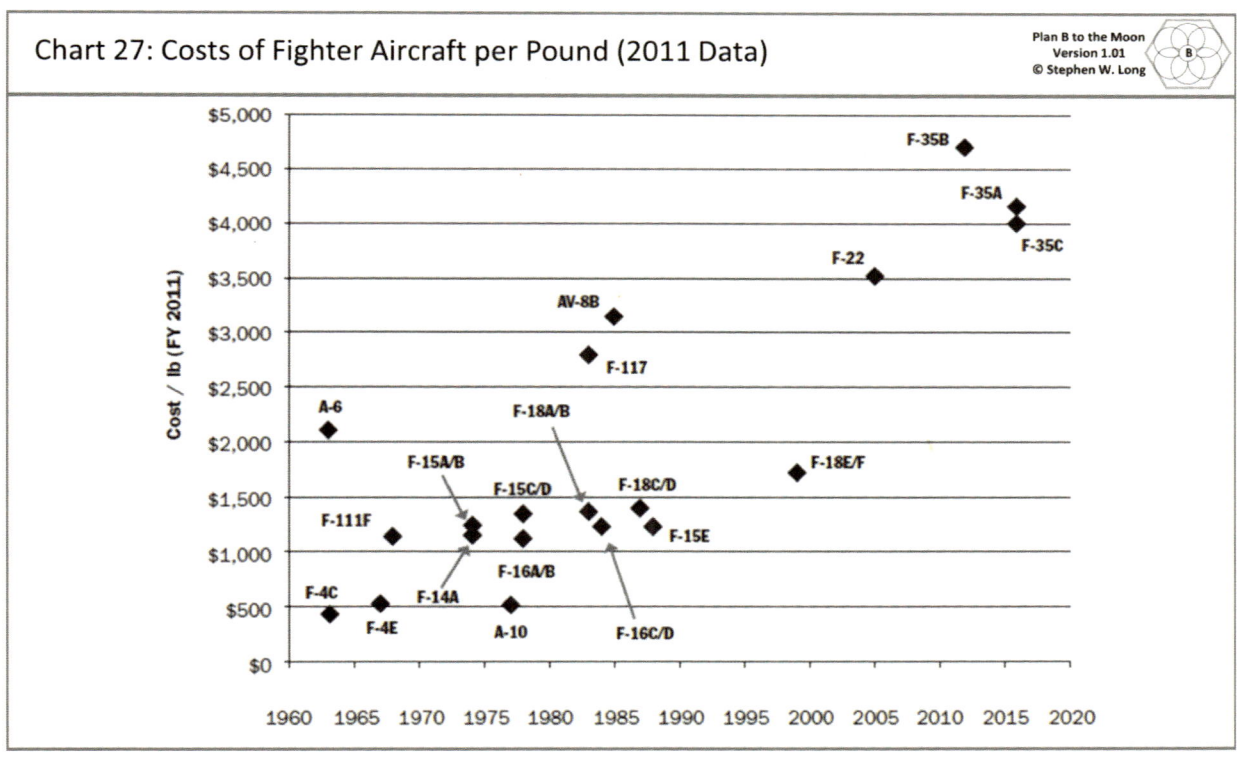

Plan B to the Moon: A How-to Guide to Return to and Colonize the Lunar Surface

It is a **Plan_B Precept** to calculate Plan_B costs using $5000 a pound for cargo mass spacecraft and $10,000 a pound for human rated spacecraft. There is something worthy to ponder at the $5000 per pound price point. The SpaceX FH Rocket launch vehicle costs $1000 to launch a pound. It costs $5000 a pound for aerospace contractors to build the thing sitting on top of the rocket. Plan_B suggests it is possible to trade more *thing* mass for lower production costs. Consider doubling the weight (a net of $2000 a pound to launch), but lower the build-it costs to say $2000 a pound. This would save 20% of total program costs. A hypothetical 3888 cubic foot space CONEX container weighs 26,000 pounds, and it would cost $26 Million to put this CONEX on-orbit. Compared to historical spacecraft launch costs (100's of millions up to billions), $26 Million is cheap. At $2000 per pound, a low-cost space-CONEX would cost $53 Million to build. At $5000 per pound, a space-CONEX would cost $130 Million to build. We offer another **Plan_B Precept**: For Aerospace Vehicle Components that cost as much as $5000 per pound to build, cost savings trade spaces are possible. It is possible to reduce vehicle component construction costs through the use of non-exquisite, heavier materials.

As described in Section 4.3, the SPACE-X standard launch faring weighs 4200 pounds. Earlier mission profiles sent this structure 90% of the distance and speed necessary to reach Earth orbit. If you double that weight (8400 pounds) and round up to 10,000 pounds, it should cost less than $20 Million to build a space-CONEX derived from launch faring materials. At $5000 per pound, it would cost $50 Million to build the space-CONEX. SPACE-X reports it costs $6 Million to build their existing launch fairings, which calculates to $1500 a pound. It would be reasonable and defendable for Plan_B to use a cost planning figure of merit of $2000 a pound for bulk volume SPM based spacecraft. However, to forestall potential naysayers, Plan_B cost calculations in Chapter 10 use $5000 a pound for cargo mass and $10,000 a pound for human rated spacecraft. Even at these high costs per pound, the overall Plan_B costs are still less than the Plan_A costs.

Space-Poly-Modules leverage these lower-cost precepts. SPMs are mass-produced components, with significantly lower costs per unit of manufacturing. SPMs are a much better financial deal than exquisitely manufactured, one-of-a-kind traditional space components. One demonstrable example for the professional cost estimators to pursue to lower overall Plan_B program costs would be to examine space-rated solar cells and panels. Consumer grade solar panels only cost $2 a watt. You can install 504 kW of consumer solar panels in a dirt field for $1 Million dollars. Compared to commercial rooftop solar panels, today's exquisite space-rated solar panels are expensive ($875 per watt), where 504 kW of space-rated panels cost $441 Million dollars. Consumer panels will not survive the rigors of launch and space environments. But is there a middle ground? Can we manufacture thousands of space-survivable solar panels that cost less than $1000 a pound?

It is a **Plan_B Precept** to lower the costs of spacecraft by assembling on-orbit structures from Standardized, High Quantity, Lower-Cost Modules: X-Polyhedron-Modules ™ (XPMs).

Enabled by XPMs, Plan_B divides the payload masses required to reach the Moon (and beyond) into smaller Payloads that each fit within the FHR 130,000 lb. launch mass limit. Plan_B offers

XPMs as a breakthrough approach (a set of tools) to sub-divide on-orbit vehicles to fit within the FHR 130,000 lb. launch limits.

This premise of using mass manufacturing / production to lower spacecraft costs is a fundamental **Plan_B Precept:** Build-to costs of $5000 a pound. As later seen in the MISSION MATRIX, using costs-per-manufacturing-pound provides a useful program budgeting tool. If you don't agree with the premise of $5000 per pound for SPMs, you only have to change a single variable to change the Plan_B estimated summary costs. Choose any cost number you can defend. Plan_B Chapter 10 calculates the Nominal program costs and also calculates variations on the total costs by varying the allowed cost per pound of spacecraft manufacturing.

4.2 X-Polyhedron-Modules

Chapter 4.2 defines and describes X Polyhedron (or Poly) Modules, where assembled combinations of different X's become Space-Poly-Modules (SPMs) (Chart 28). Plan_B also defines families and grouped modules (multiple, coupled SPMS) to create larger, configurable spacecraft, space habitats, moon habitats, Mars habitats and more (Chart 29). The key design advantage—breakthrough approach—is that XPMs allow launch payloads to be sub-divided to fit within the FH Rocket launch mass capabilities.

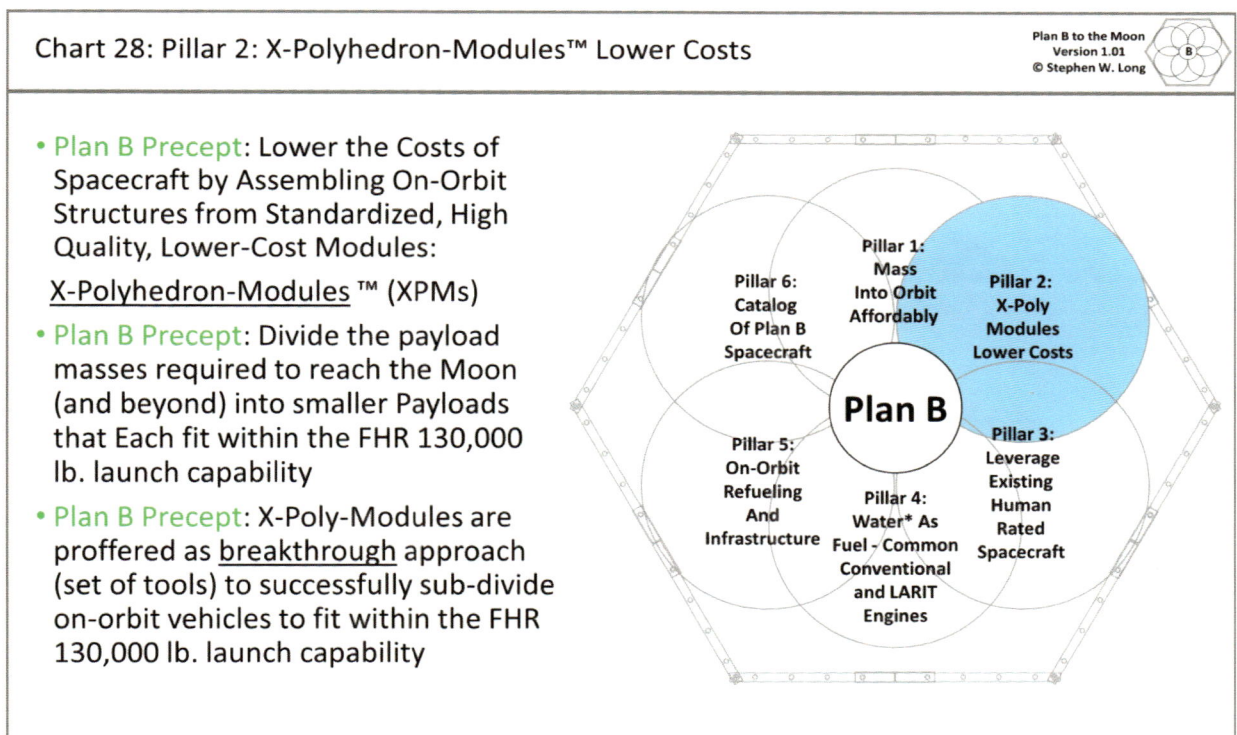

Chart 28: Pillar 2: X-Polyhedron-Modules™ Lower Costs

There is a subtle but profound consequence of establishing and enabling standardized SPMs and derivative spacecraft and habitats. Standardized modules allow many people / organizations / governments / international partnerships to build compatible and interoperable modules, and through competition decrease costs and promote innovation. In much the same way that today's ubiquitous ISO CONEX Boxes revolutionized world-wide shipping, SPMs will revolutionize space structures and exploration.

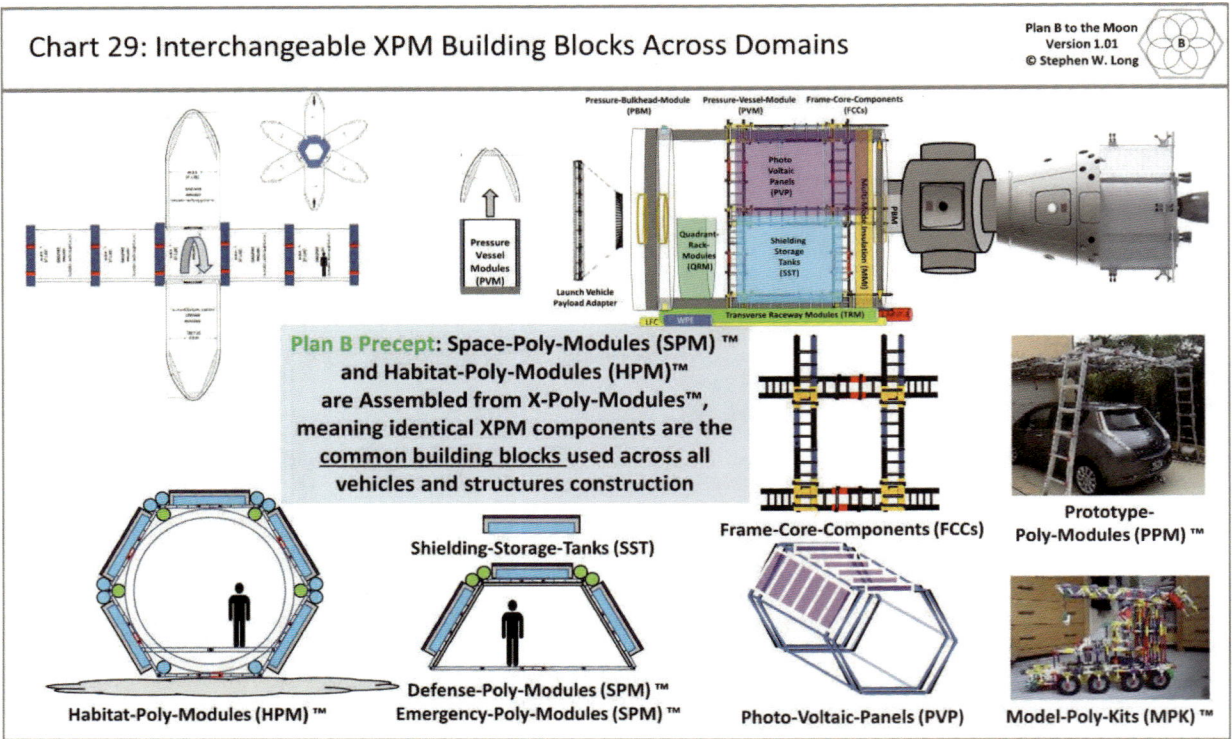

4.3 Pressure-Vessel-Modules

Launch fairings protect ~delicate satellites during the stressful launch through the Earth's atmosphere. When launch vehicles have limited spare thrust to reach LEO, a common practice is to eject (throw away) the Launch Fairing before reaching LEO to save weight. The SpaceX FHR fairing weighs 6700 pounds, and earlier launch profiles carry the fairing ~90% of the way to low earth orbit. Chart 30 asks, what if we could leverage this launch-weight investment and derive mission value?

It is a **Plan_B Precept** to replace the traditional throw away fairing with a revised (new) Fairing shell structure design, where the new shell has a dual purpose:
- The new shell will continue to provide protection during launch, serving as the universal cargo carrier for Plan_B mission payloads.
- After reaching orbit and after cargo (mission payload) extraction from the shell (Chart 31), Plan_B proposes to use the shell as a spacecraft structural component. We call this shell the Pressure-Vessel-Module (PVM). The PVM shells are assembled on-orbit with other X-Polyhedron Modules, creating Plan_B Spacecraft or Habitats on-orbit or on the Moon.

We note that the Apollo Command Module was its own Pressure-Vessel-Module. It protected the contents (humans) during launch and served as the outer structure of the craft on-orbit.

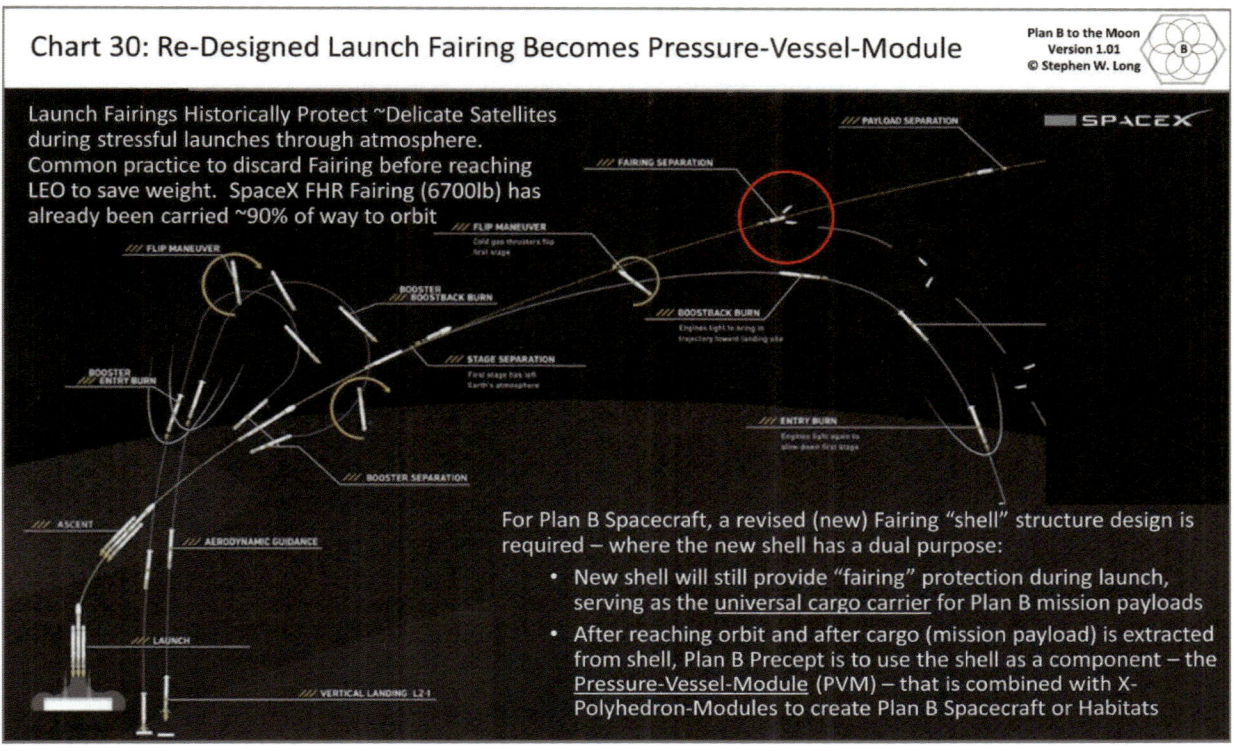

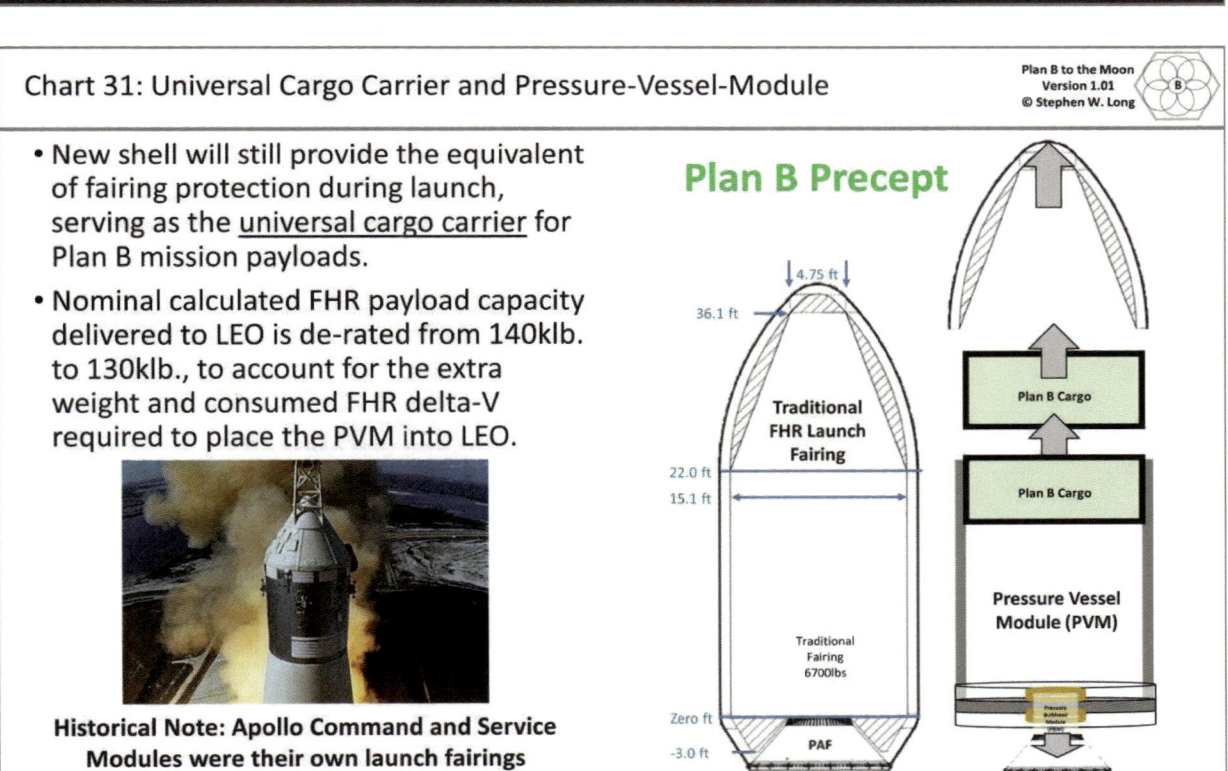

Chart 32 depicts more details of the dual roles of the PVM. Ground construction seals one end. After orbit delivery, the opposite end opens to space and the cargo contents slide out. This leaves an empty shell that once closed (sealed) is suitable to become a habitable Pressure-Vessel-Module on-orbit. Chart 32 includes a notional package of water storage and structural XPMs. What was once a cargo delivery shell is now a pressure-vessel suitable for space habitation.

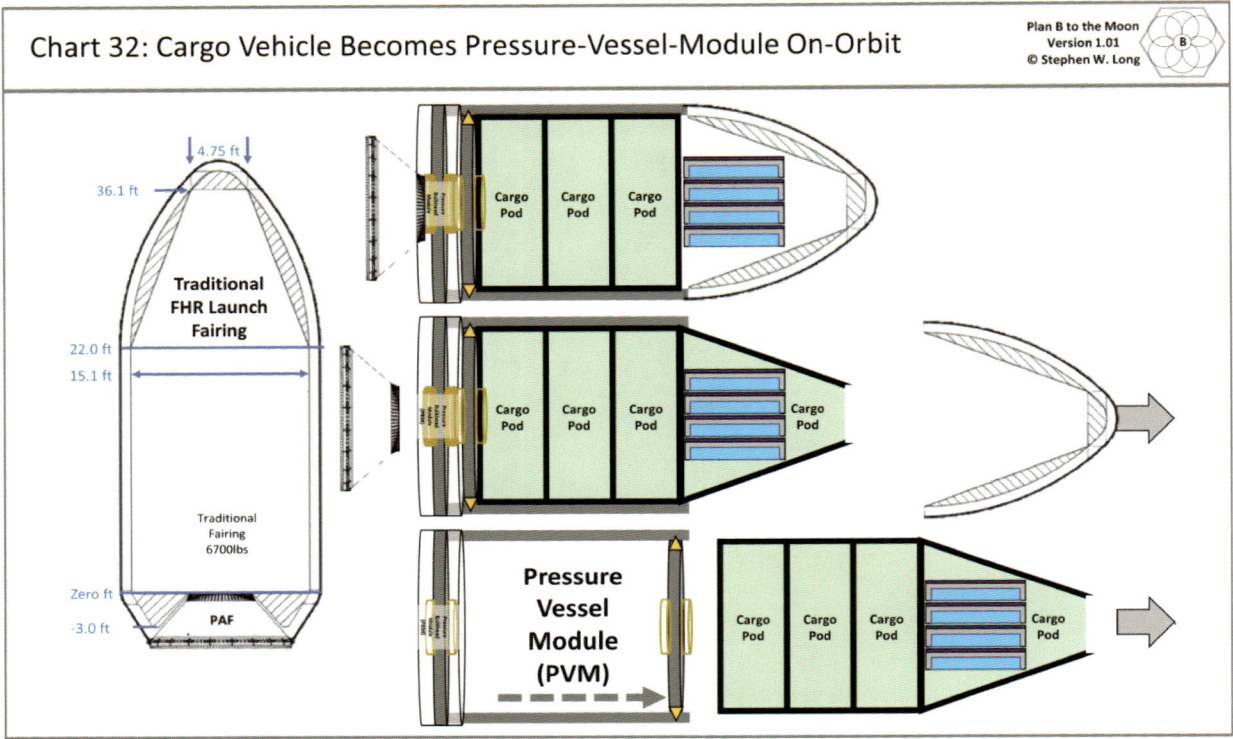

Chart 33 depicts the notional premise of a PVM surrounded by XPM structural components. There is historical precedent for using re-purposed spacecraft components to become different spacecraft. NASA built the 1970s Skylab space station using a re-purposed Saturn V launch rocket shell. The entire premise of having a Pressure-Vessel-Module is for the module to contain the atmospheric pressure and become a living environment for the astronauts. Therefore, **Plan B spacecraft** designs do not allow penetrations of the pressure shell.

Once you open the end of the PVM to remove the cargo, how do you reseal the tube to become a pressure module? Chart 34 introduces PVM designs as an analog to a classic medication delivery syringe. The plunger starts in the closed position. As you withdraw the plunger, you create a sealed space behind the plunger. Plan_B offers the design concept to use a movable floor that extends forward from the rear bulkhead (the permanently sealed end). As this floor moves upwards via tracks/rails to the open end, it locks into place. For the PVM, the end caps incorporate crew air locks and all other plumbing/electric/atmosphere penetration fixtures.

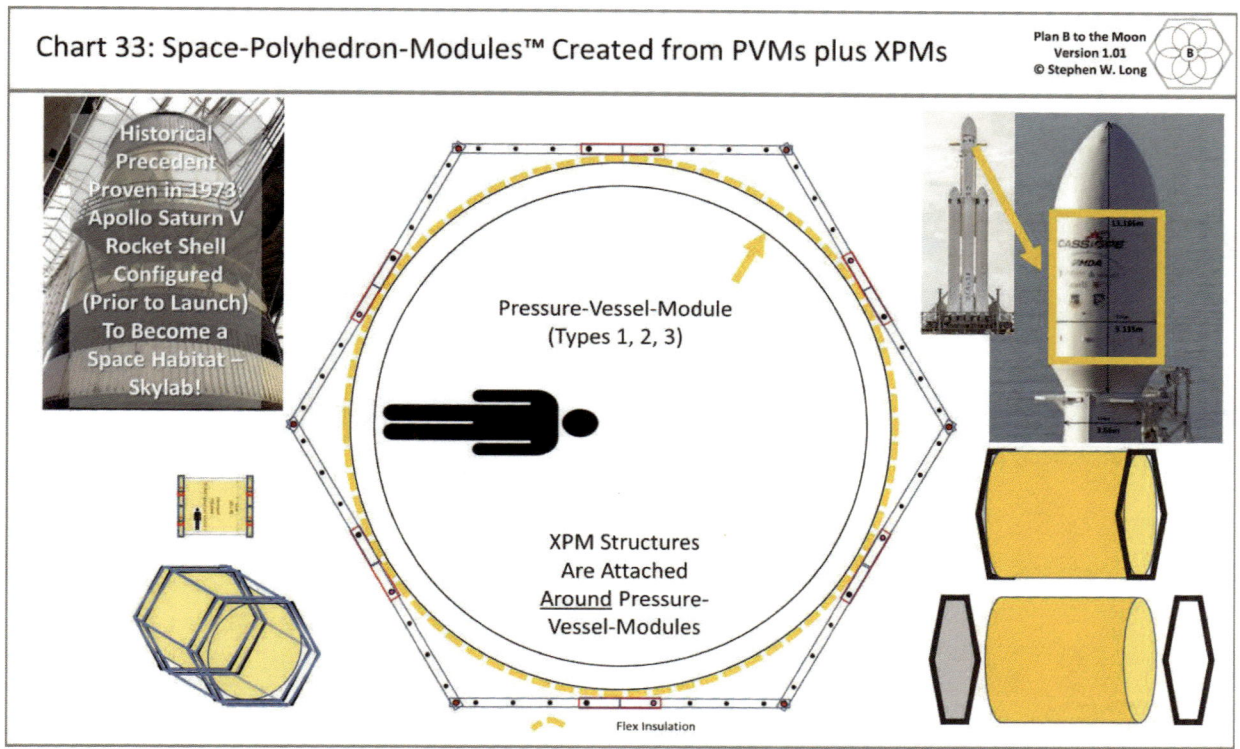

Chart 33: Space-Polyhedron-Modules™ Created from PVMs plus XPMs

Chart 35 illustrates the **Plan_B Precept** of three types of Pressure-Vessel-Modules. PVM Type 1 delivers a full cargo-only load. The extracted cargo elements are the building blocks for spacecraft. The remaining pressure vessel becomes the structural tube used in SPM construction. Pressure-Vessel-Module Type I is a cargo carrier configuration that will yield 3888 cubic feet of pressurized space.

PVM Type II is a pre-built Pressure-Vessel-Module (no removable cargo). Habitat modules will need kitchen facilities, plumbing facilities, research components, equipment racks—elements that are typically pre-built on Earth because of their complexity. This is comparable to the existing practice of launching a habitat module for the International Space Station. Plan_A also proposes building a habitat module for the lunar deep space orbital gateway. Plan_B proposes because of its common XPM/PVM components, these Plan_B Type II PVMs will be less expensive to build than ISS or Orion habitat modules.

Type III PVMs will support future mission (Plan_B Version 2.0 / 3.0) plans. For Type III designs, the tail opens instead of the head. Once closed with a syringe-like sliding floor, the Type III design provides an additional thousand cubic feet of habitable volume. Retaining a PVM nose cone adds 1000 feet to join the 3888 cubic feet of the PVM Type I shell. This design gives you a cargo carrier and large pressure vessel used for the gravity modules explained in Chapter 8.

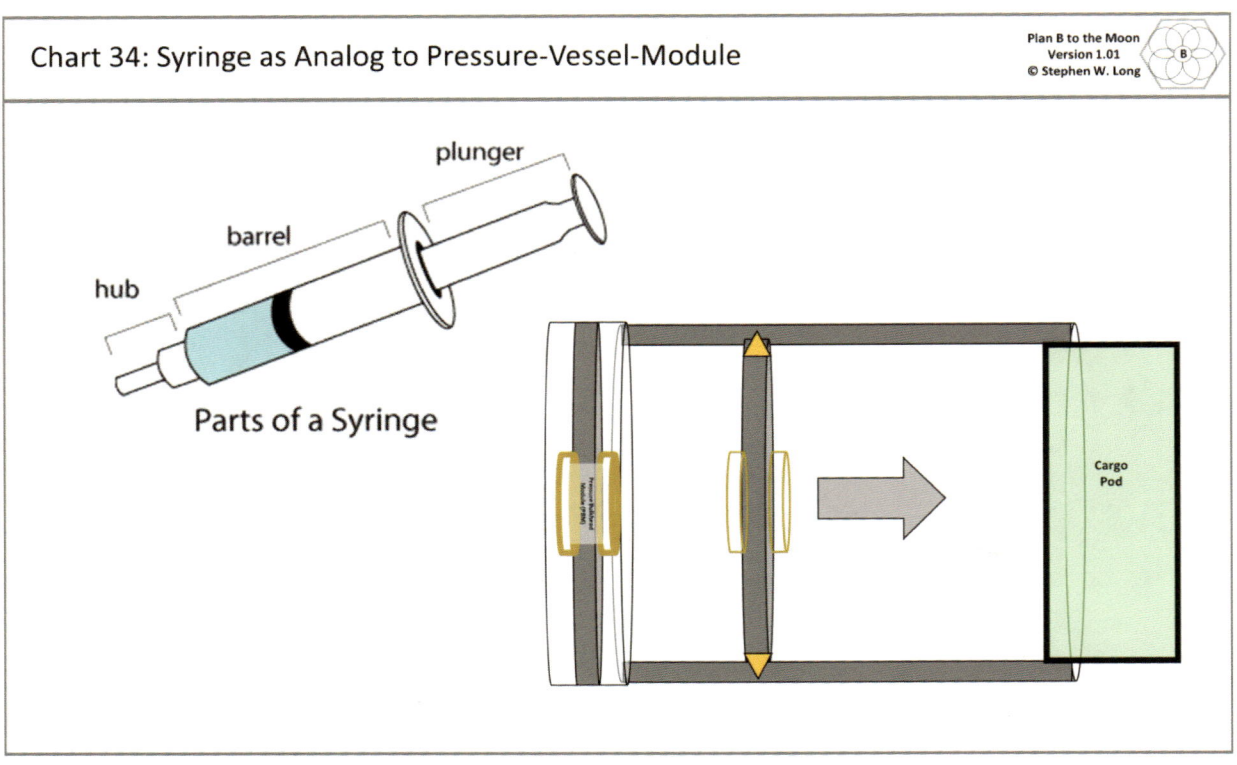

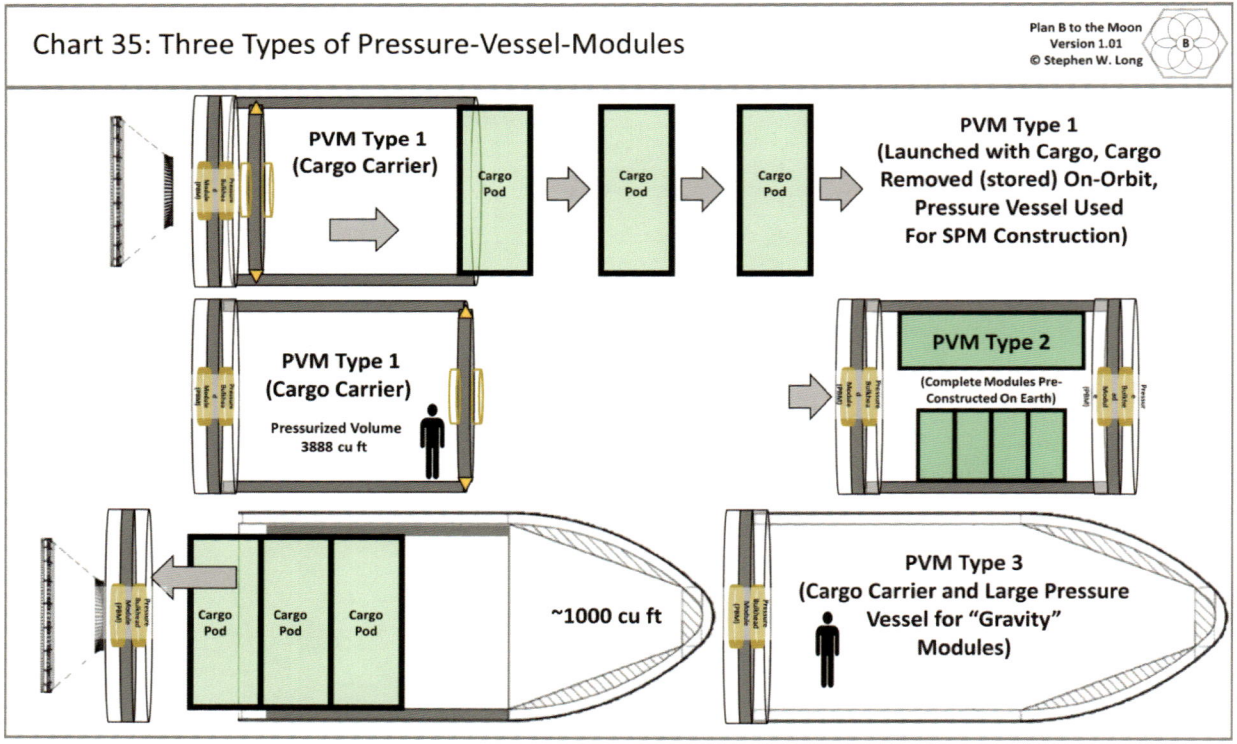

4.4 Frame-Core-Components

The simple PVM shell by itself is not a complete spacecraft. Frame-Core-Components (FCCs) wrap around the basic PVM shells, starting the process of converting simple pressure vessels into functioning spacecraft called Space-Poly-Modules (SPMs). SPMs = PVMs + XPM 1 + XPM 2 + XPM 3, adding modules as necessary to assemble a complete spacecraft.

When FCCs are fitted and locked together (Charts 36 and 37), you can create large longitudinal and transverse structures within the conforming structural strength of the aluminum or other components. It is a **Plan_B Precept** that FCC combinations will produce large space structures.

FCC elements include Cores, Longitudinal and Transverse modules. These modules as shown in Chart 38 allow extension and placement anywhere within the lockable structure. Chart 38 shows a sample of Frame Core Component sections assembled together into a basic girder structure. Combinations of interchangeable standard components—assembled on-obit, on the Moon, on Mars will form spacecraft, habitats, and support defense/emergency response needs.

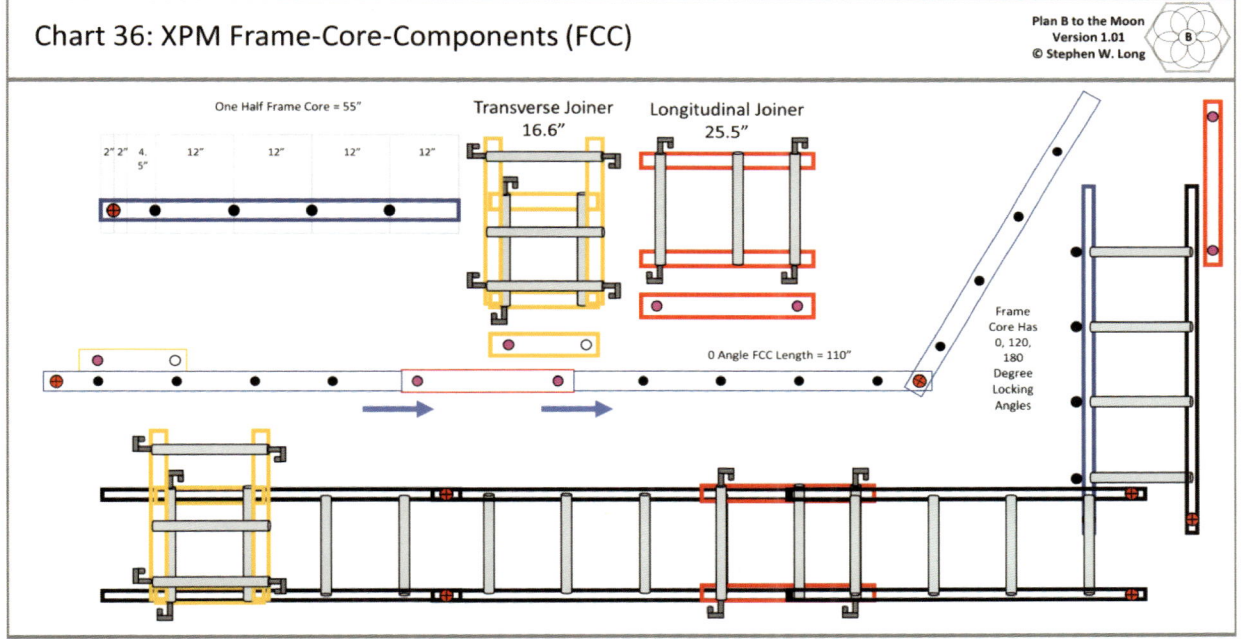

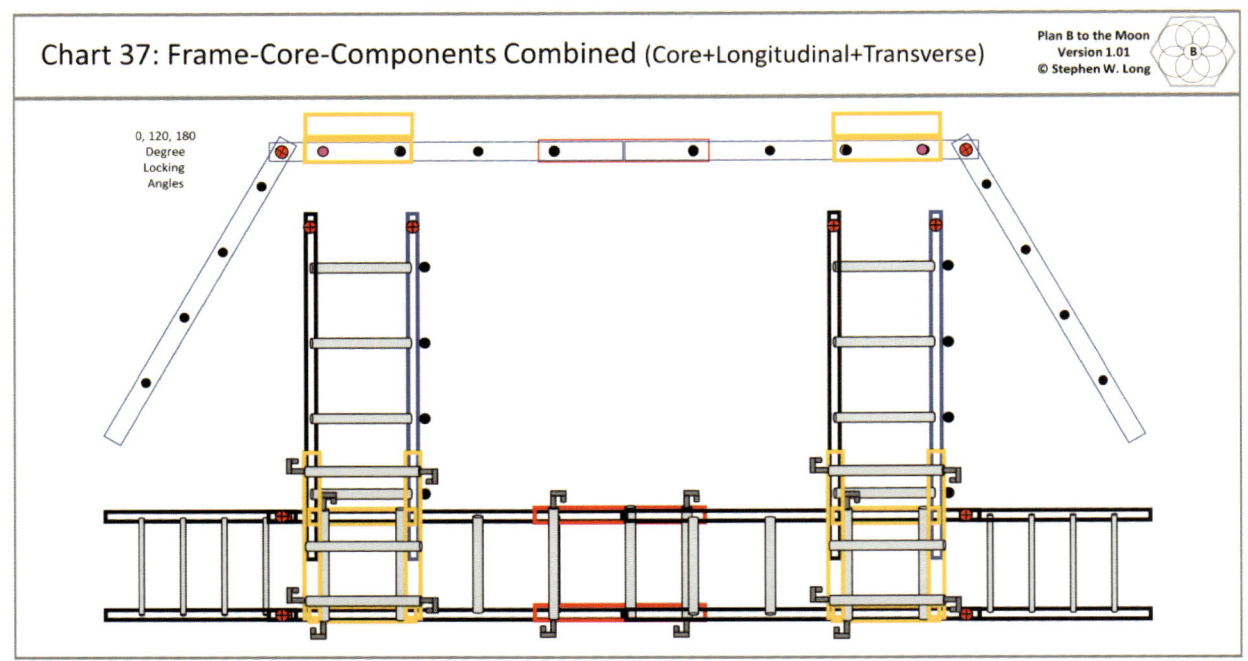

Chart 37: Frame-Core-Components Combined (Core+Longitudinal+Transverse)

0, 120, 180
Degree
Locking
Angles

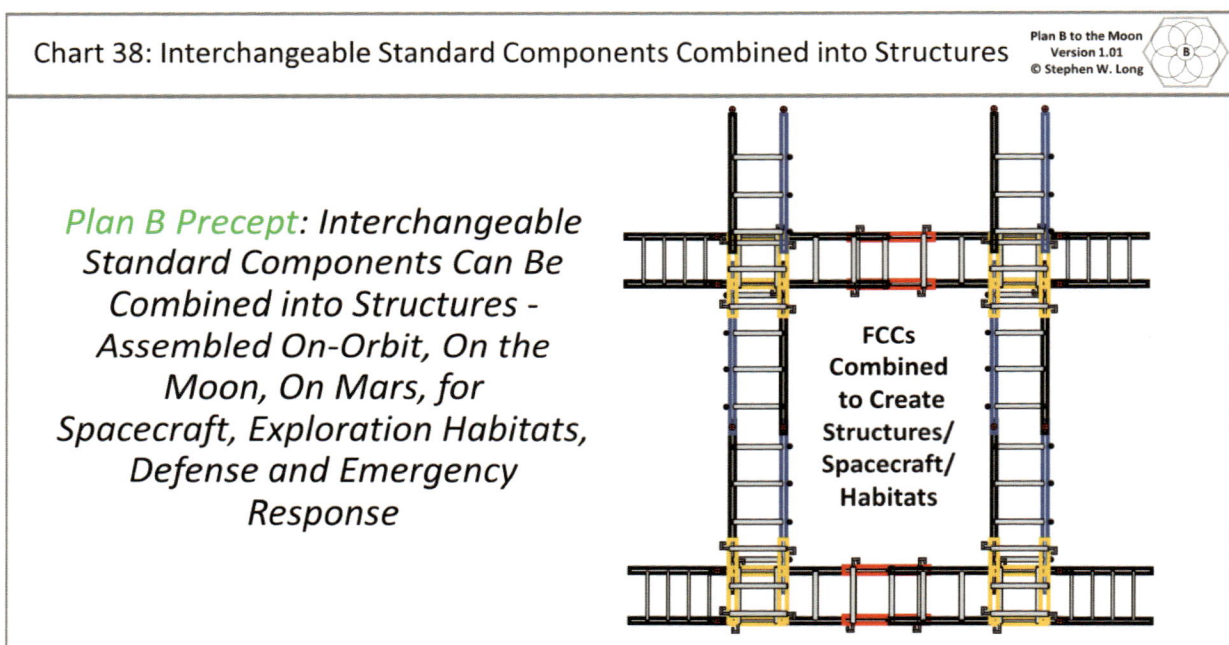

Chart 38: Interchangeable Standard Components Combined into Structures

Plan B Precept: Interchangeable Standard Components Can Be Combined into Structures - Assembled On-Orbit, On the Moon, On Mars, for Spacecraft, Exploration Habitats, Defense and Emergency Response

FCCs Combined to Create Structures/ Spacecraft/ Habitats

4.5 XPM Hexagons

Chart 39 depicts FCC / XPM structural components formed into hexagons. FCCs fold for transport. When six FCCs are unfolded (expanded), locked at 120° increments and joined, a structural hexagon emerges. The folding and locking connectors support rapid assembly of FCC based structures.

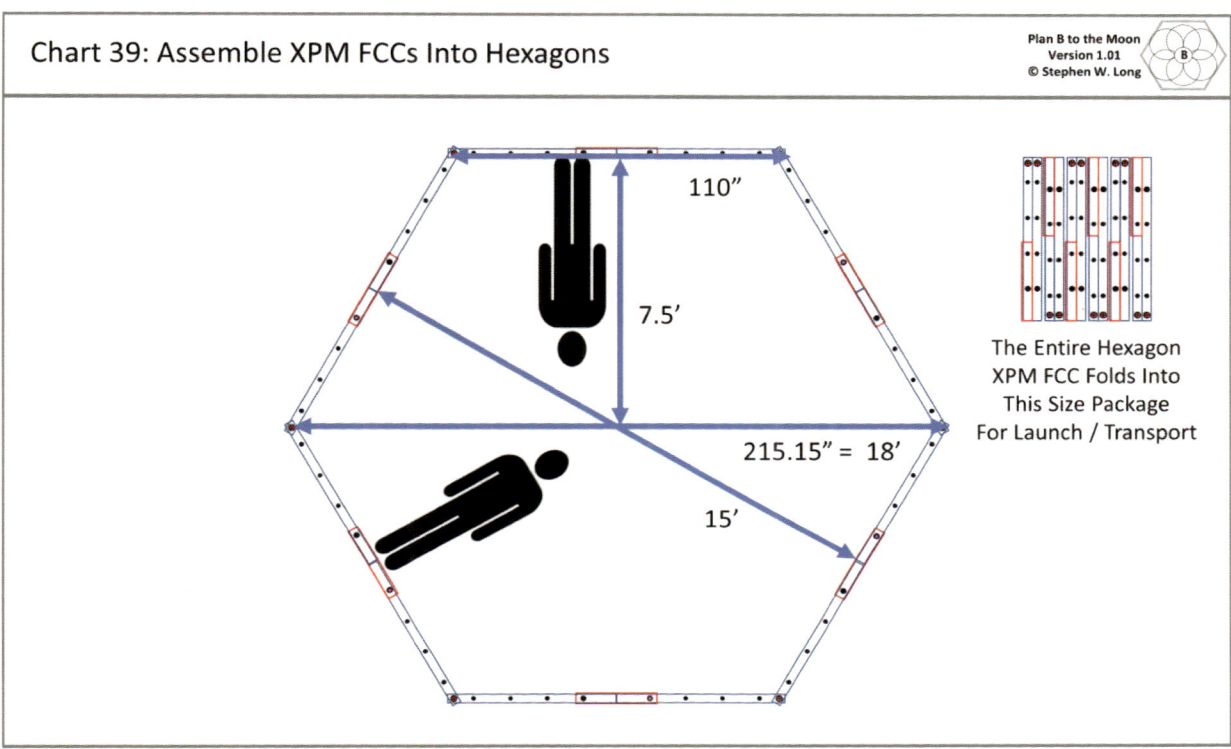

For a designed linear element length of 110 inches (plus transverse and longitudinal modules), the resulting sturdy hexagon shape has a 15-foot Hex flat-to-flat measurement. The Hex has an 18-foot corner to corner measurement. This yields an estimated center height at 7.5 feet. Chart 39 depicts a six-foot human shape to portray the large structure size enabled by such a hexagon structure.

By design, these Hexagons precisely fit the circle outer diameter of the standard Pressure-Vessel-Modules. This engineering/design choice enables the PVMs to stay round (a shell). Round shapes provide the best pressure containment while the Hexagon FCCs provide straight-line/girder elements to support other XPMs—without compromising the pressure structure of the PVM. The Plan_B Team had a true eureka moment when we first assembled the prototype FCC hexagons. The first prototype yielded a sturdy structure that fit within inches of the published outer diameter of the SpaceX launch fairings. You don't have to believe in such things, but the Plan_B team took the eureka moment as a sign that we were on the right track. It was proof we could design and eventually construct large spacecraft from foldable (compact) building blocks (XPMs) that would fit around existing space faring designs (PVMs).

4.6 Space-Poly-Modules

Combinations of standard components–X-Poly-Modules plus Pressure-Vessel-Modules create Space-Poly-Modules (SPMs). Chart 40 depicts an FCC frame wrapping around a Pressure-Vessel-Module (Type 1, Type 2 or Type 3), assembled into a hexagon shape.

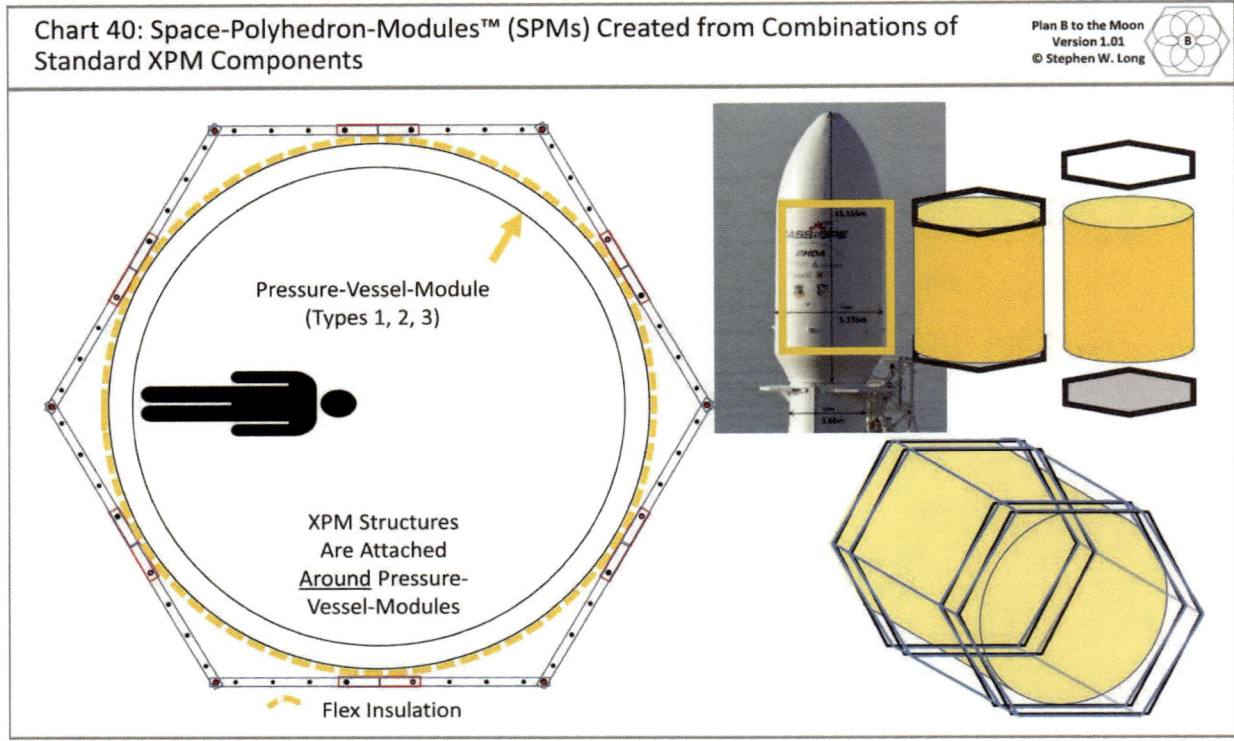

Chart 40: Space-Polyhedron-Modules™ (SPMs) Created from Combinations of Standard XPM Components

Plan B to the Moon
Version 1.01
© Stephen W. Long

Pressure-Vessel-Module
(Types 1, 2, 3)

XPM Structures
Are Attached
Around Pressure-
Vessel-Modules

Flex Insulation

The FCC structures attach OUTSIDE (around) the Pressure-Vessel-Modules. There is higher risk (and expense) when spacecraft walls have penetrations. Any penetration is a potential source of leaks. Never penetrate the PVM is a **Plan_B Precept**. This important cost savings feature keeps PVM shells intact. There are no penetrations or only limited end cap penetrations. The straight FCCs also simplify structural designs where all subsequent XPM structures attach to the straight structural girder FCCs around the outer skin of the curved pressure vessels.

Chart 41 depicts the Skylab space-station exhibit at the National Air and Space Museum (NASM). Skylab used a flight-ready, re-purposed Saturn V rocket shell for the pressure vessel of the space-station. In these images you can see the large Saturn V outer shell. The Skylab shell features habitat components built into the inner surface walls. The NASM exhibit has 1973 video footage available at the exhibit showing astronauts floating inside a weightless environment inside the large Skylab open area (Chart 41 right half). This giant / open area provided the astronauts with a very large space volume in which to live.

Chart 42 shows creepy manikins sitting in the Skylab shell. The interior of the Skylab appears to be utilitarian and Spartan. The Plan_B team posits that over months, or years of continuous use any such utilitarian environment would become very wearisome to long-term inhabitants. Plain beige walls and lack of creature comforts are not conducive to long duration mission human needs.

4.7 Space-Poly-Modules Built from X-Poly-Modules

Chart 43 further introduces concepts where combinations of XPM components assembled together on-orbit form human-habitat-capable Space-Polyhedron-Modules (SPMs). Yes, this is an alphabet soup of acronyms. Plan_B assembles SPM from FCCs plus PVMs plus PBMs plus MSMs plus TRMs plus SSTs plus MMIs plus PVPs plus LFCs plus QRMs plus future X options (such as X partridges in Y pear trees). Each of these XPM design concepts are further explained below.

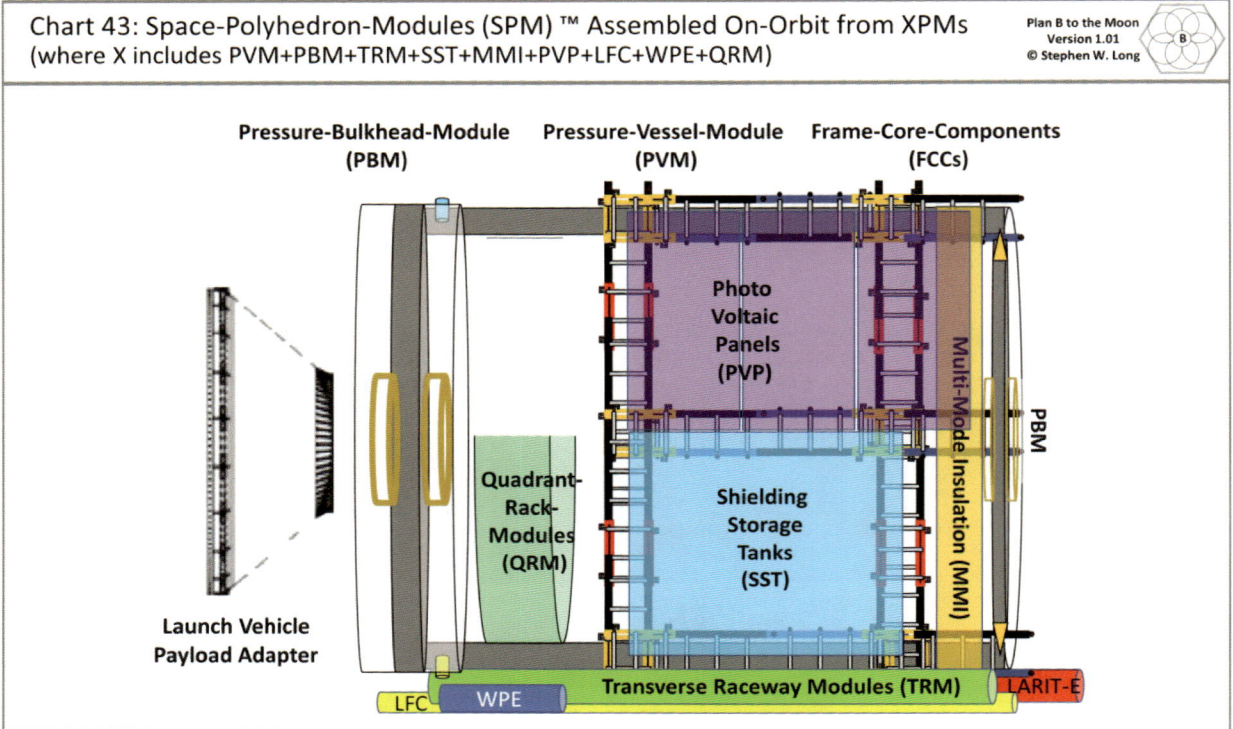

4.8 Pressure-Bulkhead-Modules

We started with Pressure-Vessel-Modules and added Frame-Core-Components. The next components are the Pressure-Bulkhead-Modules (PBMs). PBMs are toroidal (disc-shaped) structures with a pressure hatch (nominal 48 inches) built into the center to allow human access through the bulkhead (Chart 44). The key premise is that the PBM seals the end of a PVM, making the PVM airtight and therefore becomes suitable for use as a spacecraft human habitat.

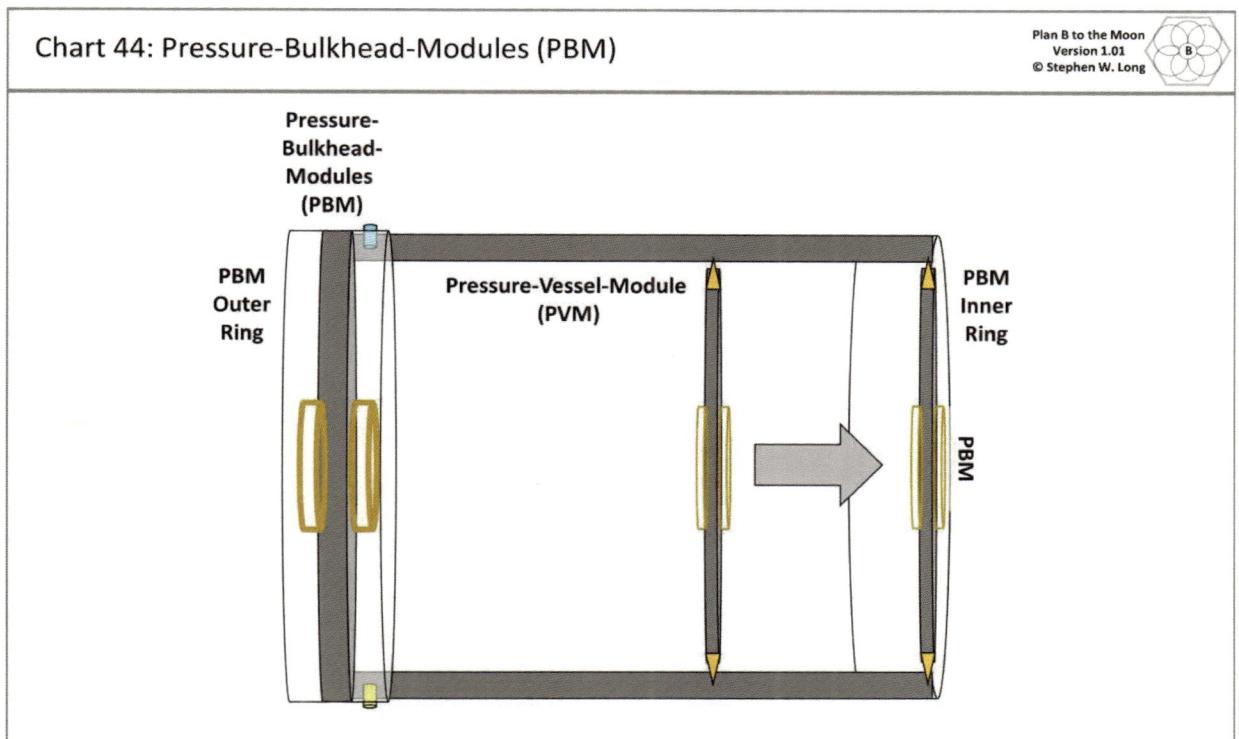

Plan B to the Moon
Version 1.01
© Stephen W. Long

Chart 44: Pressure-Bulkhead-Modules (PBM)

4.9 Transverse-Raceway-Modules

Every vessel or structure or modern building uses the concept of conduit. Conduit is an enclosing metal tube or pipe into which wires or fluids flow from one location to other locations. Chart 45 depicts a PVM with an external FCC hexagon structure and Transverse-Raceway-Modules (TRMs). TRMs are large conduit tubes that precisely fit between the 120-degree corner gaps of the FCCs and the surface of the PVM. The TRMs provide conduits supporting wiring, plumbing and other SPM system needs. The TRM engineering design premise keeps the PVM shell from being penetrated—only the Pressure-Bulkhead-Modules at the ends of each of the PVMs have penetrations. TRMs attach outside the PVM and run the length of the PVM.

Separate TRMs will support different system routing functions. One TRM would hold power and data cables. Another transverse raceway would route water piping. Another raceway moves fuel gases or fuel fluids. Note that the Plan_B Team designed the FCC hexagon corners to allow the transverse raceways to fit between the outer circumference area of the PVM and the FCC frame. Chart 46 depicts how wires etc. will enter the PVM via the bulkhead pressure modules at the ends of each of the PVM. Transfer raceways will have pre-engineered and constructed pressure bulkhead plugs, allowing rapid assembly on-orbit (think ball and socket pressure fittings).

Chart 45: Transverse-Raceway-Modules (TRM)

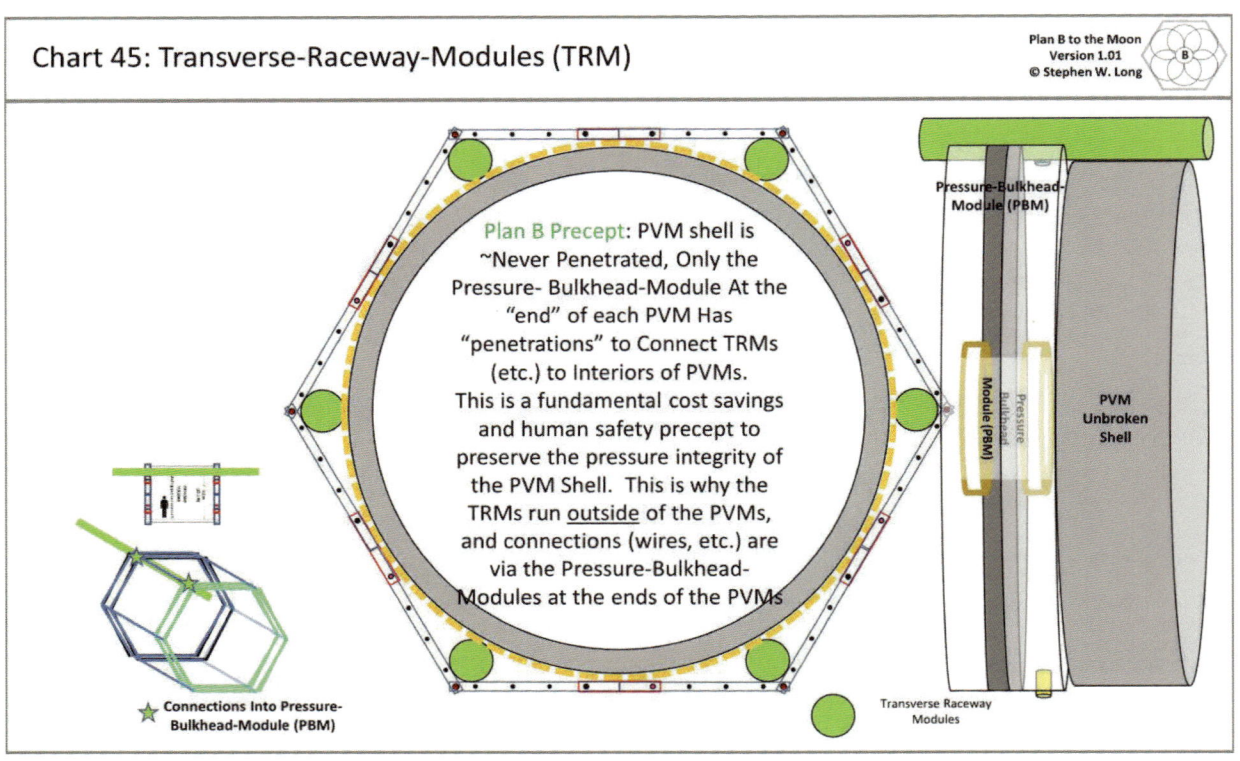

Plan B Precept: PVM shell is ~Never Penetrated, Only the Pressure- Bulkhead-Module At the "end" of each PVM Has "penetrations" to Connect TRMs (etc.) to Interiors of PVMs. This is a fundamental cost savings and human safety precept to preserve the pressure integrity of the PVM Shell. This is why the TRMs run outside of the PVMs, and connections (wires, etc.) are via the Pressure-Bulkhead-Modules at the ends of the PVMs

Pressure-Bulkhead-Module (PBM)

Pressure Bulkhead Module (PBM)

PVM Unbroken Shell

Connections Into Pressure-Bulkhead-Module (PBM)

Transverse Raceway Modules

Chart 46: Transverse Raceway Connection Details

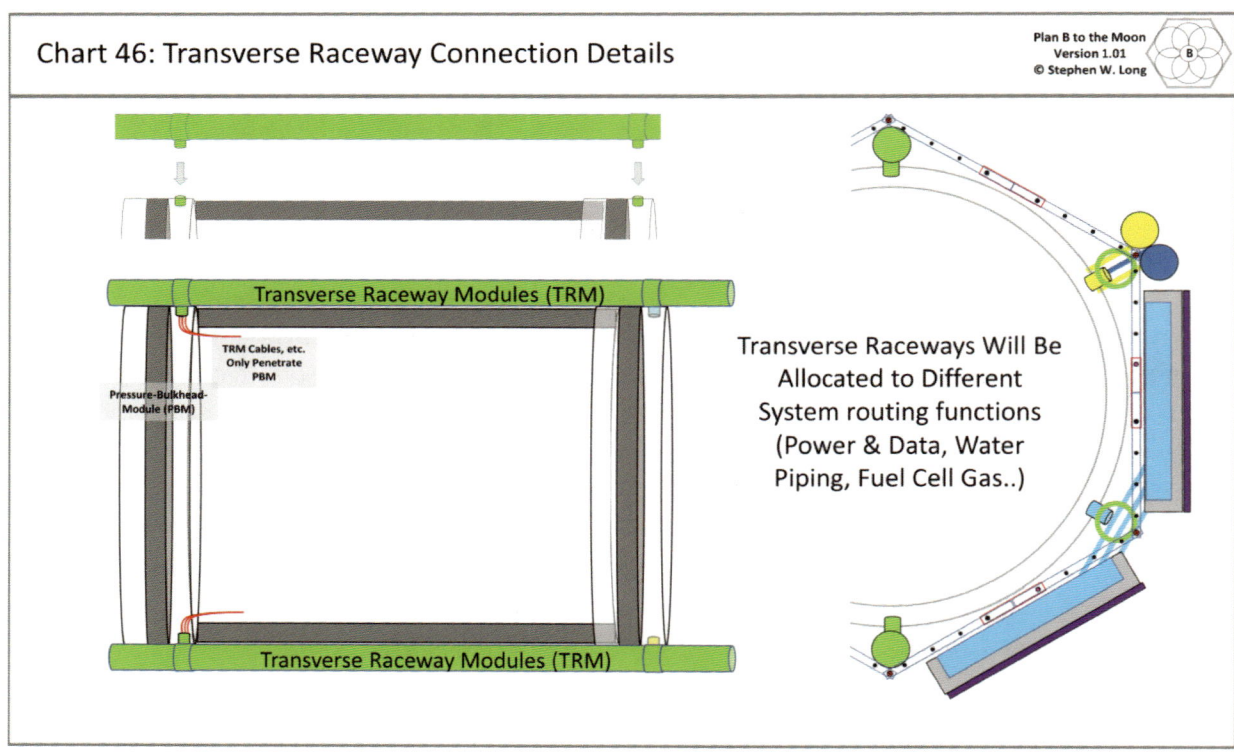

Transverse Raceway Modules (TRM)

TRM Cables, etc. Only Penetrate PBM

Pressure-Bulkhead-Module (PBM)

Transverse Raceway Modules (TRM)

Transverse Raceways Will Be Allocated to Different System routing functions (Power & Data, Water Piping, Fuel Cell Gas..)

4.10   Shielding-Storage-Tanks

Plan_B Team believes future spacecraft must provide extensive protective shielding for human inhabitants. Humans in space receive much higher exposures of cosmic rays than humans living on the earth's surface. Humans living on-orbit also face the constant danger of micrometeorite strikes which could penetrate the pressure vessel and other spacecraft components.

Serendipitously, six months into Plan_B Team research, NASA published a study (see Chart 47) that warns of the significant danger to human gut bacteria on long duration space voyages.

Chart 47: Thick Water Shield May Be Critical Spacecraft Design Choice

Plan B to the Moon
Version 1.01
© Stephen W. Long

A NASA-funded study says long trips in space could destroy astronauts' stomachs and cause cancer

By Lauren Kent, CNN
Updated 3:42 PM ET, Tue October 2, 2018

Plan B Included the Concept of Extensive Crew Shielding 6 months before this radiation report released – Plan B Ahead of its time...

The NASA spacecraft Orion is designed to allow astronauts to journey to asteroids and Mars.

(CNN) — Astronauts may not be able to stomach long voyages into space -- literally speaking.

A new NASA-funded study reveals that exposure to space radiation on long trips, like a voyage to Mars, could permanently harm astronauts' intestines and lead to stomach and colon cancer.

The study, published by cancer researchers at Georgetown University Medical Center, used mice to test exposure to heavy ion radiation, which mimics the galactic cosmic radiation found in deep space. If that sounds complicated, essentially researchers compared "space" radiation to X-ray radiation and found its effects to be much more dangerous.

It is a Plan_B Precept to use water (liquid or ice) as shielding to protect human space habitats. The PVM and its associated FCC frame design holds external Shielding-Storage-Tanks (SST), as shown in Chart 48. These shielding tanks get launched as cargo elements (empty tanks) and are on-orbit attached to the external FCC frame. Crews use TRM fittings to fill the tanks with conditioned water (or other suitable fluids). Crews fill the tanks with liquids, and the fluids freeze into hard ice, with ice as a nominal storage state. These shield tanks will provide a minimum of 12 inches of effective shielding to the interior habitants of the PVM. As depicted, these outer water shells do not completely cover the PVM. But the limited surface area / outer areas of the PVMs not protected by the tanks have protection from other structures—namely the TRMs and other raceways. Please note that the Chapter 10 MISSION MATRIX calculations account for the mass of the tanks and the mass of the shielding water.

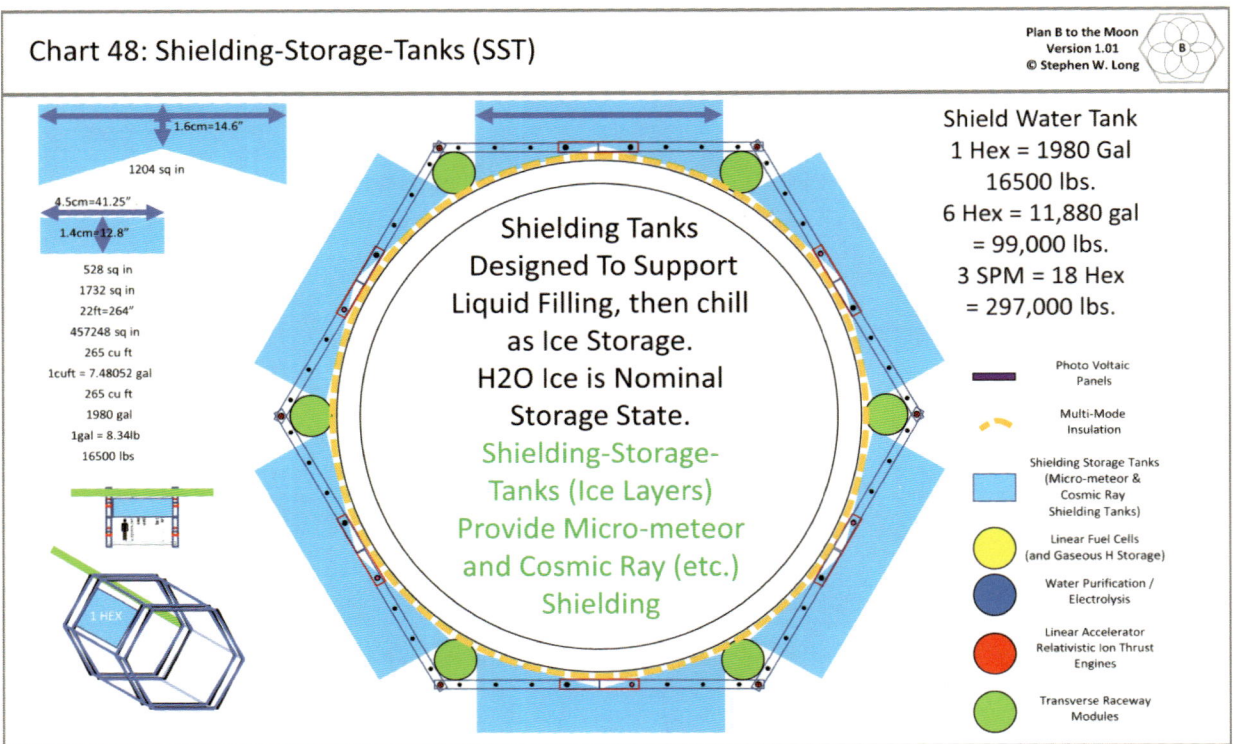

Shielding Storage Tank design details include:

- One *Hex* is one flat surface of the hexagon. One Hex shell holds 1980 gallons of water, which is 16,500 pounds of launch mass. Six Hex around the PVM hold 11,880 gallons of water or 99,000 pounds of launch mass. Engineering studies will determine the optimum depth of an enclosed ice shield to protect the crew from cosmic rays and micrometeorites. Based on our Plan_B Team research, studies say 12 inches of liquid water will stop a 50-caliber rifle round. This stopping-power value informed the 12.8 inches thick design assumption of the SST at the thinnest portion of the water tank.

- Future trade studies may identify lower mass SPM shielding, requiring less weight (and saving money) than 99,000 pounds of water mass. But one benefit of using water as shielding is that this water can also serve as an emergency reservoir. Such a reservoir would provide emergency water for crew needs. It may be unpleasant to imagine, but such external storage tanks may best serve as organic effluent (septic) tanks of the SPMs. Such effluent tanks with the correct engineering recycling system could become compost tanks to create the soil and fertilizer for 0.1 G green-houses (see Chapter 8.21).

- It is likely that the Shielding-Storage-Tanks will be multi-compartment tanks. The outer half of the tanks get filled with water (or organic slurries), which freeze into ice. The inner tanks would hold potable water, which when mixed with O2 (forming highly oxygenated water) provide the crew with an emergency O2 supply. Perhaps the inner half of the tanks have heated water which moderates SPM interior temperatures (like an old-fashioned home radiator). The inner half of the SPM would use potable water heated to remain liquid. This design becomes a combination water storage tank and radiant heating system for the PVM.

4.11　Linear-Fuel-Cells and Water Electrolysis Modules

Chart 49 depicts concepts for Linear-Fuel-Cells (LFCs) and Water Electrolysis Modules (WEMs) tubes attached to the gaps in the FCCs / WSTs. The external-to-the-pressure-vessel LFC and WPE tubes do not waste interior PVM human habitable volume. The WEMs produce electrolyzed water (H and O2) as fuel for subsequent use by the LFCs to make fuel-cell electricity. Water bi-products of the LFCs are available for crew use. The Plan_B Team designed the Photo-Voltaic-Panels (see 4.13) to create surplus electricity. This electricity enables the WPEs to create and store Hydrogen and O2 during surplus electricity production periods. LFCs consume the H-O2 and make power during low PV power production (such as during orbital shading). PVPs and LFCs work together as a closed system. The water becomes H and O2, the fuel cells use the H and O2 to make electricity and water. PVP electric power electrolysis converts water back into Hydrogen and O2. LFC tubes store the H and O2 gases and associated system components. SSTs are good candidate fuel cell water storage tanks.

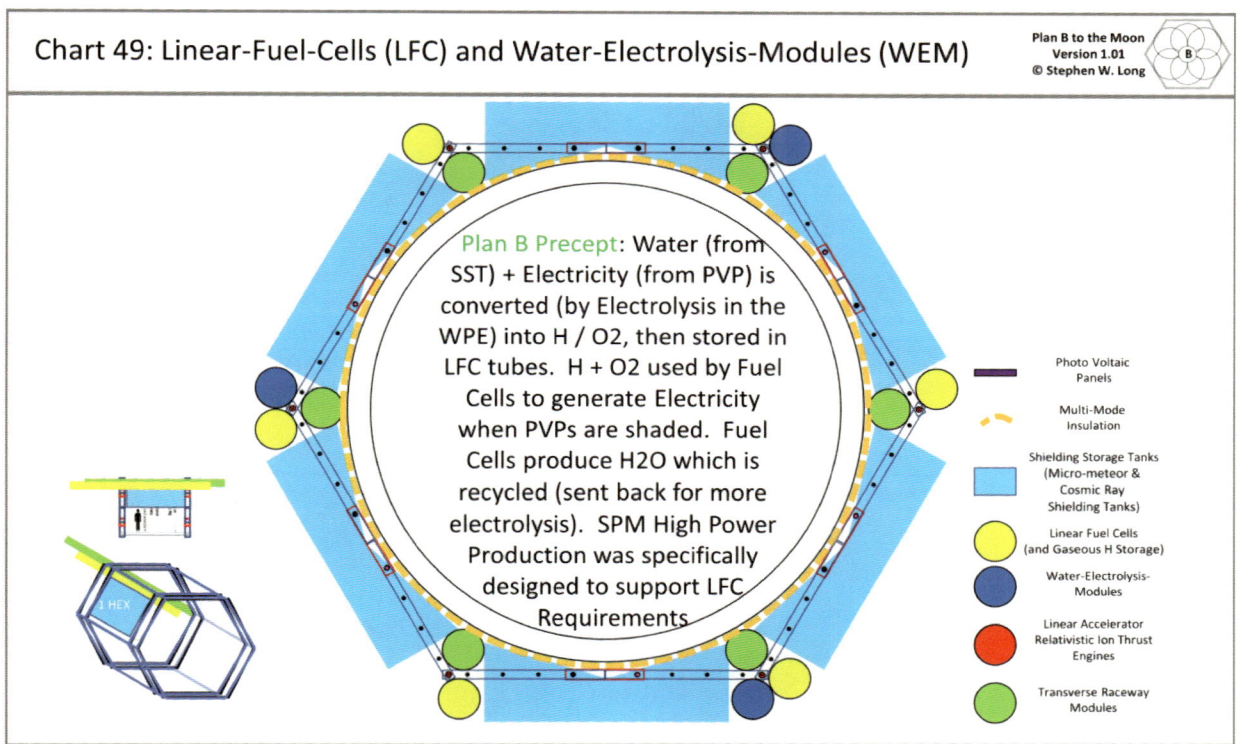

Chart 49: Linear-Fuel-Cells (LFC) and Water-Electrolysis-Modules (WEM)

One key premise behind creating standardized tube/same-shape XPMs such as the LFCs and WPEs is to provide opportunities for purposeful redundancy and global competition. Interoperable but different LFCs built by different organizations (competitors) will provide opportunities for side-by-side evaluations in real-space conditions. Newer and better designs will swap in and out over time, affording the SPMs and their constructed habitats opportunities to improve. This also enables full component tube swap-out maintenance opportunities.

4.12   LARIT Engines

Chart 50 introduces Linear Accelerator Relativistic Ion Thrust (LARIT) Engines. Plan_B Version 1.0 does not require LARIT Engines to support mission objectives. However, Plan_B reserves core SPM design space for the LARIT Engine tubes. Chart 50 shows how the LARIT tubes fit outside the PVM shell. This is the same concept as reserving space for Linear-Fuel-Cells and Water Purification and Electrolysis modules. The 120-degree equal spacing of the three LARIT modules around the PVM will enable a uniform linear accelerator thrust vector.

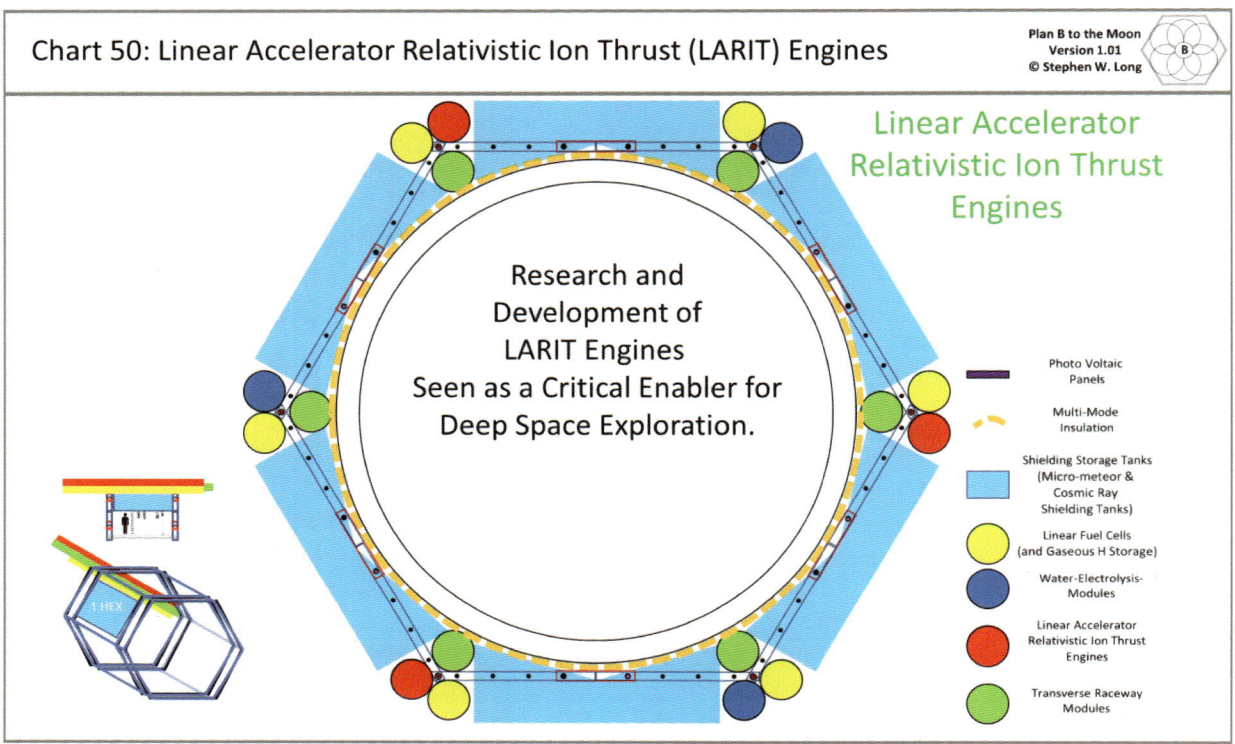

4.13   Photo-Voltaic-Panels

Chart 51 depicts Photo-Voltaic-Panels (PVPs) attached to the FCCs over-top of the SPM XPMs. The P-V Panels form the outermost shell of the entire SPM structure. They have a minor structural (rigid) function plus their actual photo-voltaic electricity producing layer, forming the outmost and replaceable shell layer around the inner SPM layers. With suitable engineering controls, such an outer P-V layer will also serve as the micrometeorite strike warning skin. Strike-caused interruptions of a specific cell's electricity output would inform a detection matrix showing the locations of micrometeorite strikes. The PVP design has designed structural strength (a rigid panel). This panel design allows the PVPs to be attached to FCCs. But the PVPs do not always need an associated Pressure-Vessel-Module base to form a space structure. This is the underlying design concept that enables assembly of the very large solar panel arrays for the Space-Fuel-Terminal (see Chapter 7.2). This dual-use of PVPs is another example and premise of building-block XPMs—different and useful on-orbit and on-planet structures emerge from the building blocks.

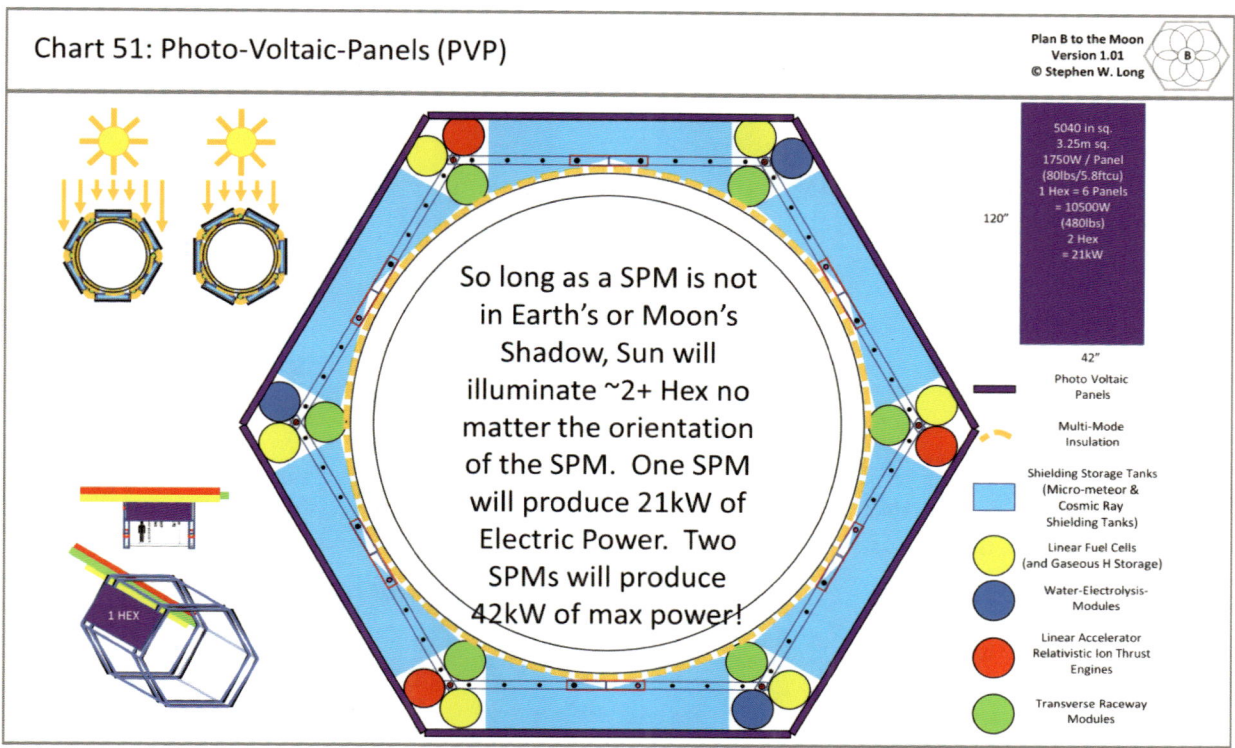

As depicted in the upper left of Chart 51, the sun will illuminate two hexagon face equivalents no matter the orientation of the SPM. Yes, there will be instances where the Earth's or Moon's shadows shade the SPM structure. Constant illumination of two hexagons will produce 21 kW of electric power (basis for calculations captured in the upper right of Chart 51). We note the Plan_A lunar orbital gateway design produces 40 kW of maximum electric power and those solar arrays also sail into the Moon's shadow. Therefore, Plan_B assumes two SPM's (covered with PVPs) will produce at least the same instantaneous solar electric power as the Plan_A lunar gateway array. Chart 51 depicts a designed panel size of 120 inches by 42 inches which yields 540 square inches or 3.25 square meters of surface area. This produces 1750 W of electric power per panel. The estimated weight is 80 pounds of launch mass and 5.8 cubic feet of launch volume. Each SPM hex face can hold six panels, calculating each hex will produce 10,500 W and weighs 480 pounds. Two Hex therefore produces 21 kW of power.

4.14   Quadrant-Rack-Modules

Chart 52 depicts how Space-Polyhedron-Module interiors are populated as necessary with Quadrant-Rack-Modules (QRMs). It is a common practice in commercial cargo and especially military cargo aircraft designs to build / create an empty outer shell. Pre-fabricated modules of seats or cargo racks or other equipment roll on or roll off the aircraft via a large hatch. In this approach, the empty shell aircraft (or spacecraft) assumes the mission flavor of the technology installed via the Quadrant-Rack-Modules. Chart 52 hints at one design for the rail system built into the inner wall of the PVM, allowing the slide-in insertion of QRM components.

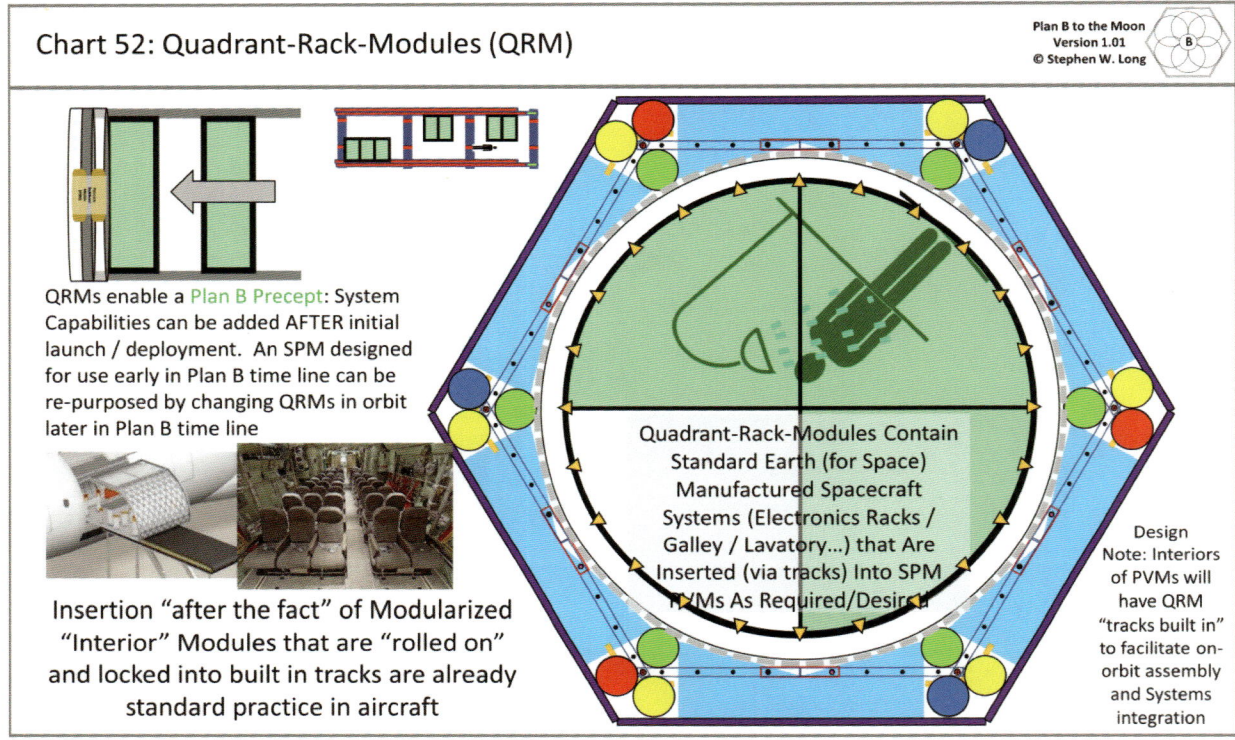

One farther future premise is that flexible SPM shells enable yet-to-be-determined capabilities. Early Plan_B mission time lines may only need to launch empty SPM shells and later new QRM launches populate the shells. This enabling design premise means we don't have to define every mission capability before we start initial mission launches. We can add future unplanned-for-today capabilities during later phases or launches. For instance, in the farther future, perhaps SPMs used as long-distance spacecraft structures get later re-configured to become distant planet on-orbit habitats.

It is a **Plan_B Precept** to not lock down future end-states early in the Program. The Plan_B Program Director often reminds the Plan_B Team that everyone is smarter at the END of a project development cycle than they were early in the project cycle. With gentleness, we note large PoR teams appear convinced they must lock down even minor PoR system component designs years if not decades before flight or launch. There is near religious fervor to stabilize the technical baseline, denying smaller self-contained components years of technological advancements while waiting for completion of larger more complex systems. In this smarter-at-the-end mindset, Plan_B asserts it is unnecessary to design all Plan_B XPM and QRMs components up-front. Instead, we need only define the interface specifications (form, fit, power, data) to enable future QRMs to mix with old SPM configurations based on future needs. This premise, as depicted in Chart 52, illustrates the **Plan_B Precept** that the outer shell of a spacecraft is just the shell. Instead of building a new spacecraft every time you need a new mission capability, Plan_B advocates re-using and re-using common shells. Just add, just slide in new interior modules for new missions. There is also great value to this approach for designing new mission components. We do NOT have to up-front design all the QRM components that will eventually populate the PVM. We do need to define the core specifications, such as locking down the PVM size and the TRM wiring and plumbing. Well-defined standards enable multi-vendor plug-and-play systems. Such plug-and-play QRM systems may include: pre-assembled electronics racks, a galley, a water closet, an additive manufacturing module, a bunkhouse; and any yet-to-be-thought-of system humans need to live and work in space.

An earlier Chapter introduced concepts where Type II PVMs launch as pre-configured, functional modules—a full habitat for instance. Several Mission Circles in Chapters 9 and 10 include these Type II PVMs. Such habitat modules are by far the most expensive habitat modules conceived of under Plan_B (SPM-9 in Mission Circle 323 costs $394 M).

It is a **Plan_B Precept** that SPMs add QRMs to add future system capabilities. Plan_B sends pre-built habitats (Type 2 PVMs) to space early in Plan_B phases, so the astronauts have a place to live in orbit. Later phases fill launches with tools and construction supplies. Once supply (cargo) deliveries reach the moon, those assembled-on-site components will support human colonization. As additive manufacturing (3D printing) technologies become ever more mainstream, future SPMs will arrive with tools-and-seeds (see Section 1.2). After delivery, the tools will create components to populate the PVM shell with on-site printed parts that become workshops, factories, green-houses or habitats.

4.15   Built in Extensibility

It is a **Plan_B Precept** that all XPMs are modular. They must also be separable. Inspired by the premise of the OSI Seven Layer stack for networking, each XPM connects to other components (up-stack or down-stack) by a defined interface specification. Each interface design supports re-use. Any given XPM must be changeable over time to include improvements without changing the other associated up-stack or down-stack XPMs.

XPMs borrow a premise from pods on military aircraft (Chart 53). The military adds new capabilities to an aircraft—such as a new targeting sensor—without changing the flying parts of the aircraft. New mission capabilities are easy to add by just uploading a new pod to a standard interface (attachment, power, data, cooling), which connect to the existing onboard systems.

For XPM designs, if at all practicable, the outer diameter sizes of the various tubes (TRM, SST, WPE, LARIT) should be interchangeable if not identical. Using the same materials (carbon fiber or stainless or aluminum tubes) will facilitate larger scale manufacturing cost reductions. For instance, prototype testing may determine that the Linear Fuel Cell (LFC) requires 30% more hydrogen gas storage volume. Such a storage volume shortfall could be a program crisis if learned late in the program development cycle. Instead, replacing an XPM tube becomes a straightforward engineering decision. It becomes an engineering trade to substitute LARIT engines with LFC tubes, without having to change the entire SPM / spacecraft design.

Chart 53: XPMs are Inherently Modular and Separable

Plan B to the Moon
Version 1.01
© Stephen W. Long

• **Plan B Precept**: All XPMs (XPM + PVM + PBM + TRM + SST + PVP + LFC + QRM) are inherently modular, they must also be separable:

  • Borrowing from the premise of the OSI Seven Layer Stack for Internet (etc.), each XPM connects to other components by way of a <u>defined interface</u>, so each interface must be carefully designed and maintained, but within any given XPM (literally inside of a M – in the shell or tube or...) it is allowed to change over time (improvements, etc.), which can be made without changing other XPMs.

  • This is the premise behind "pods" on military aircraft – where new capabilities can be added to an aircraft (such as a new targeting sensor) without changing the aircraft – the new "pod" is simply attached to a standard interface.

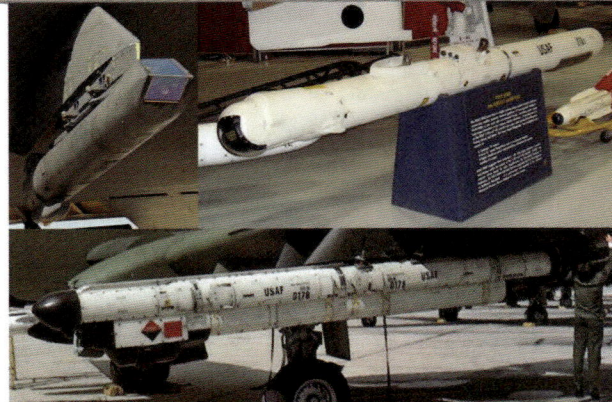

If practicable, the various "tube" X's (TRM, SST) should use comparable exterior tube materials (same diameter, etc.), to facilitate cost reductions

---

4.16 Combine SPMs to Build Larger Structures

Since by design SPMs are modular and extensible, they can also combine to form large space structures. Chart 54 depicts strings of SPMs connected end-to-end to form a larger structure. Chart 55 shows a preliminary conceptual view of 10 combined SPM's forming a large tubular V habitat structure to support human colonization on the Moon.

Chart 54: SPMs Connected to form Larger Structures and Vehicles

Plan B to the Moon
Version 1.01
© Stephen W. Long

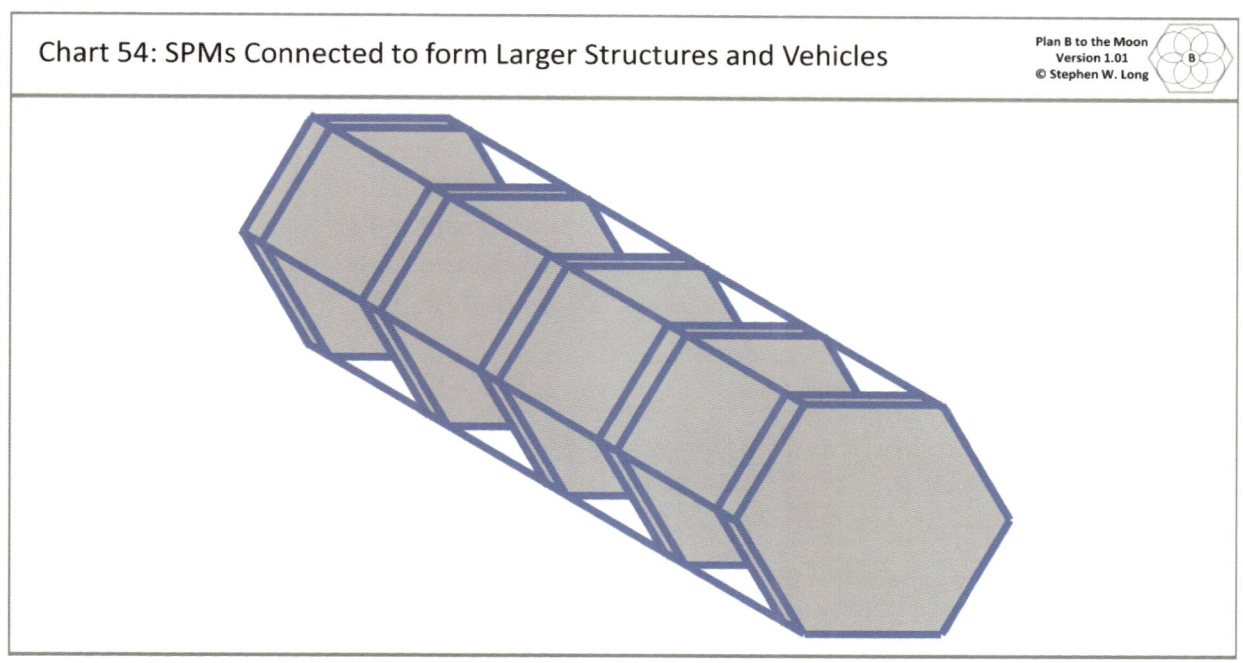

Chart 55: Plan B to the Moon: Lunar Surface Operations

Plan B to the Moon
Version 1.01
© Stephen W. Long

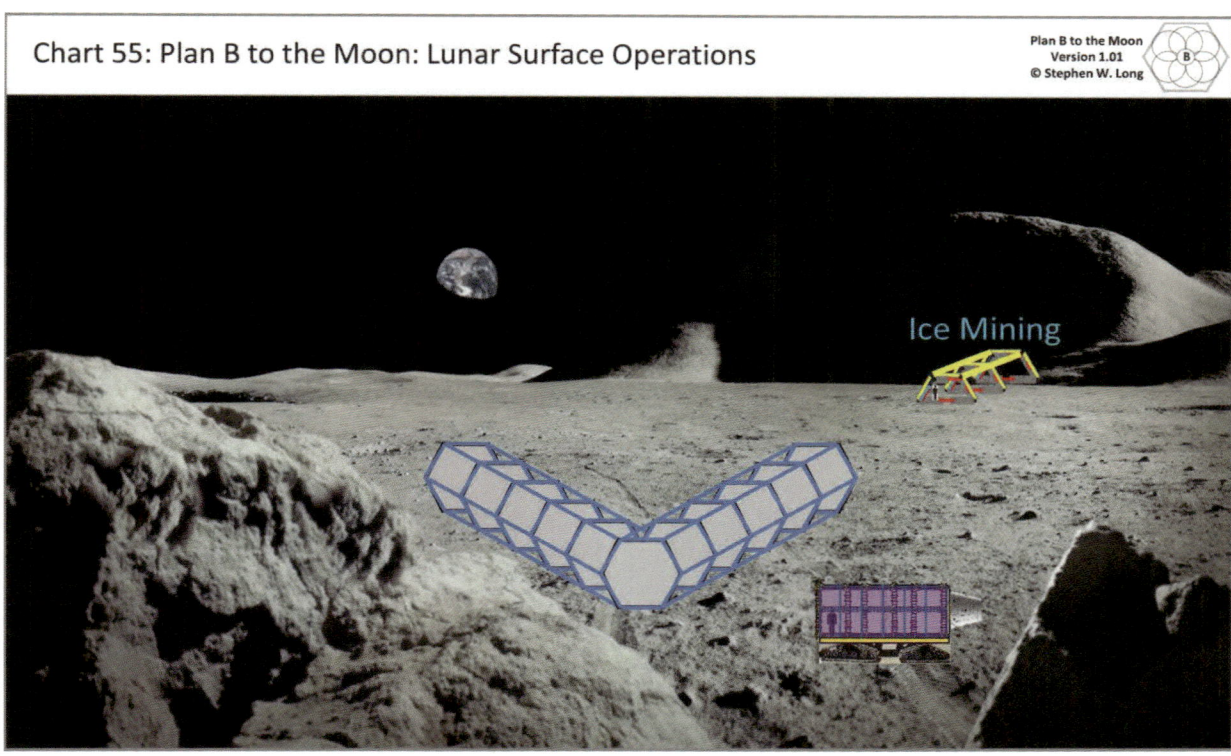

4.17   XPM Prototyping Early Experiments

XPMs came from humble origins.   The Plan_B PD needed to solve a different real-world residential structural engineering problem.   The PD discovered it is possible to combine low-cost interchangeable aluminum components–movable structures–to form large and inexpensive habitat shapes.   The act of experimenting with human-sized-building-blocks lead to a Plan_B Team eureka moment.   We realized that these low-cost interchangeable aluminum components were the structural analogs needed for low-cost structural components for Plan_B spacecraft and habitats.   These aluminum structural components became the inspiration and earliest prototypes of Frame-Core-Components (Section 4.4).

It is sometimes amazing (and or exasperating) to experience how different people react to different situations.   While we assembled the earliest FCCs, someone called the first structure the Moon Hut, and the name stuck.   Well-intentioned observers said the low-cost aluminum components would never survive space environments.   These good people could not see beyond the literal components.   They could not see that the low-cost components were only analogs of what appropriate material space structures could be.   This learning experience inspired new classes of Plan_B XPM components: Prototype-Poly-Modules (PPMs) and Model-Poly-Kits (MPKs).   Neither the PPMs nor the MPKs use space-rated materials—because they are not for space.   PPMs and MPKs are inspirational components, not space-flight components.

Chart 56 depicts some very early Prototype-Poly-Modules assembled by the Plan_B Team.   This picture documents a very early experiment using a version 0.1 XPM as a garage sun shelter (the moon hut).   Frame-Core-Components are first unfolded and locked into place.   Next add insulation panels, add patented design folding solar panels on top–providing shelter and electric power charging capabilities for an electric vehicle (a Nissan Leaf–her name is Fig).   We later used these same components in practical structural strength tests.   As a tropical storm threatened the homes of Plan_B Team members, we used the modules to climb up to second-story windows to hang hurricane shutters.   These early analog Frame-Core-Components exceeded all of our expectations.

Over the course of 2018, as Plan-B concepts matured, the Plan_B Team designed and built very much more realistic Frame-Core-Component modules.   Chart 57 depicts the first assembly of V0.5 FCCs into Space-Poly-Module sized hexagon shapes.   These hexagons precisely fit the outer dimensions of the SpaceX launch fairings flying into near LEO today.   Chart 57 also includes images of a prototype Solar Trailer.   This trailer validated concepts for folding solar panel designs and the PD used the trailer for many years as a portable (transportable) electric power source for electric vehicles.   The solar panels fold up into a compact space on a trailer and the electric car tows the trailer to no-electric-power-available destinations.   At the destination the solar panels are unfolded and they provide power to charge the electric car, extending the effective driving range of the electric car.

Chart 56: Prototype-Poly-Modules™ and Model-Poly-Kits™

Plan B to the Moon
Version 1.01
© Stephen W. Long

- **Plan B Precept**: Design and build working prototypes of XPMs to inform space vehicle design decisions and simultaneously develop new Earth-bound capabilities (such as Emergency Habitats) – at consumer price points

- **Plan B Precept**: Design and create "toy" kit versions of XPM components, to inspire millions of children and adults around the Earth to build their visions of XPM derived Spacecraft and Lunar/Martian Habitats and Colonies.

- **Plan B Precept**: Provide a MPK™ with every Emergency/Defense XPM-E / XPM-D system, where the users can built their structures as a "model" first, then assemble their actual habit module.

Gen V0.1 XPMs – Prototype Frames + solar panels, providing "shelter" and electric power charging for an Electric Vehicle

Gen V0.2 XPMs

KNEX™ toy components depicted here: Inspirational Examples for future Model-Poly-Kits (MPK) ™

The first XPM moon-hut structure inspired additional concepts based on the premise of XPM interchangeable parts. Chart 58 introduces multiple derivative concepts sharing heritage with XPMs and SPMs. Plan_B describes these as Prototype-Poly-Modules, Model-Poly-Kits, Defense-Poly-Modules, Emergency-Poly-Modules, Habitat-Poly-Modules & Space-Poly-Modules.

Chart 57: Proof of Concept XPM Systems: Hexagon Outer Frame for SPM

Plan B to the Moon
Version 1.01
© Stephen W. Long

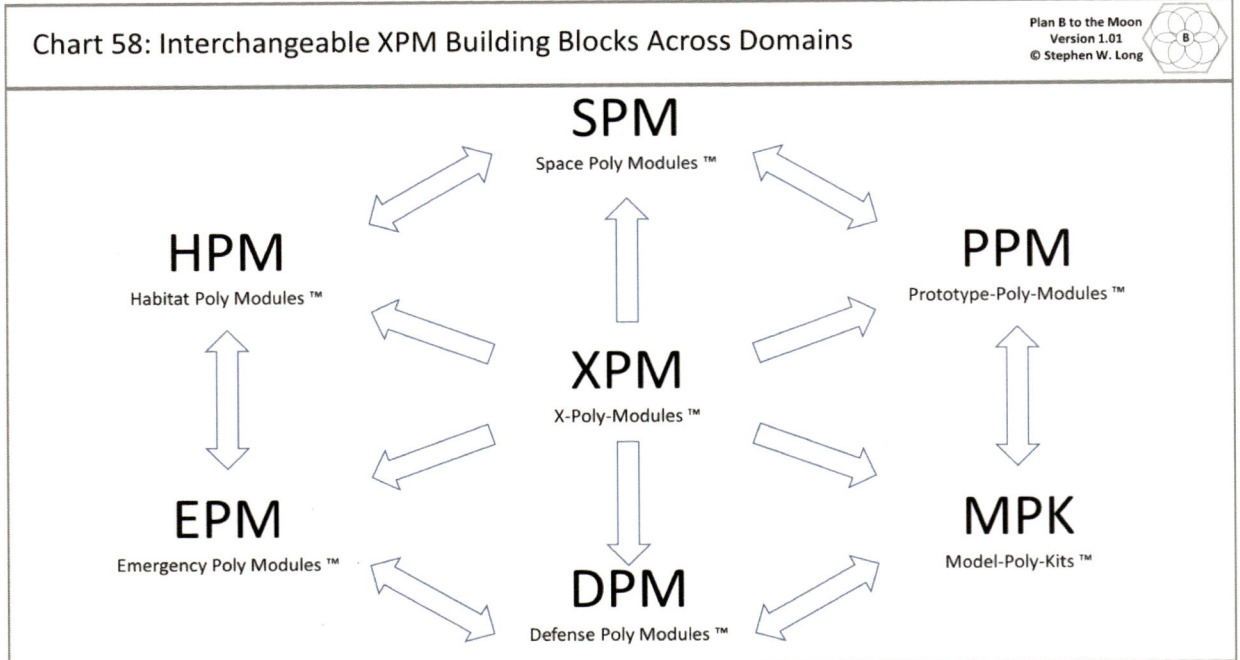

Chart 58: Interchangeable XPM Building Blocks Across Domains

4.18   Model-Poly-Kits

It is highly likely many of the scientists and engineers of today had hours of entertainment and enrichment as children playing with building blocks or similar three-dimensional construction toys. Lincoln Logs, Tinker Toys, Erector Sets, Legos®, KNEX® and similar components were profoundly entertaining toys. When you listen to an engineer discuss the toys they grew up with, you can guess the approximate age of the engineer. These 3D block toys feature combinations of standardized components that fit together into near infinite combinations.

Model-Poly-Kits (MPKs) are XPM literal scale model toy pieces (similar to KNEX®) that will allow the children (and uninhibited adults) of the world to dream in 3D. MPKs let you build space vessels, habits, even scale-model miniature cities of real Plan_B XPM parts for the Moon and Deep Space. Plan_B advocates distributing Model-Poly-Kits around the world, inspiring millions of children and adults to build their dreams and visions of XPM-derived future spacecraft and habitats. It is a **Plan_B Precept** to have toy versions of XPM components manufactured in mass. The Plan_B Program Director started negotiations to develop XPM toy components with a major toy manufacturer. The company expressed profound non-interest in the premise of MPKs. Plan_B V1.0 publication changed their minds. For uninhibited thinkers, one can think of XPMs as the adult-sized versions of KNEX® parts.

Modern computational artificial biology research predicts new biological capabilities emerge from trying millions of combinations of basic biological building blocks. Imagine a million new moon hut experiments built out of building-block-toys. One can guarantee some of those designs will out-compete committee-designed traditional construction structures. If you can build a dinosaur out of LEGOs®, you can build a toy model exploration ship mimicking real SPMs.

4.19   Prototype-Poly-Kits

The Plan_B Team learned-by-doing, experimenting with life-sized modular components. We discovered another **Plan_B Precept**: Better engineering decisions emerge from low-cost trial-and-error designs that become working analog prototypes. These prototype XPMs help inform space vehicle design decisions and inspire new earthbound capabilities, such as emergency habitats at consumer price points.

The early prototype FCC component experiments inspired the Plan_B Team to look for prior-NASA-funded research papers. We sought answers from prior work to create space structures from stick assembly components (see Chart 59). The Team found published papers describing ready-to-assemble aluminum tube (rail) elements. Just as Plan_B postulates XPMs combine to form SPM structures, these stick-together-rails form larger space structures. It was beyond the bank account balances of the Plan_B team to buy and experiment with the space-rated components described in the NASA papers. But as shown previously in Chart 57, we could afford and we built proof-of-concept analogs of eventual space structural FCCs using common aluminum and steel parts.

Chart 59: Prior NASA Research to Develop Modular Spacecraft Components

Plan B to the Moon
Version 1.01
© Stephen W. Long

4.20   XPMs for Defense and Emergency Applications

During concept development efforts for XPMs and the Frame-Core-Components, non-space concepts for Defense and Emergency application XPMs emerged. The Plan_B Team realized XPMs would meet Defense and Emergency requirements, leveraging the core XPM premise of lower-cost and highly extensible building block modules configured as the need arises. XPMs do not always have use space-rated materials. Commodity (low cost) components enable useful-for-Earth applications. Excitingly so.

Chart 60 defines new XPM components known as Defense-Poly-Modules (DPMs) and Emergency-Poly-Modules (EPMs). A big idea here is to understand and consider that XPMs distribute a common design heritage across other domain modalities. When XPMs are mass produced, less expensive materials designs are affordable as Earth-based Emergency habitats, or as Defense habitats (better than a tent). The experience gained from manufacturing ~millions of copies of Earthbound XPMs will influence how to better build space rated XPMs at far lower costs than traditional exquisite space component designs. Any money spent on research and development for one domain of XPMs further advances other modules. Interchangeable XPM building block designs apply across many domains.

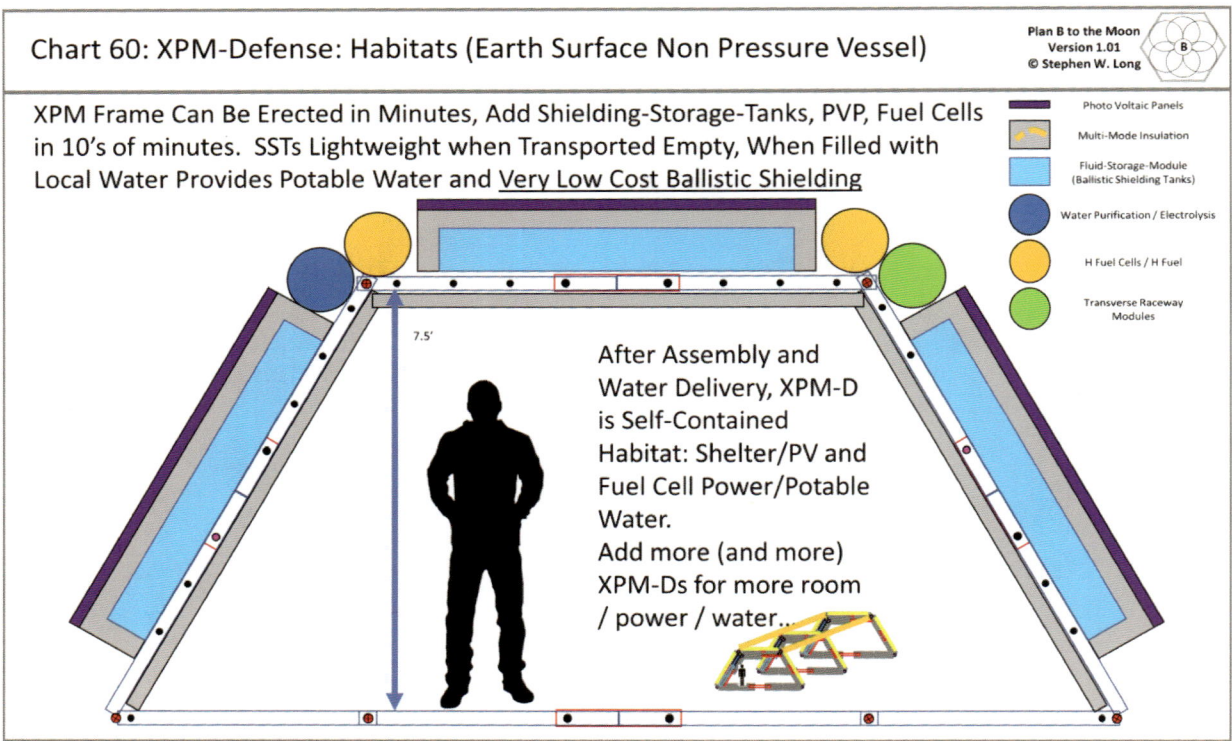

As depicted in Chart 60 an XPM-D (for defense) is a habitat segment designed for use on Earth's surface. As designed for original Plan_B structures in space, XPM-Ds erect in minutes. Fluid storage tanks are lightweight when transported empty but when attached to the FCC frame and filled with local water, they serve multiple purposes. Integral storage tanks provide a ready source of potable water. Encompassing tanks provide temperature insulation. You heat or cool the water to heat or cool the interior of the shelter. When filled with water or ice, the tanks provide effective ballistic protection against mortar or rocket shrapnel. We also recommend adding Kevlar blankets into XPM-D designs. Ground systems Plan_B space technologies (Linear-Fuel-Cells, Water Purification and Electrolysis systems, and Photo-Voltaic-Panels) create potable water and reliable power with no need for external fuel-hungry generators. Chart 60 depicts a notional ceiling height of 7.5 feet. The final interior height can be taller or shorter, depending on the lengths of the FCCs used for the structure. Each XPM component folds for transport by pickup truck class vehicles and can be unfolded in minutes to create a sturdy structure.

The Plan_B Program Team designed and built prototype XPM-D modules in the hot Florida sun during the summer of 2018. The earliest frame cores used commercial-off-the-shelf foldable aluminum components. We designed and fabricated Click-in joining structures at a local machine shop, and the Team stopwatch-proved we could assemble a sturdy structure in tens of minutes. The FCC frame goes up in just a few minutes. It takes more time to mount the water tanks and the photo-voltaic modules.

The Plan_B team believes a recursive improvement cycle will eventually prevail. Ideas for Space-Poly-Modules will inform concepts for Defense Poly Modules. Defense Modules will inform Emergency shelter modules. Defense and Emergency modules will create a large manufacturing base that will lower the costs to build Space capable modules. Space technologies such as hydrogen fuel cell modules, water purification modules, and communications infrastructure modules will also be useful for Defense and Emergency applications.

4.21 Thinking Differently—The Power of XPM and SPM Thinking

The Plan_B Team offers some additional concepts for future XPM developments:
- Create a competition (similar to X Prizes™), to foster creation and proliferation of new XPM modules and XPM derived structures. With the right incentives, dozens if not hundreds of XPM concepts can arise—leading to even more benefits to the entire XPM family of technologies and components.
- Via competitions or straightforward engineering development, make the XPMs / SPMs printable using additive manufacturing systems (3D printers). Maybe the lowest cost Model-Poly-Kit toys get stamped out of plastic. But we build space-ready SPMs from aluminum or beryllium or carbon fibers or spools of wire. Once we deliver suitable printers on-orbit, all later missions ship up printer ink, and print any new XPMs in space.
- The XPMs/SPMs are mission recyclable. Perhaps the SPMs are first used to build XPM manufacturing facilities on-orbit. Once the factory completes XPM printing, the SPM factory modules get disassembled and reassembled into new configurations. The SPMs may first serve as spacecraft to travel to Mars. Once the spacecraft arrives at Mars, the reassembled modules become Mars-orbiting space stations or habitats.

- One key premise behind XPM/SPM designs is that you don't have to have the end-design completed for on-orbit structures before you launch building blocks. Start with simpler launch-today modules and add on and improve the on-orbit structures as we go.

- Unlock the creative potential of humans from across planet earth. If people believe there are no limits to what we can build, NASA roles change. Instead of doing it all (funding, building and flying), NASA shifts their focus to develop module standards and interfaces. Perhaps NASA's role is to provide oversight for launches, in a manner similar to how the FAA supervises the National airspace. The FAA manages the airspace without operating all the aircraft that use the airspace. With this newly matured framework, the rest of humanity can help build and pay for the future of on-orbit structures.

- Consider the premise of significant ready-power produced on any given SPM covered with PVPs. If you need 100,000 W of power for a new orbiting factory, you snap together five 20-kW SPMs to build a 100-kW enabled on-orbit or on-the-Moon factory. This means you don't have to invent new power production systems every time you travel to orbit. You only have to assemble previously designed power-panel-covered SPMs with the new tool QRMs to build whatever new factory you desire. See Chapter 7.2, Space-Fuel-Terminal, for specific examples of combinations of QRMs building new space capabilities.

- The ultimate expression of the power of XPM and SPM thinking is to either travel to a comet or asteroid (see Chapter 8 Comet or Asteroid Rodeo). The mission would *lasso and tow* the asteroid to the earth's orbit. From the asteroid's raw materials, print and assemble XPM and SPM components. The asteroid provides the raw materials used for the 3D printers.

Chapter 5:   Leverage Existing Human Rated Spacecraft

5.1 CCtCap Vehicles

Building and testing Human Rated (HR) Spacecraft is extraordinarily expensive. It is a **Plan_B Precept** to leverage the already paid for, NASA in-progress human-rated rocket and spacecraft plans. It is not a Plan_B priority to spend additional resources to develop new human rated launch spacecraft. Chart 61 introduces the **Plan_B Precept**: Do not burden space cargo delivery missions with the extra safety requirements of human rated launches. Also, do not burden lunar surface cargo delivery missions with the extra safety requirements of human rated landing craft.

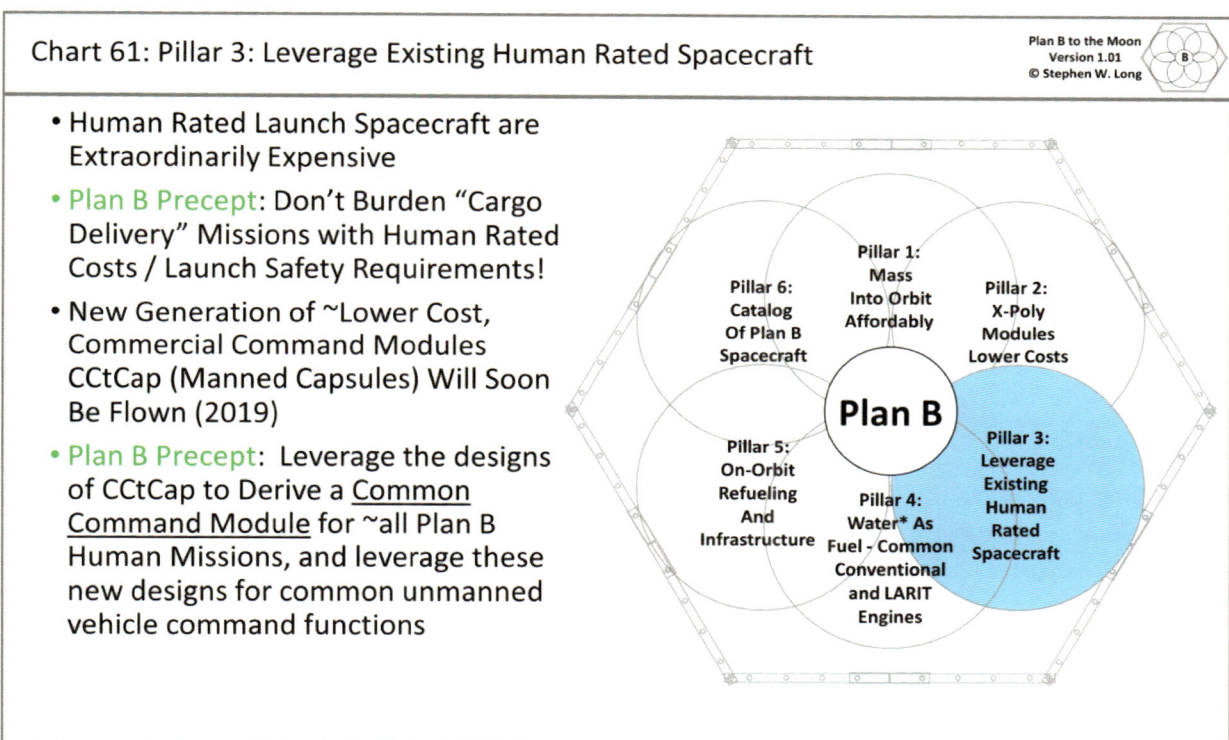

NASA has fiscally programed two competing systems for human rated launches from American-soil, what NASA calls the Common Crew Transportation Capability (CCtCap) vehicles. Chart 62 depicts both the Boeing CST-100 CCtCap and the SpaceX Dragon Capsule CCtCap.

Plan_B incorporates CCtCap vehicles throughout its planning (Chapters 9, 10, 11) to launch astronauts into LEO, and return them from LEO to the Earth. Plan_B uses the CCtCap vehicle published launch weight of 21,300 pounds. The Plan_B Team could not find a published list-price to purchase CCtCap vehicles. We therefore used a planning/budgeting factor of $10,000 per pound to calculate the purchase costs of CCtCap vehicles. Plan_B therefore budgets $213 Million per CCtCap vehicle in the Chapter 10 MISSION MATRIX mission plan.

Chart 62: Common Crew Transportation Capability (CCtCap)

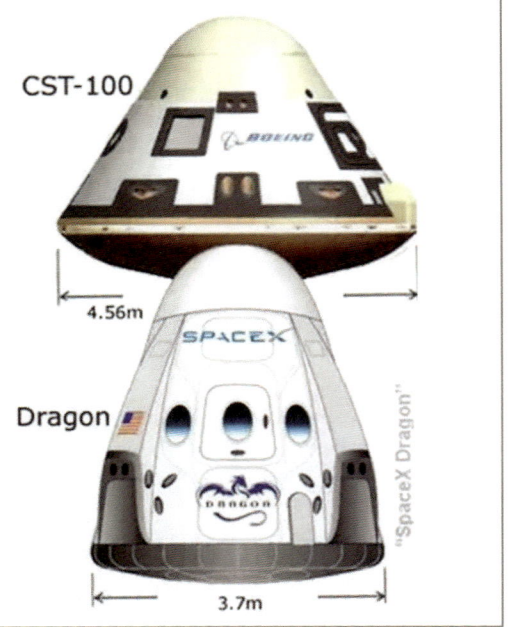

- Plan B Precept: Human Launches (Earth to LEO and return to Earth) will be done using CCtCap Modules
  - In 2014, NASA selected Boeing CST-100 and SpaceX Dragon V2 for the Commercial Crew Transportation Capability (CCtCap) program, with an award of $4.2 billion. Craft will fly in 2019. CCtCap Modules will return American Astronauts to space on American rockets (instead of Russian rockets) for first time since the US Space Shuttle
  - CCtCap modules will transport astronauts from Earth to LEO, Rendezvous with LEO-Stations, and return to Earth. Plan B Precept: CCtCap Modules will serve as Plan B Earth to LEO to Earth Return Modules. CSCMs will be used for all other Plan B missions (LEO and beyond).
  - CST-100 Starliner / Dragon Command Modules (or parts thereof) can/should also be considered as the design starting point for use as the CSCM. Orion components can be considered if costs can be constrained.

5.2 Common-Space-Command-Module Introduction

It is a Plan_B Precept to develop a derivative Common-Space-Command-Module (CSCM). The CSCM will leverage the designs, systems and components used by the to-be most successful of the two CCtCap designs. The CSCM is a human-rated cockpit—a place to sit and fly a vehicle once delivered to orbit or to the Moon. But as a thinking-differently premise CSCMs will also serve as the control hub of autonomous spacecraft. Even if a spacecraft is un-crewed, you still need a vehicle (a space rated box) that has computers and communications and power and thrustors and other components for missions.

Chart 63 depicts the relative sizes of the historic Apollo Command Module, the Orion capsule, the CST-100 CCtCap vehicle, the Dragon CCtCap vehicle, and a notional CSCM.

As an exemplar, the Plan_B Team notes that Airbus Industries made design decisions decades ago to use common cockpits across all their aircraft types. Chart 64 depicts the modern common glass cockpit used on Airbus aircraft. A common (identical) cockpit enables new aircraft development cost savings and significant crew training savings. Once pilots get trained in one cockpit that crew member is pre-trained for all other vehicle cockpits.

Chart 63: Common-Space-Command-Module (CSCM)

Plan B to the Moon
Version 1.01
© Stephen W. Long

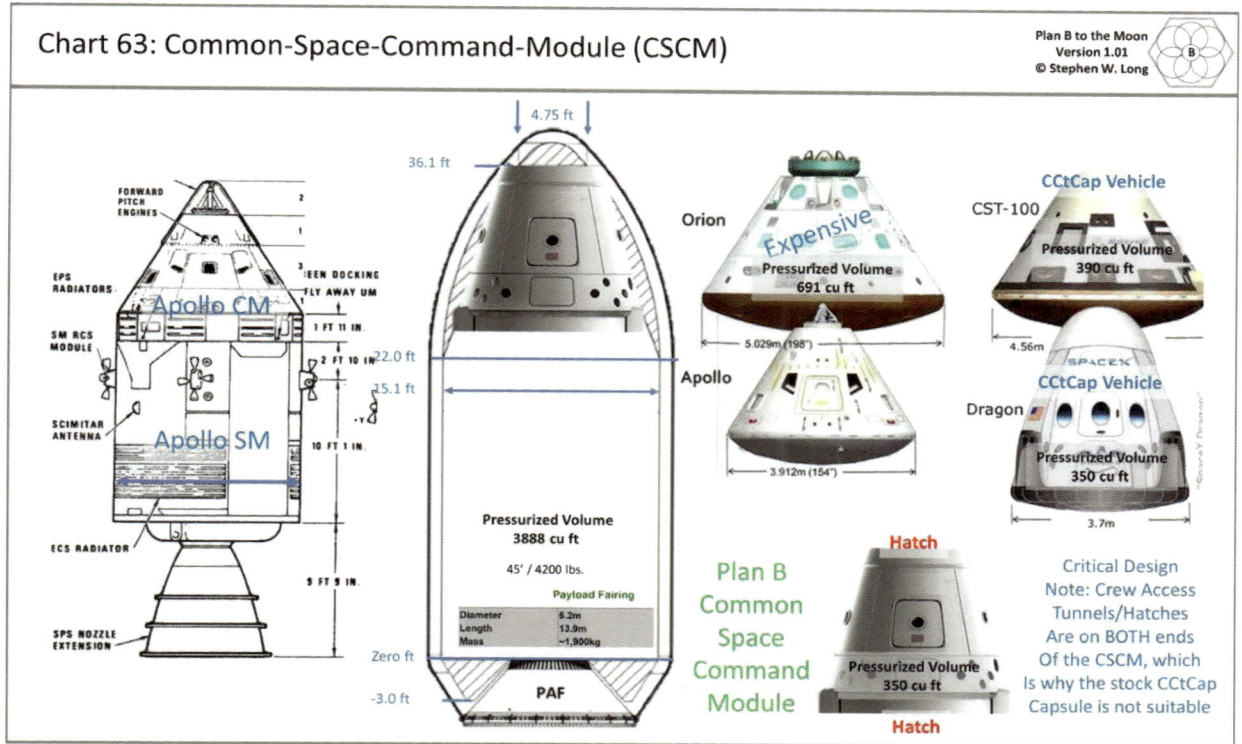

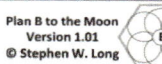

Chart 64: Common Cockpit Components

Plan B to the Moon
Version 1.01
© Stephen W. Long

- **Common Components Exemplar:** Airbus Industries made a Design Decision to use a <u>Common Cockpit</u> across ALL of their aircraft types. Provides development cost savings as well as significant crew training savings – Once trained in one cockpit, crew members are pre-trained for all vehicle cockpits

As another example, consider that one of the more curious forward-thinking concepts in the various Star Trek television episodes was the ubiquity of common controls (Chart 65). Somehow automagically, every crew member (no matter their uniform shirt color) knew how to fly every ship. The touch screen controls on the bridge of the Enterprise D (left front seat) were the same controls as the right front seat of a Star Trek NG shuttle craft. The Plan_B Program Team does not advocate using touch-screen-only controls for CSCMs. As an old stick and rudder pilot, the Plan_B Program Director finds a hand controller (center stick or arm rest controller) comforting. For intricate 3D flight paths (such as fighter jet maneuvers) a control stick (center or side) has for 100 years provided maximum flight control responsiveness for airplanes. But large, heavy planes use yokes (steering wheels) instead of sticks (that's ok). Foot activated rudder pedals are no longer mechanically necessary in space, but sometimes having four limbs providing directional / velocity inputs are better than just two limbs.

Chart 65: "Star Trek Next Generation" Common Control Panels Across Vehicles

Plan B to the Moon
Version 1.01
© Stephen W. Long

All-the-above is the likely best flexibility approach. Control functions using 2D touch screens are OK (providing they work well with gloved fingers). But there should be the option for a plug-in controller (arm rest OK) for more sensitive axis response flight controls. The tried-and-true fighter jet cockpit has a center stick or right-side attitude controller (stick) and the left side has a throttle. Radio push-to-talk uses left thumb clicks and weapons (or other time critical functions) use right thumb (pickle button) or right index finger (trigger finger) clicks.

Therefore, the CSCM design must allow a single pilot, from the left or center or right seats, to access common control inputs. The Apollo LM used a fold down/fold up control arm.

It is a **Plan_B Precept** to build a common CSCM (based on CCtCap or similar core) from which humans can pilot (command) a space craft / vehicle. When humans are not in the seats, the same computers autonomously control the spacecraft or vehicle. The Plan_B Team notes that at least one major Aerospace corporation, Northrop Grumman, produces aircraft they market as Optionally Manned. Meaning, there are controls (and seats, etc.) from which humans may fly the craft, but as automation, sensors, computers, and software become more capable, autonomous operations are possible (and sometimes preferred to save weight, etc.). Therefore, Plan_B assumes and plans for the rapid development of CSCM shells for all spacecraft command functions—for Crewed and Autonomous missions.

5.3 Crewed Versus Autonomous Flight Controls

The Plan_B Team notes considerable confusion (lack of published clarity) regarding the terms manned and crewed versus unmanned versus Remotely Piloted Aircraft (RPA) versus autonomous. Plan_B uses the terms Crewed and Autonomous. Plan_B defines Crewed as human pilots with literal hands-on flight controls—be they left-hand throttles and right-hand sticks or touchscreen and joystick type controllers. To illustrate how confusing this topic becomes, how do you describe a vehicle flight control system when a human pilot engages the auto pilot? How do you describe computers that operate flight controls while the human pilot observes and is ready to override the autopilot and resume control?

The US Defense Department has decades of experience with Earth based, non-crew-on-board-aircraft (called unmanned aerial vehicles). Combat missions have proved keeping humans-in-the-loop to control (fly) the vehicle during critical flight phases is a key to operational flexibility. In the lunar exploration context, one only has to recall the drama of the Apollo 11 landing sequence. Neil Armstrong had to take over the automatic landing to steer the LM to a landing spot free of giant boulders. An autonomous automatic landing attempt would have likely crashed Apollo 11.

Plan_B defines Crewed as meaning maximized human control, using appropriate levels of flight automation support and Autonomous as meaning maximized computer control with occasional human pilot inputs. Vehicle flight control options span a spectrum of capabilities. In some not-to-distant future, full autonomous-control will be possible—no human pilots required at all. Until that time, the Plan_B Team notes a middle ground that the USAF has successfully deployed–their concepts for Remotely Piloted Aircraft (RPAs). RPAs mean the pilot (crew) is sitting in a land-cockpit (a room that looks and feels as if it were in an airplane). But the aircraft (to date an unmanned aerial vehicle or UAV) has no cockpit–no place at all for a human to sit. A human pilot within the literal line of site of the UAV runway observes and remotely flies the aircraft during takeoff roll and initial climb out. Once the aircraft is safely away from the ground, UAV control switches to another pilot, often physically sitting on the other side of planet earth. Via satellite communications the remote pilot either manually flies (stick and rudder) the aircraft or uses standard aircraft autopilot systems to control the UAV. When it is time to land, the UAV flies to within the line-of-site environment of the landing field. A pilot on the ground assumes control of the aircraft and safely lands the aircraft. Pilots hand-fly the UAV during the most difficult flight phases–takes offs, landings, and weapons employment. The pilots engage the Autopilot for routine straight-and-level flights. Plan_B adopts this same flight control premise–where human pilots (crew) can and will provide spacecraft flight control during the most difficult phases (docking, undocking, landings). Automation will provide spacecraft flight control during the space equivalent of straight-and-level flights.

There are other UAVs, such as the USAF Global Hawk, that take off and land autonomously with no human pilot touching the flight controls. But such autonomous operations take place from well understood permanent landing facilities–such as commercial-airline-grade airports or surveyed military base runways. Such airfields have hyper-accurate runway coordinates for the Global Hawk flight control system to pilot itself autonomously during takeoffs and landings. Having human-in-the-loop control over Lunar-Long-Boats (LLBs–see Chapter 8) for landings is the lowest risk / most affordable LLB option.

In recent Expert System, Narrow Artificial Intelligence (AI) projects, humans in the loop teach AIs how to perform tasks. Whereas Plan_B intends the CSCM for both manual and autonomous operations, a trainable AI-pilot-in-the-loop would be a good early program investment. As human pilots (in simulators and in space) fly tens, hundreds, and eventually thousands of space maneuver operations, the AI can learn from such pilots. The AI will become capable over time of autonomously performing similar operations.

5.4 Common-Space-Command-Module Design Notes

The reusable CSCM (Chart 66) will become the command module for LEO Station, LLO Station, ELTV, Long Boat, Lunar-Lander, and Lunar Habitat Transfer Vehicles. While the CSCM can and should borrow from proven CCtCap or similar designs, the CSCM will never return to earth. CSCM designs can reduce costs and weight by removing Earth Return components such as heat shields and parachutes.

The CSCMs will host the computation and communications needed by Plan_B vehicles. We save time and money if we do not reinvent these systems for each different craft design. Since the CSCM designs build upon previously designed CCtCap vehicles, they are by definition human-rated. The CSCM will provide the very important role of a life boat on Plan_B Vehicles if there are larger-craft (SPM) life support problems.

The Plan_B Team learned a painful engineering lesson during space station rack-and-stack maneuvers. If there is only one-entrance into a vehicle (a single hatch), it becomes VERY difficult to manage stacking of vehicles and components on-orbit. No matter how well you think you plan out a sequence of vehicle attachments (dockings/undocking's in space), you eventually paint yourself in a corner. You have no empty hatch upon which to dock later missions. Chapters 9 and 10 (Mission Circles and MISSION MATRIX) often ran into such painted corners–eventually leading (in some frustration) to the design decision and need for multi-hatch vehicles. We learned the CSCM design must include both forward and aft hatches. Plan_B drawings depict triangular cones (for aesthetics, spaceships always have pointy front ends). An elongated toroid (think a soda can with a donut hole running down the middle) may be a more practicable dual-docking-port design for the CSCM.

Chart 66: Common-Space-Command-Module (CSCM)	Plan B to the Moon Version 1.01 © Stephen W. Long

• Plan B Precept: Reusable capsule will serve as the Common-Space-Command-Module (CSCM) for ALL Plan B vehicles (LEO Station, LLO Gateway, LEO/LLO Transfer, Long Boat, Lunar Lander, Lunar Vehicles...)

  • CSCMs never return to Earth, therefore CSCM costs/weight can be reduced by removing Earth Return components (such as heat shield, parachutes, etc.)

  • CSCMs will host computation and communication systems required by ALL Plan B vehicles, and will also serve as the "life boat" on Plan B spacecraft if there are larger-craft life support problem

Common-Space-Command-Module (CSCM) [Notional]

As documented in Chapter 10, the CSCM has a design weight of 10,650 pounds, with a design manufacturing cost $10,000 per pound or $106.5 Million per unit. This is TWICE the **Plan_B Precept** spacecraft costs of $5000 per pound. We doubled the budgeted costs for CSCMs to capture the expected higher costs of building human rated spacecraft components.

5.5 Common-Space-Service-Module

Plan_B on-orbit operations require use of a Common-Space-Service-Module (CSSM) (Chart 67). NASA's Plan_A reports contracting with the European Space Agency (ESA) to build the ESA Service Module (ESASM) to provide the Orion Space capsule thrust, power, etc. The Orion Space capsule plus the ESASM creates a complete and functioning spacecraft. Public access records do not provide the full technical capabilities (or the costs) of the ESASM. One can conjecture that the ESASM provides sufficient Delta-V to transport the Orion Space capsule from LEO to Lunar Orbit.

The ESA has already completed RDT&E for the ESASM. The ESASM may be a cost-effective candidate to serve as the Plan_B LEO primary orbital maneuver and boost motor for LEO operations (see Space Refueling Chapter 7). The ESASM may be able serve as the trans-Lunar boost rocket engine (the rocket to send mass from LEO to Lunar Orbit and back to LEO). However, based on ESASM drawings and press photos, the ESASM appears to be physically smaller than the Apollo era Saturn V Third Stage and the Service Module rocket engine. Based on size, it is unlikely the ESASM has sufficient Delta-V to transport the 100,000-pound Plan_B lunar cargo deliveries. On this basis, future versions (after V1.0) of Plan_B will examine use of the ESASM as the Plan_B CSSM. Plan_B develops the concepts necessary for a new (under development) Common Conventional Engine (CCE). The CCE will provide the Delta-V necessary to transport Plan_B payloads from LEO to LLO. See Chapter 6 for more details about the CCEs.

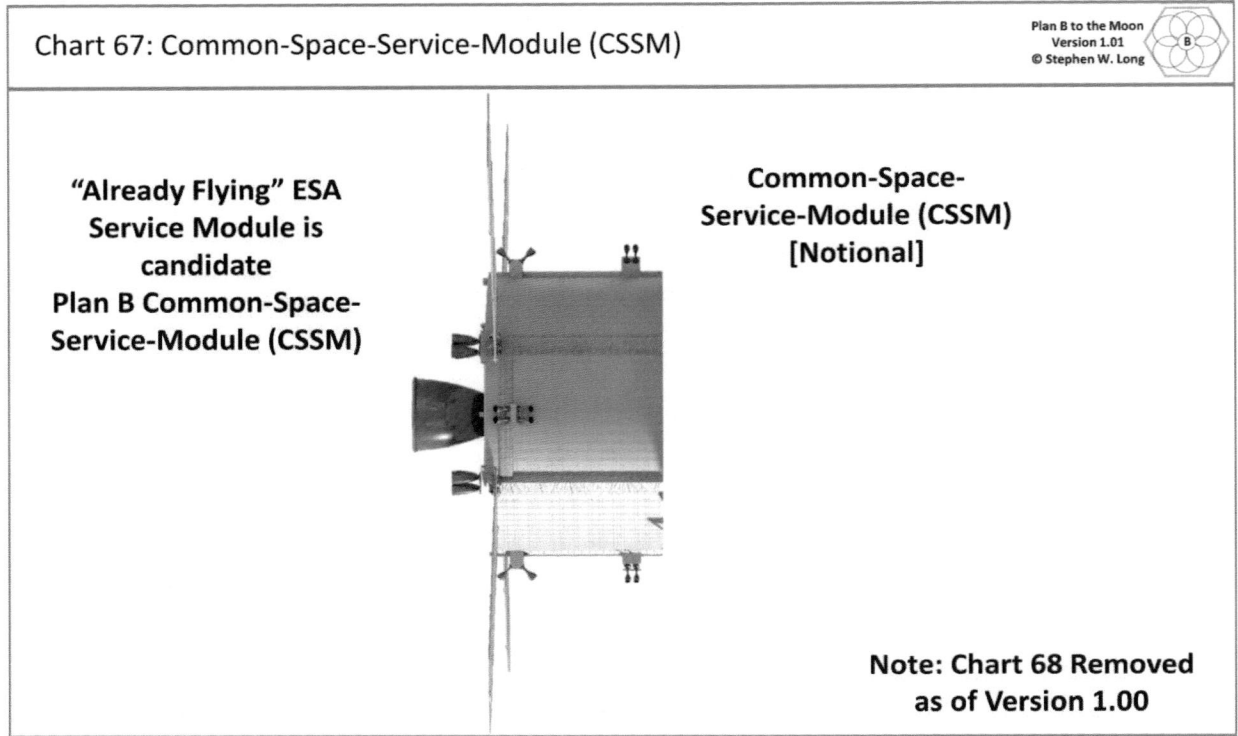

Chart 67: Common-Space-Service-Module (CSSM)

Plan B to the Moon
Version 1.01
© Stephen W. Long

"Already Flying" ESA Service Module is candidate Plan B Common-Space-Service-Module (CSSM)

Common-Space-Service-Module (CSSM) [Notional]

Note: Chart 68 Removed as of Version 1.00

Chapter 6: Water* as Fuel for Conventional and LARIT Engines

6.1 Water* as Fuel

It is expensive to have to bring all the fuel you need for deep space exploration from the Earth's surface. Even with the cost savings of using the Falcon Heavy Rocket as the launch vehicle, it still costs $1000 per pound (see Chapter 3). To send 500,000 pounds of mass to the Moon to begin construction of a Lunar Colony, you need 1,000,000 pounds of fuel, which in launch-mass-costs-alone is $1 Billion. Is there a less expensive long-term Colonization approach for spacecraft fuel? Yes, there is.

It is possible to create rocket fuel from the simple elements found in water (H_2O), Carbon Dioxide (CO_2), and lots of electricity. H_2O plus CO_2 plus cooling produces Liquid Methane (LCH4) and Liquid Oxygen (LOX) (Chart 69).

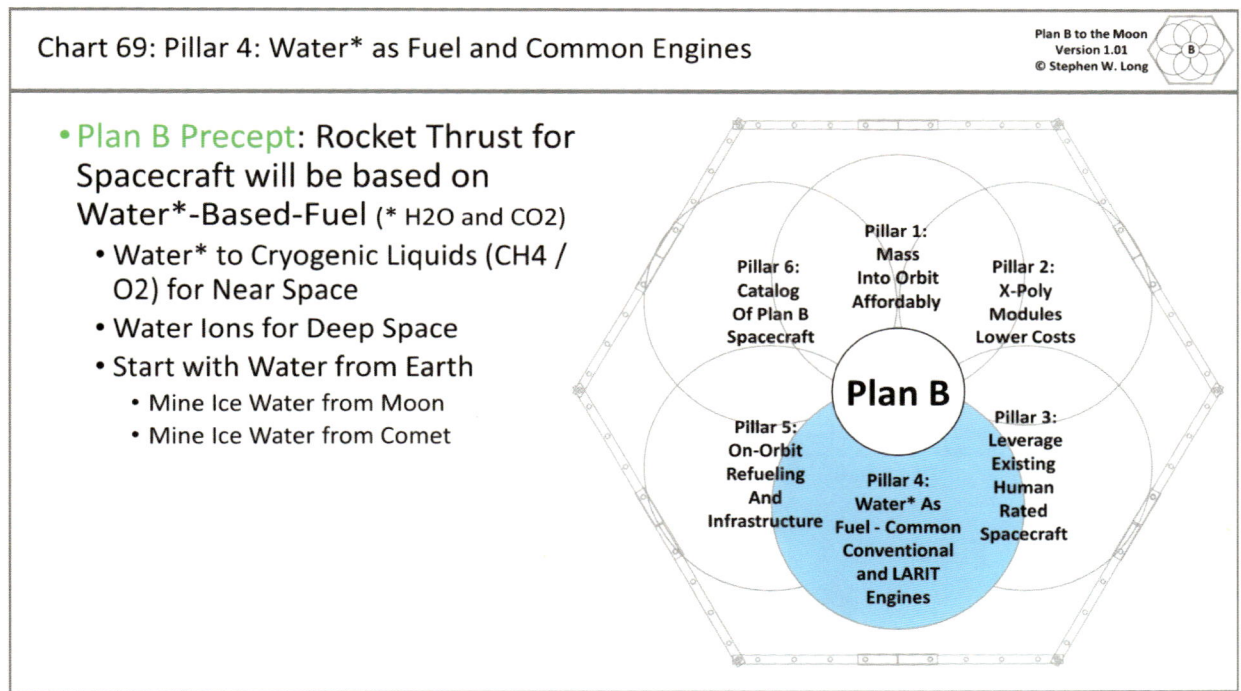

Chart 69: Pillar 4: Water* as Fuel and Common Engines

The Moon (and comets) have ICE (frozen H_2O). This concept of mining Ice (frozen H_2O) on the moon to create rocket fuel has been a staple of serious-fiction-based-on-science for decades. It is a **Plan_B Precept** to use Water* as Fuel (* Water plus CO_2). If you intend to use Water* as fuel, you must first have rocket engines that can ~burn the water derived fuel—liquid Methane (LCH4) and liquid oxygen (LOX) engines. Sections 6.3 and 6.4 will describe the Plan_B rocket engines what will ~burn Water*.

Plan_B notes there is an even better potential rocket fuel than LCH4 and LOX—liquid hydrogen combined with liquid oxygen. The Plan_B Team found no designs or technology to STORE liquid hydrogen for months or years in space. Liquid hydrogen and liquid oxygen were the fuels of the US Space Shuttle main engines launched from Earth. Other launch rockets continue to use Liquid hydrogen and liquid oxygen. However, none of those cryogenic fuel engines had to store super-cold hydrogen for long periods of time. In contrast, we have existing cryogenic technology to store warmer liquid methane and oxygen for extended periods. Many corporations ship Liquefied Natural Gas (which is methane) across Earth's oceans in giant tanker ships. If a modern genius invents how to store liquid (or solid) hydrogen in space for years on end, we can / should move from liquid methane to liquid hydrogen fuel. Until that future technology invention occurs, LCH4/LOX rocket fuel is our best option moving forward for space exploration that lasts longer than hours or days.

6.2 The Science Behind Water* to Rocket Fuel

As depicted in Chart 70, the science behind Water* as Rocket fuel is straightforward:
- The electrolysis of water (passing electricity through water) produces gaseous Hydrogen (H) and gaseous Oxygen (O2). This is a common experiment in 8th grade science classes (the PD became interested in science because of this 8th grade class).
- When gaseous Hydrogen and Carbon Dioxide combines within a nickel catalyst chamber (other catalysts possible) at 300-400 degrees Celsius, the process produces gaseous Methane and water vapor. This is the Sabatier reaction, discovered by Paul Sabatier in the 1910s.
- Gaseous Methane will condense to liquid Methane at 112 Kelvin, or -161.6 Celsius.
- Oxygen will condense to liquid Oxygen at 90 Kelvin.
- Gaseous Hydrogen will convert to liquid Hydrogen below 33 Kelvin where 20 Kelvin is optimum. Understand that 20 Kelvin is devilishly cold, and it is difficult to get down to and maintain this super cold temperature.
- Liquid Methane and Liquid Oxygen create a very potent rocket fuel. We note NASA, SPACE-X and others are intensely developing (and have already test fired) LCH4/LOX rocket engines.

One of the seldom admired and free attributes of outer space is the ready availability of super cold temperatures. Near-Earth space is already cold to about 175 Kelvin (minus 100 Celsius). Research shows passive cooling is possible from native near space (175 Kelvin) down to 112 Kelvin (for liquid methane). 90 Kelvin (for liquid O2) cooling is reachable using passive sun-shade cooling and available/space-rated multi-stage cryo-coolers. The James Webb Telescope team has published details of their development of passive cooling techniques that will cool down to 37 Kelvin. As good and cold as 37 Kelvin might be, it is still not cold enough for liquid hydrogen (20 Kelvin). But 37 Kelvin is cold enough for LCH4 (112 Kelvin) and LOX (90 Kelvin). Therefore, Plan_B believes available now or near-term technology will liquify methane and oxygen on-orbit. This enables the **Plan_B Precept** that liquid methane and LOX are the preferred fuel and oxidizer of choice for Plan_B missions. Readers can find more details about passive cooling solutions at https://jwst.nasa.gov/cryocooler.html.

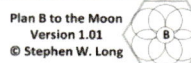

Chart 70: The Science Behind Water* to Rocket Fuel	Plan B to the Moon Version 1.01 © Stephen W. Long

- Water* plus Electricity = Rocket Fuel! (*plus CO2)

- Electrolysis of water produces gaseous Hydrogen and Oxygen (H / O2)

- Liquid Hydrogen / Liquid Oxygen would be an ideal rocket fuel – except that liquid hydrogen is created/stored at 31K (-242C) – very difficult (expensive) to do on orbit

- Hydrogen and Carbon Dioxide can be combined (Sabatier Reaction) to form Methane (CH4) (natural gas). Discovered by Paul Sabatier in 1910s
  - Reaction of hydrogen with carbon dioxide at 300-400 deg C, in presence of nickel catalyst produces Methane and Water

$$CO_2 + 4\,H_2 \xrightarrow[\text{pressure}]{400\ ^\circ C} CH_4 + 2\,H_2O$$

Item	Temp K	Temp C	Temp F
Near Earth Space Shadow	174.0	-99.2	-146.2
Liquid Methane (CH4)	112.0	-161.2	-257.8
Lunar Shadow	100.2	-173.0	-279.0
Liquid Oxygen	90.1	-183.1	-297.2
Liquid Hydrogen	31.0	-242.2	-403.6
Deep Space	2.7	-270.5	-454.5
Absolute Zero	0.0	-273.2	-459.4

- Liquid Methane (CH4) when combined with Liquid Oxygen (O2) creates <u>potent Rocket Fuel</u>

- Gaseous methane converts to liquid methane at 112K (-161.6C), Gaseous oxygen liquidation is 90K (-183C). Near-Earth Space is already very cold 175K (-100C), so if properly exploited, native cold of space can be used to reduce energy required to compress / convert gaseous CH4 / O2 to liquid cryogenic fuel

Turning Water* into fuel (turning water into CH4) needs a lot of electrical power. It is a **Plan_B Precept** to design and build space structures that generate a LOT of electric power. See Chapter 7.2, the Space-Fuel-Terminal, for an on-orbit solar array design that produces 504 kW of electric power. Water* on-orbit is convertible into fuel using the massive amounts of electricity available from photo-voltaic solar panels on-orbit or the Moon. The magic is putting Water* into orbit. The near-term money-green rabbit will use Water* from the Earth. We will find the longer-term White Rabbit in magician-hat dark holes on the Moon or on comets.

Plan_B acknowledges its Space-Fuel-Terminal concepts have higher developmental risks than just launching/ferrying LCH4/LOX up from Earth. However, producing fuel in space (on-orbit or on the Lunar Surface) is a key enabler of future deep space exploration. The sooner we tackle and overcome this engineering challenge of creating fuel from Water*, the better for humanity.

To advance this noble cause while also reducing program risk, Plan_B includes duel-fuel-source redundancy into the Plan_B timeline and budget. Plan_B includes the launches, weight and cost budgets to ferry all necessary LCH4/LOX fuel up from Earth to return to the Moon. Plan_B also includes the launch, weight and cost budgets to deliver all the components to build the Space-Fuel-Terminal on-orbit. Missions also launch into orbit large amounts of Water* for conversion into rocket fuel. Meaning, if the SFT is a developmental disappointment (does not produce fuel on-orbit in meaningful quantities), Plan_B can proceed to the Moon without additional programmed funds or schedule delays. Chapter 10, MISSION MATRIX calculations provide explicit details on the required fuel-ferry trips from Earth and on-orbit SFT fuel production.

Plan_B Version 2.0 and Version 3.0 introduce the mission plan steps required to mine ice found on the Moon. Version 3.0 factory modules convert this ice into LCH4/LOX as fuel for future space activities. As stated in Chapter 1.4, creating the ability to produce fuel outside the deep gravity well of Earth will be a critical milestone in human history.

6.3 Common Conventional Engines

Plan_B needs a Common Conventional Engine (CCE) that uses LCH4 and LOX. LCH4 and LOX fuel engines are needed for LEO operations, LEO to LLO / LLO to LEO transport, and derivative CCEs for Lunar Surface operations (Chart 71).

Chart 71: Common-Conventional-Engine (CCE) Will Power Multiple Spacecraft

Plan B to the Moon
Version 1.01
© Stephen W. Long

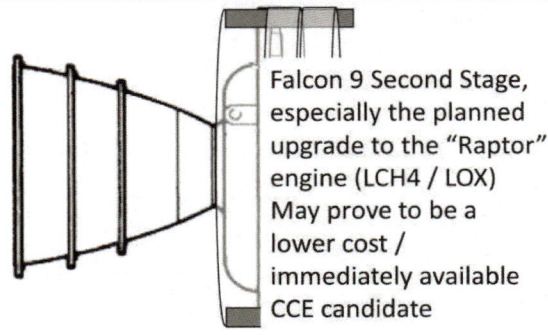

Falcon 9 Second Stage, especially the planned upgrade to the "Raptor" engine (LCH4 / LOX) May prove to be a lower cost / immediately available CCE candidate

Candidate Common-Conventional-Engine (CCE)

Plan B Precept: CCE will Have Sufficient Thrust to Transport Wet CTugRV spacecraft and 100,000lb. payload to LLO and then return to LEO ("dry" upon LEO return)

Current Falcon 9 Second Stage

NASA has developed advanced LCH4/LOX Engines for Spacecraft (not flown yet). SpaceX is Developing their "Raptor" Engine that is based on LCH4/LOX

NASA's Plan_A is developing (has developed) a rocket engine to transport the Orion capsule to Lunar Orbit. The Plan_B team could not determine from published data if the European Space Agency (ESA) Common-Space-Service-Module (CSSM) is that rocket engine. We could not determine if it has sufficient Delta-V to move larger payloads to Lunar Orbit, or if NASA is planning to use a different engine. The ESA CSSM could serve as the Plan_B CCE if it has sufficient Delta-V to transport 100,000 pounds of mass to the moon. The ESA CSSM would also have to be upgraded to burn LCH4/LOX. If the ESA CSSM engine is too small (insufficient Delta V) or does not burn LCH4/LOX, Plan_B requires a suitable new rocket engine.

For Plan_B, the most promising LCH4/LOX engine on the near horizon is the SPACE-X upgrade of their Falcon 9 Second Stage (F9SS). SpaceX calls this new LCH4/LOX the Raptor engine. It is a **Plan_B Precept** that this is best candidate to be the Plan_B Common Conventional Engine

(CCE). Preliminary calculations show that the FHR first stage (three boosters) has sufficient thrust to insert the F9SS into orbit without having to fire the F9SS to reach orbital velocities.

This enables the Plan_B design of the Common-Tug-and-Refueling-Vehicle (CTugRV) and the derivative Earth Lunar Transfer Vehicle (ELTV).

Plan_B V1.0 CTugRV designs use an F9SS, plus an SPM based fuel storage module frame, plus a CSCM, plus fuel. Even if the F9SS is not suitable for use as the verbatim CCE, the F9SS likely has most of the parts needed for the CCE. Plan_B Chapter 10 MISSION MATRIX uses the F9SS as the calculation basis for CTugRV and ELTV mass and fuel loads.

The Apollo Lunar Module Ascent Engine produced 3500 pounds of thrust (16 kN), used Aerozine 50 fuel, N2O4 oxidizer (dinitrogen tetroxide). This oxidizer is hypergolic—it spontaneously reacts upon contact with various forms of hydrazine, which has made the pair a common bipropellant rocket fuel. This Apollo engine has been reborn as the RS-18 engine, designed to use LCH4/LOX for the NASA Exploration Systems Architecture Study. This redesigned RS-18 engine is the Plan_B Version 1.00 candidate common engine for the Lunar-Long-Boat and the Lunar-Lander. The Lunar-Long-Boat will probably require multiple engines.

The Apollo Lunar Module Descent Engine produced 10,125 pounds of thrust (45.04 kN), designed for variable thrust between 10% and 60%. It used Aerozine 50 fuel, N2O4 oxidizer. Key advantages to these bipropellants are liquid storage at room temperature. Plan_B does not advocate the use of hydrazine and dinitrogen tetroxide as fuel because both substances are toxic to humans.

Given the at-present development indeterminacy of Lunar Operations engines, Plan_B can only provide engine mass and fuel figures-of-merit for Lunar vehicle Mission Circle planning. Plan_B mission costs therefore use the estimated masses of the components to calculate costs.

6.4 Linear-Accelerator-Relativistic-Ion Engines

Plan_B introduces concepts known as Linear-Accelerator-Relativistic-Ion-Thrust Engines (LARIT-Es). Plan_B Version 1.0 does not require development of LARIT Engines. Plan_B introduces these concepts to show that advanced future engines are possible for deep space exploration that will further leverage Water*-as-Fuel. We begin Plan_B missions using Water* converted to LCH4/LOX, and as LARIT Engine research progresses, switch to native Water-Ion propellent. The Plan_B team expects LARIT Engines will be a significant element of Plan_B Version 3.0. We hope to inspire current young readers to make building LARIT Engines their life's work. See Chapter 8 for additional information about LARIT Engines.

Apollo 11, Buzz Aldrin on the Moon, NASA ID: AS11405875

Chapter 7: On-Orbit Refueling and Infrastructure

The **Plan_B Precept** of enabling / relying upon space based, on-orbit refueling (Chart 72) has tremendous impacts on mission capabilities and lowering long-term costs. As defined in Chapter 6, the fuels for such refueling will be LCH4 / LOX / Water and CO2.

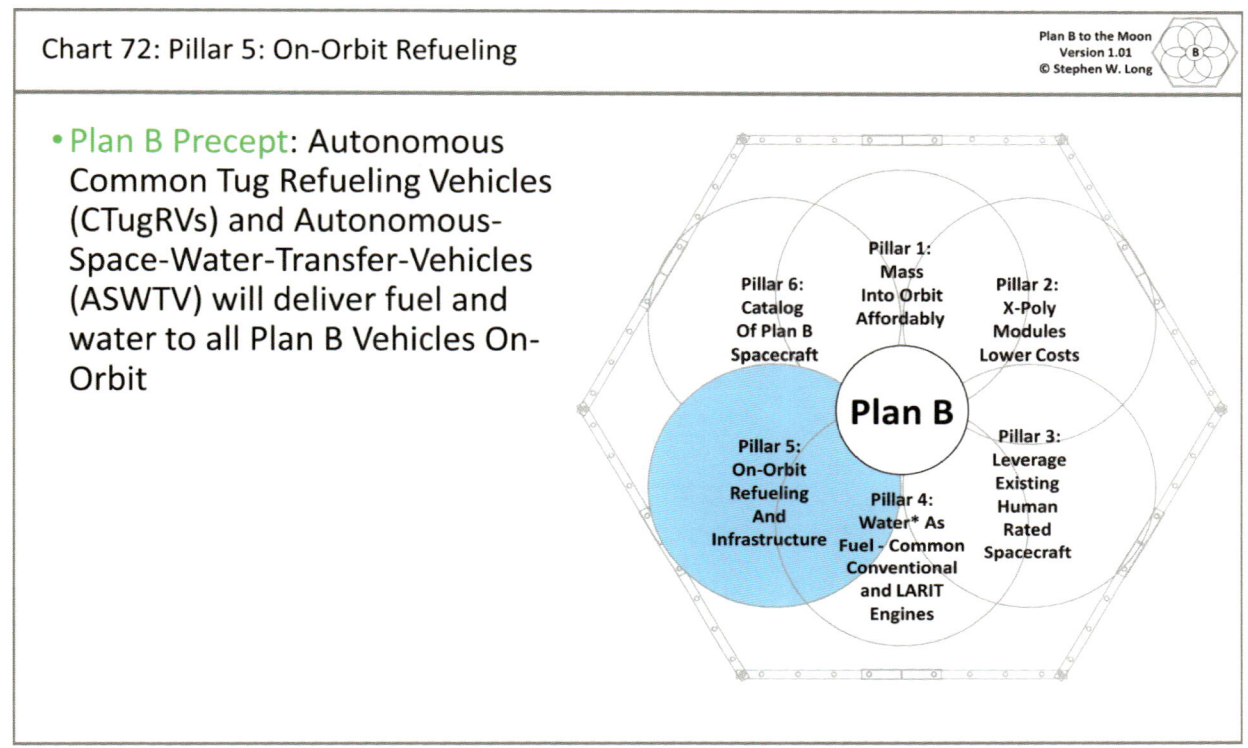

Chart 72: Pillar 5: On-Orbit Refueling

Plan B to the Moon
Version 1.01
© Stephen W. Long

• Plan B Precept: Autonomous Common Tug Refueling Vehicles (CTugRVs) and Autonomous-Space-Water-Transfer-Vehicles (ASWTV) will deliver fuel and water to all Plan B Vehicles On-Orbit

Pillar 6: Catalog Of Plan B Spacecraft

Pillar 1: Mass Into Orbit Affordably

Pillar 2: X-Poly Modules Lower Costs

Plan B

Pillar 5: On-Orbit Refueling And Infrastructure

Pillar 4: Water* As Fuel - Common Conventional and LARIT Engines

Pillar 3: Leverage Existing Human Rated Spacecraft

7.1 Refueling

To understand the power of refueling, consider that few national air forces in the world have made the investments to use air-to-air refueling. But for those that have made the investments, the modern combat-enabling results are profound. For the US military, all strategic missions and almost all tactical air force missions use refueling after launch. Air-to-air refueling is the simple but transformative design choice that enables global power projection (see Chart 73).

Nothing in Plan_B is more important than making this design decision—where re-fueling a run-dry space vehicle refreshes the vehicle to do another mission. This goes to the core of reusability—where a spacecraft becomes more like an aircraft. This therefore is a fundamental difference between Plan_A and Plan_B. Plan_A builds rockets big enough to launch and carry all the fuel required for all lunar mission phases. It is a **Plan_B Precept** to deliver lunar mission spacecraft into orbit (LEO or LLO) with near-empty fuel (called Dry). We add fuel (called Wet) via space-refueling to enable the craft to do its next mission phase. The US military has proven autonomous aircraft to autonomous aircraft airborne refueling is possible and will soon become routine.

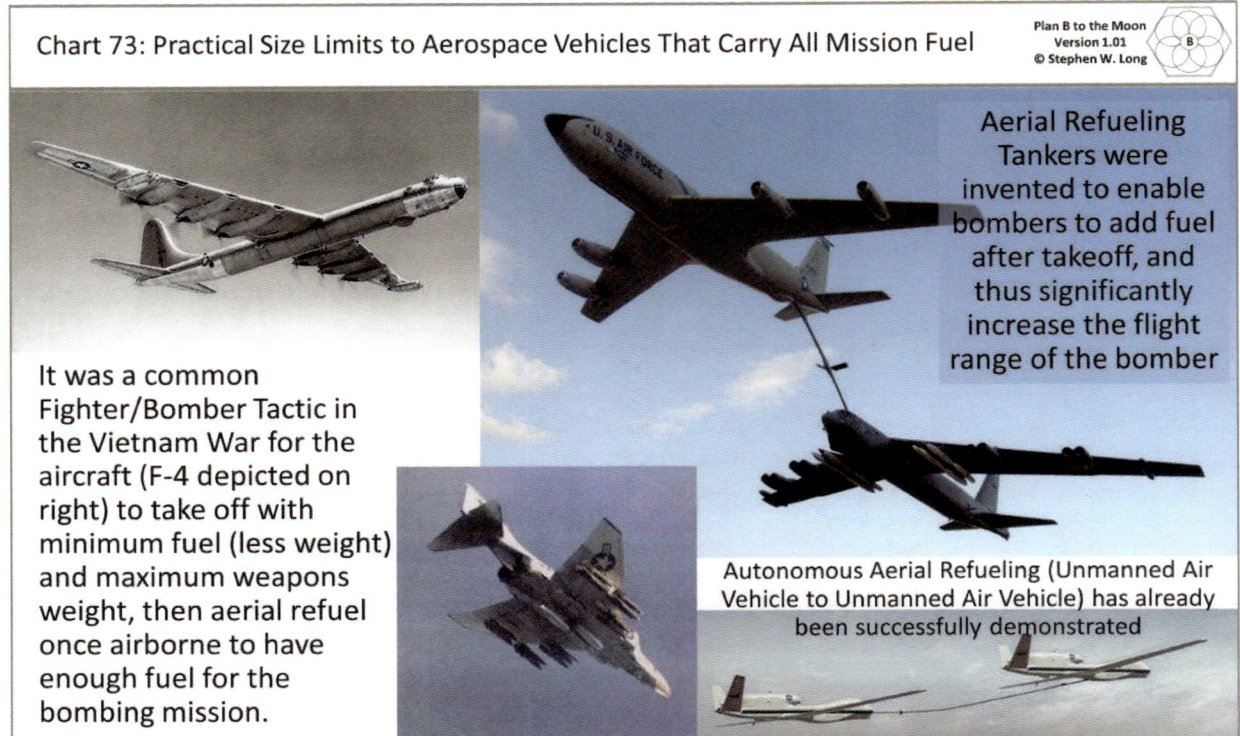

Chart 73: Practical Size Limits to Aerospace Vehicles That Carry All Mission Fuel

Plan B to the Moon
Version 1.01
© Stephen W. Long

Aerial Refueling Tankers were invented to enable bombers to add fuel after takeoff, and thus significantly increase the flight range of the bomber

It was a common Fighter/Bomber Tactic in the Vietnam War for the aircraft (F-4 depicted on right) to take off with minimum fuel (less weight) and maximum weapons weight, then aerial refuel once airborne to have enough fuel for the bombing mission.

Autonomous Aerial Refueling (Unmanned Air Vehicle to Unmanned Air Vehicle) has already been successfully demonstrated

Plan_B designs and uses the Autonomous-Space Water-Transfer-Vehicle (ASWTV) to deliver Water* from Earth to LEO and to move Water* from one spacecraft to another spacecraft. See Chapter 8.10 for additional ASWTV design details.

Plan_B designs and uses the Common-Tug-and-Refueling-Vehicle (CTugRV) to deliver cryogenic fuel from Earth to LEO and to move cryogenic fuel from one spacecraft to another spacecraft. The CTugRV uses the Common Conventional Engine (CCE) to provide on-LEO refueling station thrust, and Translunar Orbit Insertion burns (travel to and from the Moon). See Chapter 8.12 for additional CTugRV design details.

7.2 Space-Fuel-Terminal

It is a **Plan_B Precept** that deep space exploration using moon-based fuel will be significantly more affordable than previous earth-delivered-fuel mission planning paradigms. Plan_B proposes development of a fuel manufacturing facility in LEO, called the Space-Fuel-Terminal (SFT). Plan_B includes specific technical risk reduction strategies to build the Space-Fuel-Terminal to make fuel on-orbit and also deliver cryogenic fuel from Earth to LEO.

7.3 Space-Fuel-Terminal Design

Chart 74 depicts the ~five core SPM shell components of the Space-Fuel-Terminal. Right to left, there are H_2O/CO_2 storage tanks filled by ASWTV delivery missions (see Chart 75). The Electrolysis Reactors convert Water* into gaseous H, O_2 and CO_2. Sabatier Reactors produce gaseous CH4 and H2O. SFT calculations use the mass, volume, production outputs and power requirements of commercial earth-based electrolysis generators and notional Sabatier reaction engine QRMs. The Liquid C4H / O2 Compression module uses the native cold of space and cryogenic coolers to convert the gaseous C4H and O2 into Liquid C4H and LOX. Cryogenic Fuel Storage Nodules (CFSN) store the now-liquid fuels. The SFT uses identical Toroidal Storage Modules (TSMs) and Cryogenic-Storage-Modules (CSMs) designed for the ASWTV and the CTugRV.

The Plan_B ELTV nominal mission fuel load of 200,400 pounds is a mix of 43,565 pounds of liquid methane and 156,835 pounds of liquid oxygen. It will take 119,804 pounds of carbon dioxide and 98,022 pounds of water as raw materials to manufacture the 200,400 pounds of LCH4/LOX. It is a mission timeline / design **Plan_B Precept** that the SFT should create 200,400 pounds of LCH4/LOX within 90 days (and every 90 days after Water* raw stock deliveries). The SFT fuel creation system calculations (see Chapter 10.6) need 500 kW of instantaneous photovoltaic power to produce 200,000 pounds of fuel in 90 days. Chart 75 depicts how Water* enters the right side of the SFT and a CTug-RV receives finished liquid C4H / O2 fuel. There are no current plans for crew members to wash the windshields during refueling operations. Chart 75 also depicts the likely approach for Lunar-Lander or Lunar-Long-Boat refueling.

7.4 Space-Fuel-Terminal Power Array

Plan_B Version 1.00, Phase 2, delivers Space-Poly-Module-based capabilities to produce 504 kW of electrical power on LEO. Note that Plan_A delivers (produces) 40 kW of electrical power as part of their Deep Space Orbital Gateway. Plan_B will audaciously build and produce ten times more solar photovoltaic power on-orbit than Plan_A. Plan_B advocates creating high-power on-orbit to turn water and carbon dioxide into liquid methane (LCH4) and oxygen (LOX) for rocket fuel. Chapter 10.6 provides the calculations for the electric power and raw materials required to make rocket fuel for Plan_B ELTV missions.

Charts 76-78 depict the overview design of the SFT and the LARGE photovoltaic array. The SFT fuel factory throughput was matched to the fuel deliveries needed for lunar missions. The Plan_B array and photo-voltaic-panel designs use low-cost XPM modular components previously defined in Chapter 4. After SFT-01 operations prove routine on-orbit fuel creation is possible, it is only a financial decision when to build additional Space-Fuel-Terminals. Derivative SFT designs would-be built-in LEO, LLO and on the Lunar Surface. Chapter 10 (Mission MATRIX), calculates it will cost $2.45 Billion to build the first on-orbit SFT.

Plan_B includes building the SFT early in the Plan_B timeline (PHASE 2) as an example of the virtue of Colonization-patience as described in Chapter 1. Other Plan_B Phases ferry all required LCH4/LOX fuel up from Earth (to reduce program risk). It is therefore not critical to build the SFT for the return to the Moon. But as emotionally argued in Chapter 1, Plan_B is not about the quick fix—it is about the patience required to return to and colonize the Moon. As written in Chapter 1, one of the key reasons to return (land) on the Moon is because of the raw materials

(Ice and Helium 3) we will find there. Helium 3 may be vital to the future of energy production on the Earth. Ice from the Moon will profoundly lower costs for deep space exploration.

Therefore, the SFT is a necessary early investment to build the future of humans beyond the Earth. The sooner we harvest this potential of turning Lunar ice into rocket fuel, the sooner we reap the cost benefits. As we lower costs, we get closer to achieving long-term exploration objectives (see Chapter 1.4). Said with some asperity, the calculated costs of the SFT are as of Plan_B V1.00 less than the cost over runs already spent for the Plan_A Space Launch System.

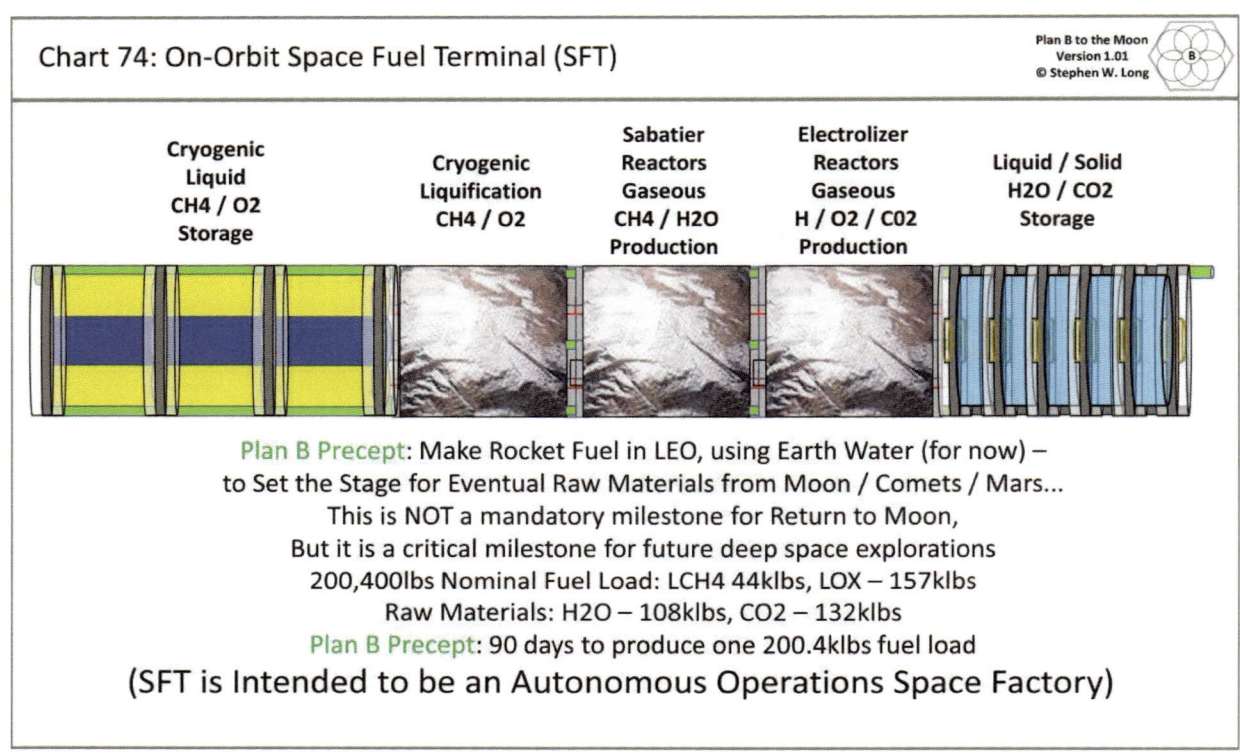

Chart 74: On-Orbit Space Fuel Terminal (SFT)

Cryogenic Liquid CH4 / O2 Storage	Cryogenic Liquification CH4 / O2	Sabatier Reactors Gaseous CH4 / H2O Production	Electrolizer Reactors Gaseous H / O2 / C02 Production	Liquid / Solid H2O / CO2 Storage

Plan B Precept: Make Rocket Fuel in LEO, using Earth Water (for now) –
to Set the Stage for Eventual Raw Materials from Moon / Comets / Mars...
This is NOT a mandatory milestone for Return to Moon,
But it is a critical milestone for future deep space explorations
200,400lbs Nominal Fuel Load: LCH4 44klbs, LOX – 157klbs
Raw Materials: H2O – 108klbs, CO2 – 132klbs
Plan B Precept: 90 days to produce one 200.4klbs fuel load
(SFT is Intended to be an Autonomous Operations Space Factory)

It is a **Plan_B Precept** to build the first SFT in LEO to develop and validate the technologies and mission concepts. Later Plan_B versions will build additional SFTs on the Lunar surface and in LLO. As depicted in the Plan_B Version 15.0-time horizon, one or more deep space missions will capture a comet and bring the comet into Lunar or Earth orbits. Comets would provide the ice– the space fuel–for future decades of deep space exploration. Until then, we will have to haul Water* up from Earth until we can mine ice on the Moon.

It is a **Plan_B Precept** to use the native cold of space to assist with the compression of gaseous CH4 and O2 into liquid cryogenic fuels. Previous NASA research has developed Advanced Solar Shades. The Shades cool spacecraft via sun-reflection and provide thermal protection down to temperatures as low as 33 Kelvin. The Plan_B Team conservatively believes passive shading can get gases down to 100 Kelvin (see Chapter 6.1). Existing space-rated multi-stage cryo-coolers will get down to LOX 90 Kelvin.

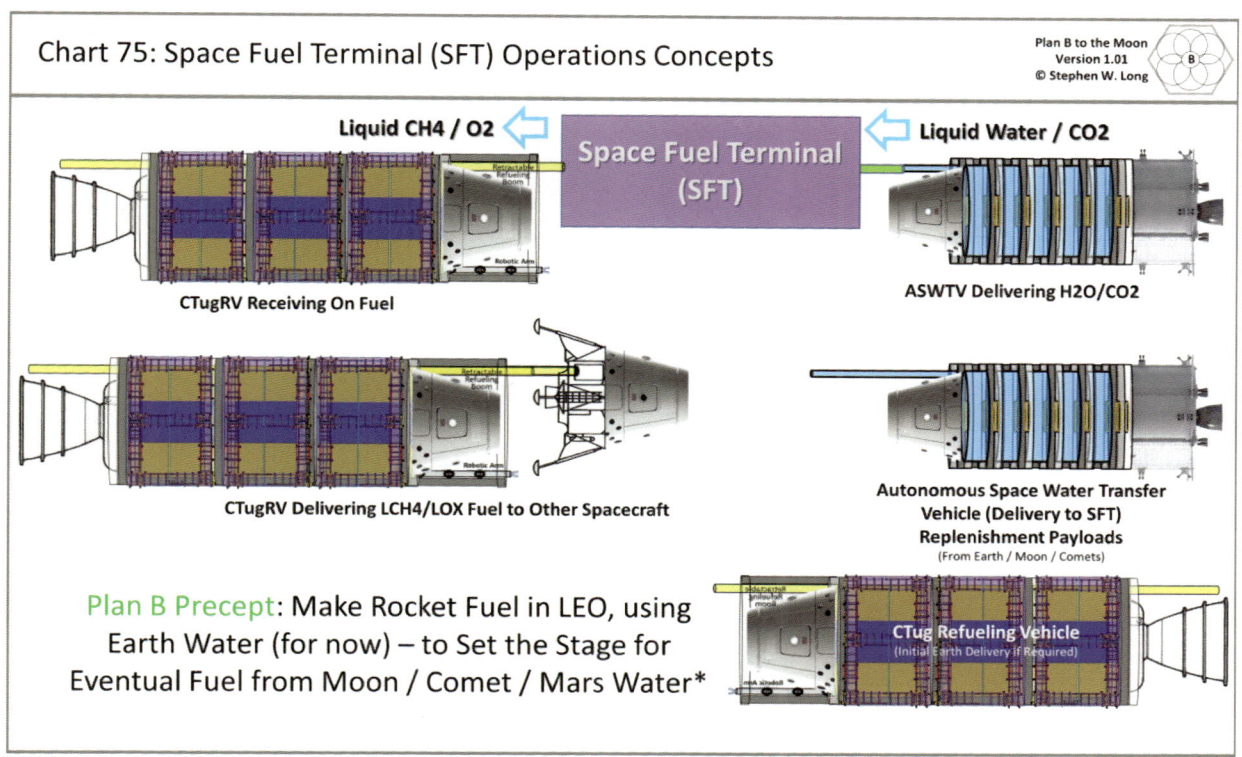

Chart 75: Space Fuel Terminal (SFT) Operations Concepts

Plan B to the Moon
Version 1.01
© Stephen W. Long

Liquid CH4 / O2

Space Fuel Terminal (SFT)

Liquid Water / CO2

CTugRV Receiving On Fuel

ASWTV Delivering H2O/CO2

CTugRV Delivering LCH4/LOX Fuel to Other Spacecraft

Autonomous Space Water Transfer Vehicle (Delivery to SFT) Replenishment Payloads
(From Earth / Moon / Comets)

Plan B Precept: Make Rocket Fuel in LEO, using Earth Water (for now) – to Set the Stage for Eventual Fuel from Moon / Comet / Mars Water*

CTug Refueling Vehicle
(Initial Earth Delivery if Required)

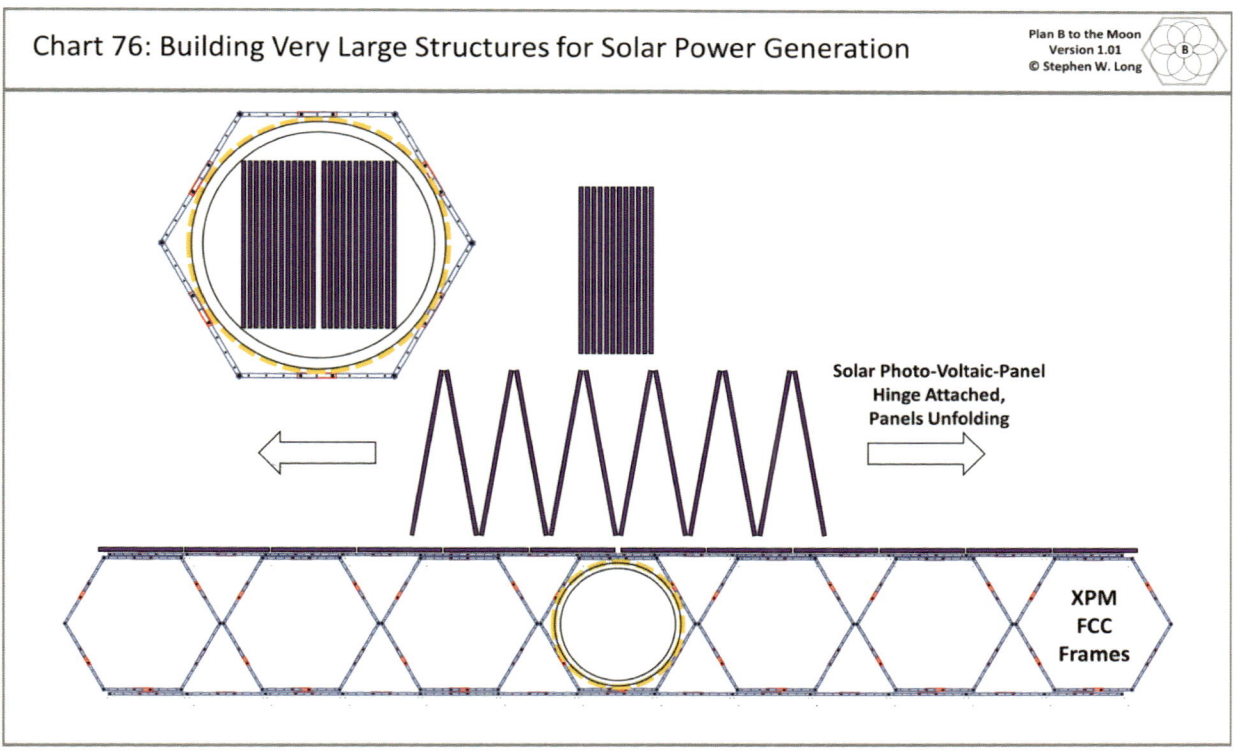

Chart 76: Building Very Large Structures for Solar Power Generation

Plan B to the Moon
Version 1.01
© Stephen W. Long

Solar Photo-Voltaic-Panel Hinge Attached, Panels Unfolding

XPM FCC Frames

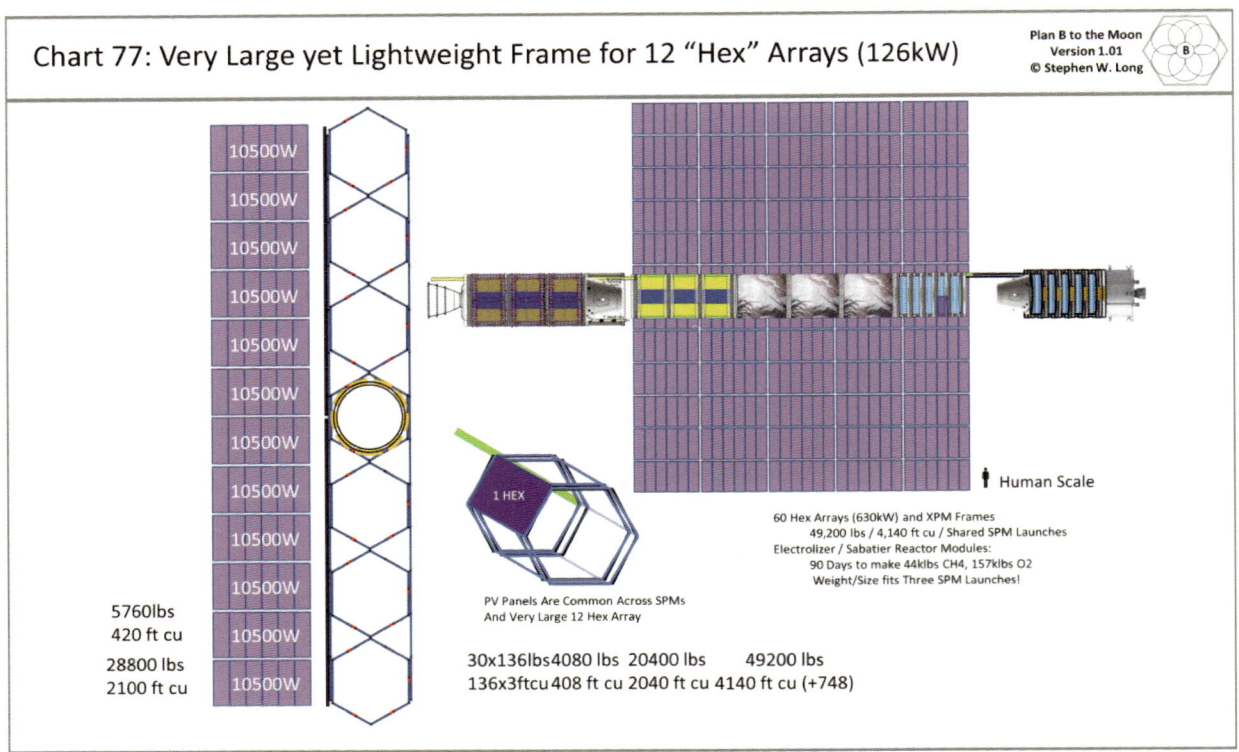

Chart 77: Very Large yet Lightweight Frame for 12 "Hex" Arrays (126kW)

10500W (×12)

5760lbs
420 ft cu

28800 lbs
2100 ft cu

1 HEX

PV Panels Are Common Across SPMs
And Very Large 12 Hex Array

Human Scale

60 Hex Arrays (630kW) and XPM Frames
49,200 lbs / 4,140 ft cu / Shared SPM Launches
Electrolizer / Sabatier Reactor Modules:
90 Days to make 44klbs CH4, 157klbs O2
Weight/Size fits Three SPM Launches!

30x136lbs 4080 lbs 20400 lbs   49200 lbs
136x3ftcu 408 ft cu 2040 ft cu 4140 ft cu (+748)

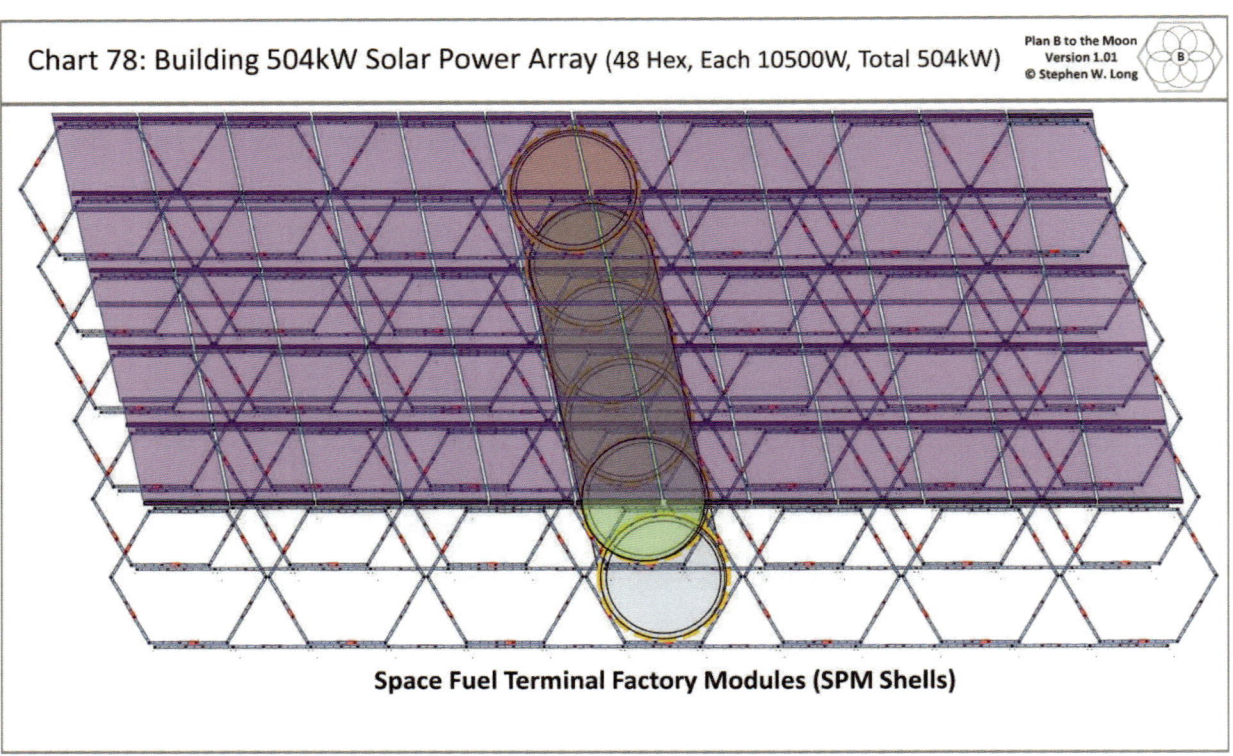

Chart 78: Building 504kW Solar Power Array (48 Hex, Each 10500W, Total 504kW)

Space Fuel Terminal Factory Modules (SPM Shells)

Chapter 8: Catalog of Plan B Spacecraft

8.1 Introduction

Chapter 8 introduces and summarizes all Plan_B spacecraft and provides details for spacecraft designs based on combinations of general-purpose SPMs (Chart 79). SPMs are in-turn assembled into larger structures, using defined connecting components.

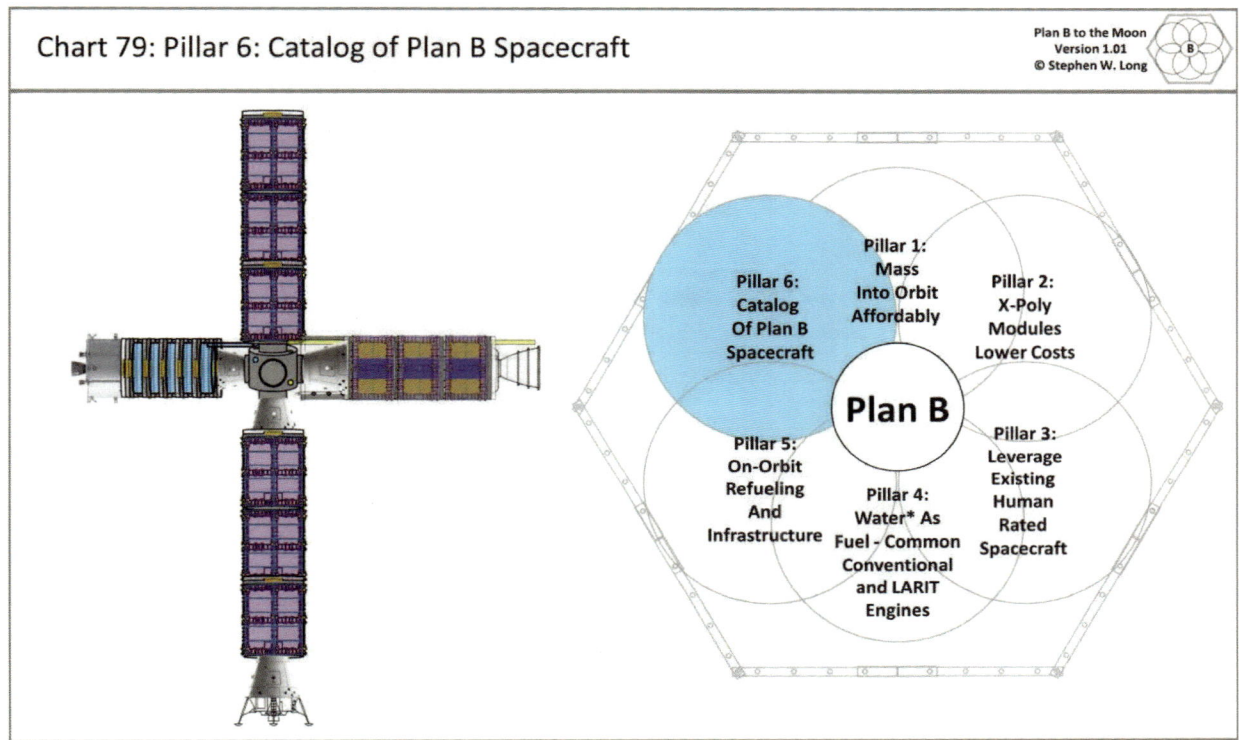

In some long-term future, organizations with large checkbooks may have ready access to a commercial catalog. From this catalog one can easily purchase spacecraft—as you can do today to choose model features and purchase automobiles. Since that spacecraft catalog does not exist (yet), the Plan_B Team had to create its own Catalog of Spacecraft, from which we developed mission plans and calculated program costs. Before Plan_B could peruse through a Catalog of Spacecraft, we had to create the catalog of spacecraft parts—the XPMs. XPM design details are in Chapters 4, 5, 6, and 7.

Plan_B Spacecraft look alike because they share common DNA. They are all built from standard XPM parts into standard Space-Poly-Modules (SPMs), grouped into standard Stations or other large structures. This is a key XPM-to-SPM construction paradigm premise. These identical building block components are also easy to assemble into Lunar-Habitats and eventually Martian and other deep space habitats. By using common building blocks, we reduce risk (human life and program cost risks) by learning to assemble and refine deep space components in the relative safety of LEO. We use those same earned skills (and components) to assemble LLO Gateway structures, and Lunar Base Structures.

Chart 80 illustrates the **Plan_B Precept** that the various common components enable different combinations and different configurations to create specialized spacecraft. Combinations of XPMs, SPMs, the Common-Space-Command-Module (CSCM), the Common-Space-Service-Module (CSSM), and the Common Conventional Engine (CCE) enable hundreds of derivative, composite structures and vehicles.

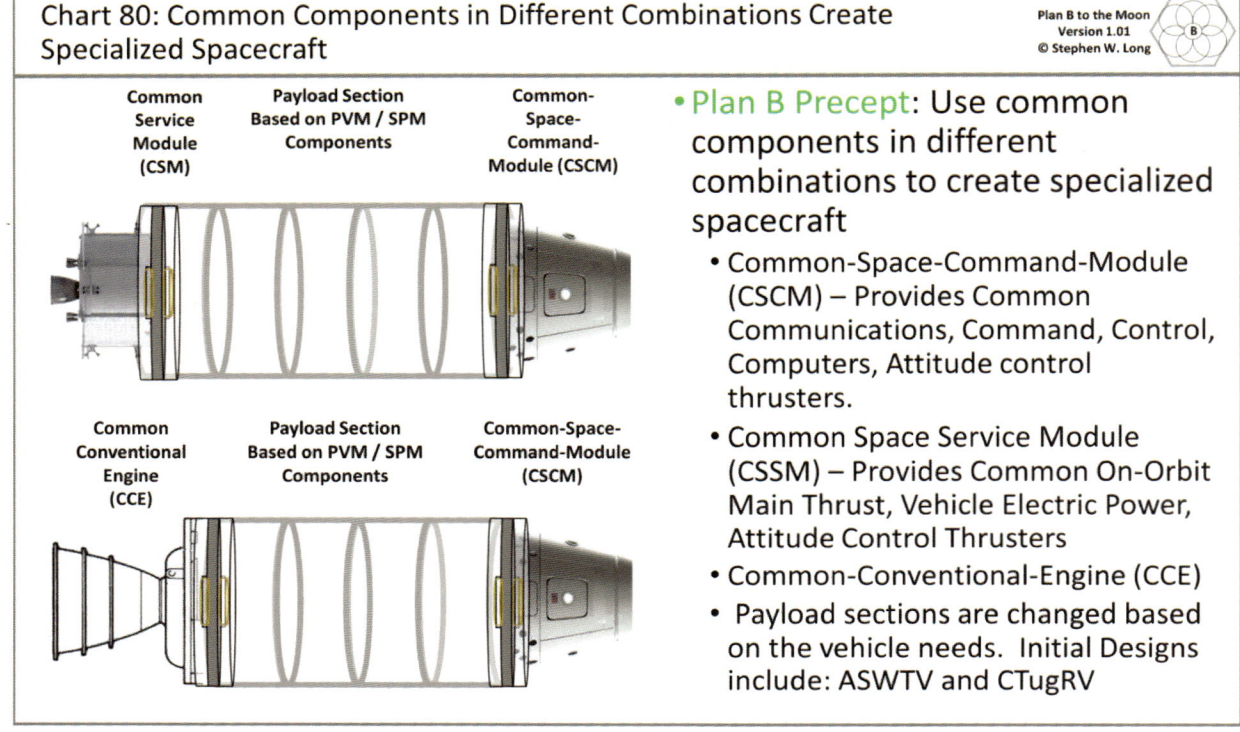

Chart 80: Common Components in Different Combinations Create Specialized Spacecraft

Plan B to the Moon
Version 1.01
© Stephen W. Long

Common Service Module (CSM) — Payload Section Based on PVM / SPM Components — Common-Space-Command-Module (CSCM)

Common Conventional Engine (CCE) — Payload Section Based on PVM / SPM Components — Common-Space-Command-Module (CSCM)

- **Plan B Precept**: Use common components in different combinations to create specialized spacecraft
  - Common-Space-Command-Module (CSCM) – Provides Common Communications, Command, Control, Computers, Attitude control thrusters.
  - Common Space Service Module (CSSM) – Provides Common On-Orbit Main Thrust, Vehicle Electric Power, Attitude Control Thrusters
  - Common-Conventional-Engine (CCE)
  - Payload sections are changed based on the vehicle needs. Initial Designs include: ASWTV and CTugRV

Chart 81 summarizes the Plan_B Catalog of Spacecraft Designs. All of spacecraft are all based on common XPM building blocks and related parts.

Some Plan_B spacecraft arrive on-orbit ready to function, they arrive on Earth-orbit assembled and fueled. Other classes of spacecraft are ready to operate after refueling on-orbit. Stations and Gateways and similar large structures will require extensive on-orbit assembly. There is the existence proof (the International Space Station) that astronauts, using space walks and remote-controlled manipulation arms can construct large space habitats on-orbit. The ISS proves we know how sequentially to launch small modules, and over time assemble the modules into large structures. The US Space Shuttle missions delivered about 50,000 pounds of cargo (such as ISS modules) into LEO per mission. It took many Space Shuttle and other vehicle launches to get the current ISS mass (900,000 pounds) into orbit to assemble the today-large ISS on-orbit. Press accounts report that the ISS cost $150 Billion (yes Billon) dollars to build.

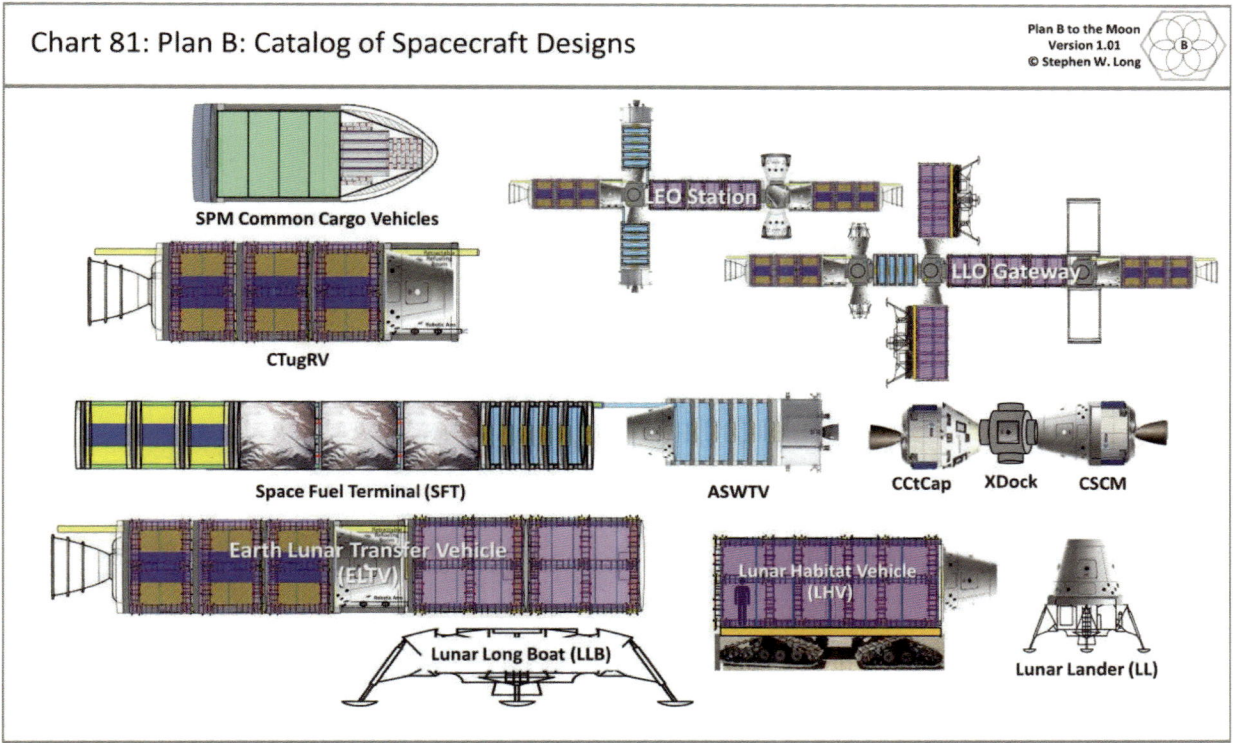

The Plan_B Team further notes highly trained astronauts have the fine motor skills even inside bulky spacesuit gloves to repair the Hubble Space Telescope. We as a species have therefore shown we have the fine manipulation skills necessary to assemble and plug-and-play together building block parts on-orbit into larger space structures.

Astronauts will assemble each spacecraft type from XPM or similar parts in various configurations. Chapter 5 includes the space-conventional components for Plan_B spacecraft– the Common-Space-Command-Module (CSCM), the Common-Space-Service-Module (CSSM) and the Common Conventional Engine (CCE). We use these components in Plan_B designs.

8.2 Unplanned but Enabled Creation

There is a **Plan_B Precept** of previously unplanned but enabled creation. Plan_B designs and constructs early prototypes of Initial Operating Capabilities (IOC) of the first spacecraft, structures and habitats for LEO / LLO / LS. We believe the eventual Lunar and beyond Colonists will reach beyond the design constraints of earthbound human imaginations.

Plan_B Versions 2.0 and 3.0 presents plans to build long-term permanent colonies from sufficient quantities of building block ~raw materials (XPMs) and tools made in situ. The right tools will enable the colonists to build the structures and infrastructure THEY need and want to build–based on the field conditions they experience at hand. It is in this mindset that the Plan_B Model-Poly-Kits (MPKs) become so important. The Plan_B team believes that highly adaptable and motivated humans will experiment with and try out dozens if not hundreds of model habitat design combinations over just hours or days. As heroically depicted in the Apollo 13 movie, the leader told the team, "We have to build a working CO2 scrubber out of the parts on the table." And they did.

If you have already built your future home as a scale model, it is much easier to build the real-thing with life-sized (more massive) components. By model-building-first you discover problems early. Early discovery affords one time to seek solutions early in the development process. You can change (measure twice cut once) the attempted design before you find yourself in the middle of false-start unusable designs.

After Plan_B Version 2.0 and later, we will only need to send tools such as additive manufacturing robots (3D printers and ~ink) into space. The colonists will print their XPM components using the delivered ink or site-available raw materials. Note that current 3D printers use large spools of thin-wire engineered plastics and metals. For instance, the equivalent to a Pressure-Vessel-Module (PVM) shell may have walls built from available lunar ice (think stereotype igloos). Printed plastic or metal matrices cover the surfaces to keep the ice from vacuum sublimating. Concepts for printable concrete homes on Earth are already in press reports. By the time the first Plan_B launches are underway in the 2020s, 3D printing using metal and plastic stock spool materials may produce all XPM components.

The Plan_B team predicts some of the greatest joys and eureka moments of the future will arise from the imaginations of highly invested/motivate humans. They will build habitats and spacecraft structure combinations never conceived by earthbound engineers/scientists. We note again that computational biology (also computational electromagnetic antenna design) experiments show that experimenting with millions of designs yields unexpected and better result designs. Committees of consensus-view limited humans classically produce ordinary designs. Extraordinary designs emerge from millions of near-zero-cost experimental trials.

8.3 Future Concepts for Deep Space–Motion Imaging Applications

Chapter 8.3 presents preliminary findings. The Society of Motion Pictures and Television Engineers (SMPTE) standard ST-2110 is a likely core enabling technology framework for motion imaging applications throughout Plan_A and Plan_B. ST-2110 moves motion imaging from the professional video SDI world to the IP world. HD, 2K, 4K, 8K–the frame size does not matter, you only need IP bandwidth to build video-and-audio-enabled networks. IP networks leverage commercial networking technology and components. This is NOT the packet-collision-prone Internet. These systems perform well over closed packet networks built inside facilities (or spacecraft). This technology also supports remote location connections. The key implication is that ST-2110 enables live streaming and control of super-high-resolution video signals from anywhere to anywhere using packet-based networks.

This is the next-step-forward / answer for NASA Motion Imaging from deep space. Instead of having to develop new technology, we can instead leverage COTS gear. We only need to get COTS technology space qualified and we will have all the parts we need. If you already require and plan to buy the network infrastructure, you no longer have to build a separate TV infrastructure. Instead you remote the IP-based cameras, etc. back to your central processing plant. The way forward will be for NASA to focus on getting the networks right after which any needed motion imaging technology will be commercially available.

8.4 Future Concepts for Deep Space Communications–Optical RF

For advanced, high-bandwidth networks, Optical RF is a very strong candidate for deep space communications. Such near-optical RF frequencies (typically around 94 GHz) have the bandwidth of laser communications without the laser problems. The technological stretch is to find manufacturers to build the ~94 GHz gear with sufficient power (and compact antennas) to reach from LEO to the Moon to LEO. From LEO, we have many options to get signals to and from the Earth's surface. Atmosphere and rain massively attenuate 94 GHz signals, so these frequency bands are unsuitable for Earth to LEO to Earth communications. In the vacuum of space, such near-optical RF systems may be perfect for Plan_B communications.

Perhaps now is the time to petition for blocks of spectrum for exclusive use in deep space. Such exclusive segregation would separate deep space communications from earth communications. This would be a stepping stone to help future-proof deep space comms–allowing space comms to grow without the typical constraints caused by high RF congestion on the Earth. It also has the side benefit of protecting deep space communications from Earth mischief. Because near-optical RF frequencies get blocked by Earth's atmosphere, space use of near-optical RF reduces chances of Earth-surface hostile powers interfering with deep space communications.

8.5 Future Concepts for Deep Space Communications–Gateway

Where is the best place for a Deep Space Communications Gateway (DSCG)? Candidate locations include Earth GEO and Moon Lagrange 1. LaGrange 1 is about two-thirds the distance between the Earth and the Moon. Any spacecraft parked there will stay there, appearing to hover over a fixed Lunar geocoordinate. It is likely we will also need a Triangle (maybe quad) of satellites orbiting the Moon to provide Deep Space relay signals. Moon/Earth rotation sometimes blocks the views of L1 from Deep Space. With tongue planted in cheek, Plan_B notes we need new descriptive languages for lunar activities—the Earth uses geo coordinates, how do we label lunacy-coordinates?

An even more profound question would be to ask if Lagrange 1 should be the orbital parking spot of the Lunar Orbiting Gateway, instead of LLO. Perhaps the Deep Space Station (DSS) needs to park at L1 and not Lunar Low Orbit. In this concept, all moon, etc. traffic would travel to the L1 station first and proceed to LLO and the Lunar Surface. Future research will calculate the Delta V to get from LEO to L1 and to the Lunar Surface. If the Delta V math works, perhaps the assembly of all Moon and Deep Space vessels will be at L1. From L1 spacecraft then proceed to the Moon or to Deep Space. Do not confuse the Earth-Moon Lagrange 1 with the Sun-Earth Lagrange 1.

8.6 XDock

Plan_B on-orbit operations will need a common docking module, called the XDock. As depicted in Chart 82, the XDock has six docking ports: Alpha, Bravo, Charlie, Delta, Echo and Foxtrot.

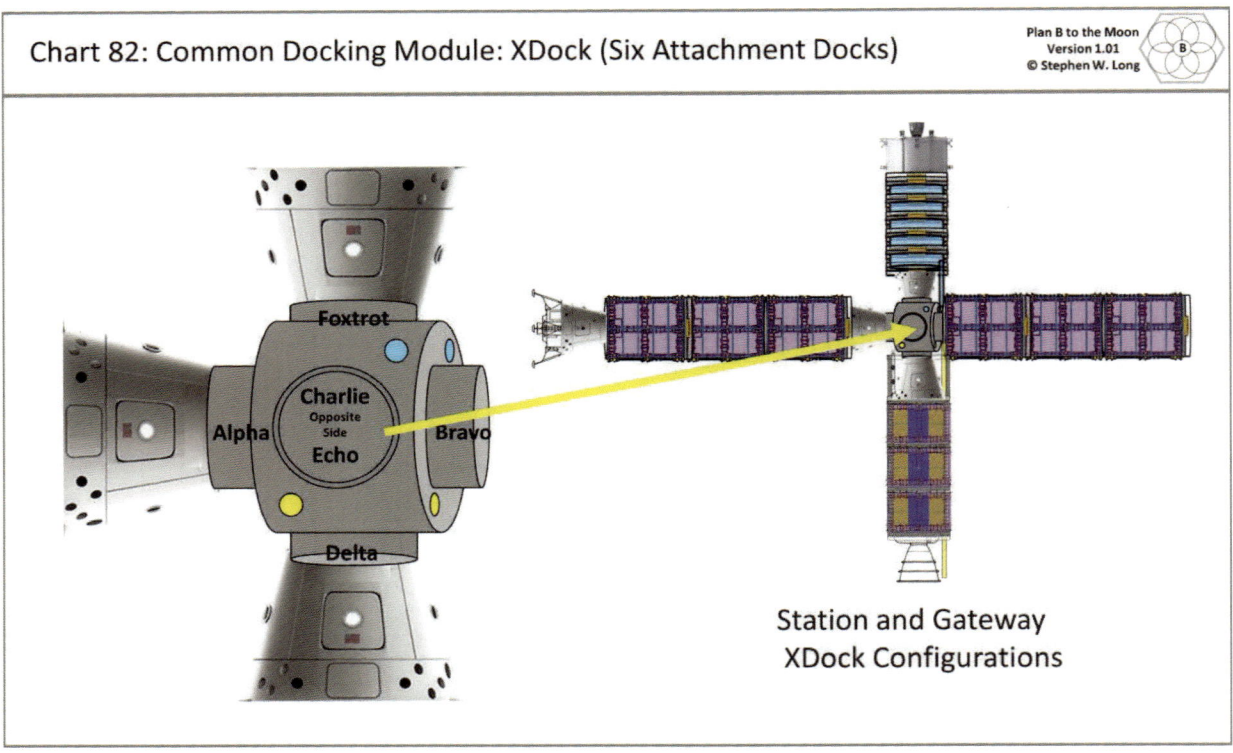

Think of the XDock as a short soda can shape with hatch / docking port openings on each end (Alpha, Bravo). The can has axially opposing docks (Charlie, Delta) and 90 degree offset axially opposing docks (Echo and Foxtrot). Alpha and Bravo docks and the XDock shell must handle the potential compression and shear-stresses of large-coupled-masses. The XDock Alpha Bravo axis will serve as the connecting bridge between two or more other spacecraft—of likely significant mass. It will also be necessary for Alpha and Bravo docks to support high torsional stresses. Plan_B missions include large mass spacecraft dockings. Thrust-pushes-through the XDock and moves the entire combined mass as a joined spacecraft stack. Alpha and Bravo docks will also need to handle counter thrusts (undocking), where opposite end spacecraft will undock and separate. Charlie, Delta, Echo and Foxtrot docks support single, smaller mass vehicle dockings. Single ship dockings include instances where a CCtCap vehicle joins LEO station, or a Lunar-Lander rejoins the LLO Gateway.

8.7 SPM Common Core Structures

Plan_B builds Space-Poly-Modules (Charts 83 and 84) from the previously defined (Chapter 4) standard XPM parts catalog of:

- Pressure-Vessel-Modules (PVM)
- Pressure-Bulkhead Modules (PBMs) (Outer Segment, Inner Segment)
- Frame-Core-Components (FCCs)
- Shielding-Storage-Tanks (SSTs)
- Transverse-Raceway-Modules (TRMs) (tubes)
- Linear-Fuel-Cells (LFCs) (tubes)
- Water-Purification-Electrolysis (WPE) (tubes)
- Photo-Voltaic-Panels (PVPs)

The XPM components get packaged together and delivered to LEO. The nose cone fairing gets ejected once out of the atmosphere. Once reaching the desired parking orbit, crew or robot arms extract the cargo XPMs. After cargo removal, the Pressure-Bulkhead-Module Inner Segment extends, sealing the open end of the PVM (see Chart 83). Frame-Core-Components (FCCs) snap into place around the PVM. Transverse-Raceway-Modules (tubes), Linear-Fuel-Cells (tubes), and Water-Purification-Electrolysis (tubes) fit within and are attached to the outer faces of the FCCs. Shielding-Storage-Tanks install between the Frame-Core-Components linear rails, and the Photo-Voltaic-Panels cover the entire outer surface. With prior planning, appropriate Quadrant-Rack-Modules are pre-loaded inside the interior of the PVM, which will allow normal cargo extractions from the open end of the PVM.

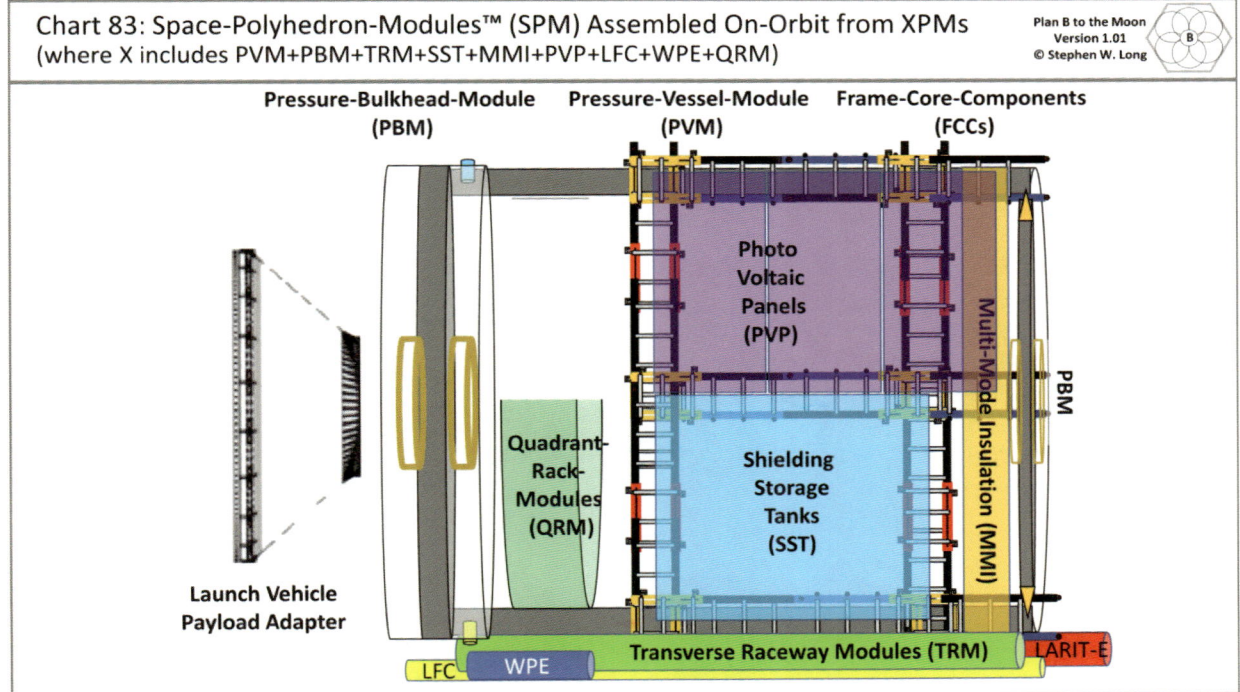

Chart 83: Space-Polyhedron-Modules™ (SPM) Assembled On-Orbit from XPMs (where X includes PVM+PBM+TRM+SST+MMI+PVP+LFC+WPE+QRM)

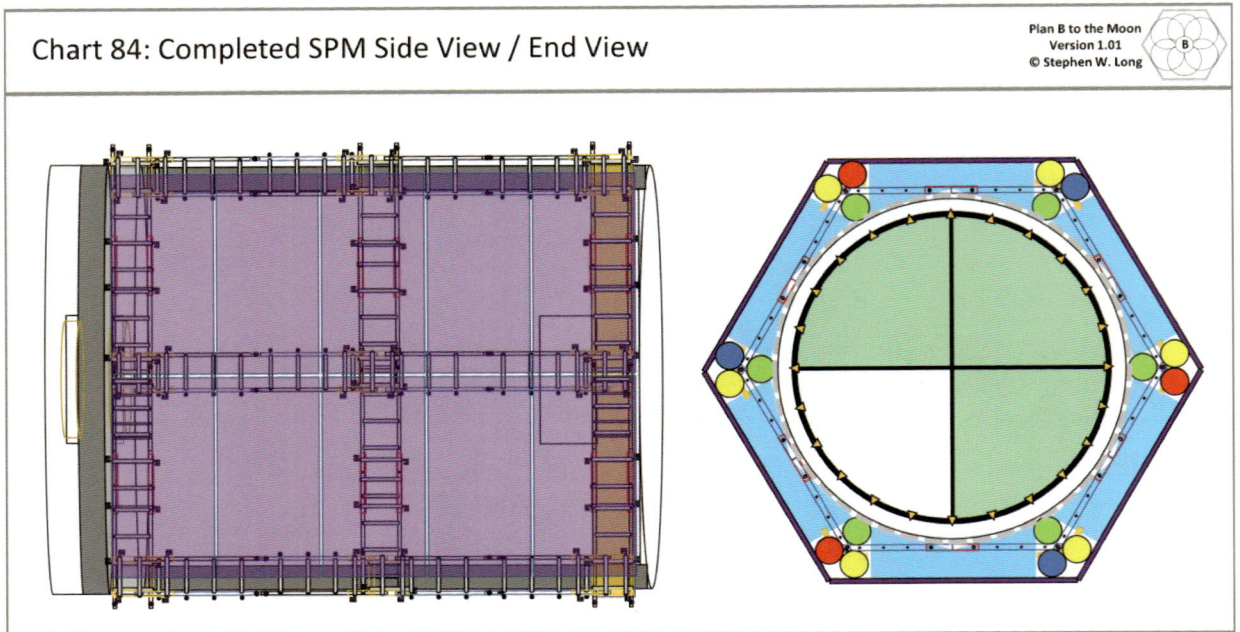

Chart 84: Completed SPM Side View / End View

Plan B to the Moon
Version 1.01
© Stephen W. Long

8.8 LEO Station-01V1

Low Earth Orbit (LEO) Station-01 Version 1 starts with PVM basic shells transformed into SPMs (Charts 85 and 86). Multiple FH Rockets launch and deliver Station component SPMs into orbit. Mission Circles 101, 128, and 134 deliver the three core SPMs (PVMs and XPM cargo) for LEOS-01 (Chart 86). Note that the Pressure-Bulkhead-Module (PBM) Outer Segment fits around the opposite end PBM Inner Segments, sealing the PVM outer shell. This approach provides a duel hatch design, meaning any SPM segment can serve at either end or the middle of strings of SPM modules. This enables incremental assembly of Station-01—adding pearls on a string. LEOS-01 becomes a much more complex station configuration as docking modules and visiting docking spacecraft Attach and Unattach over the lifetimes of Plan_B Missions.

Chapters 9 and 10 define the Mission Plans and MISSION MATRIX data for building, crewing, fueling and using LEOS-01.

8.9 LLO Gateway-01V1

Low Lunar Orbit (LLO) Gateway-01 Version 1 starts with PVM basic shells transformed into SPMs (Chart 87). LLOG-01 becomes a complex station configuration as docking modules and visiting docking spacecraft Attach and Unattach over the lifetimes of Plan_B Missions (see Chart 88).

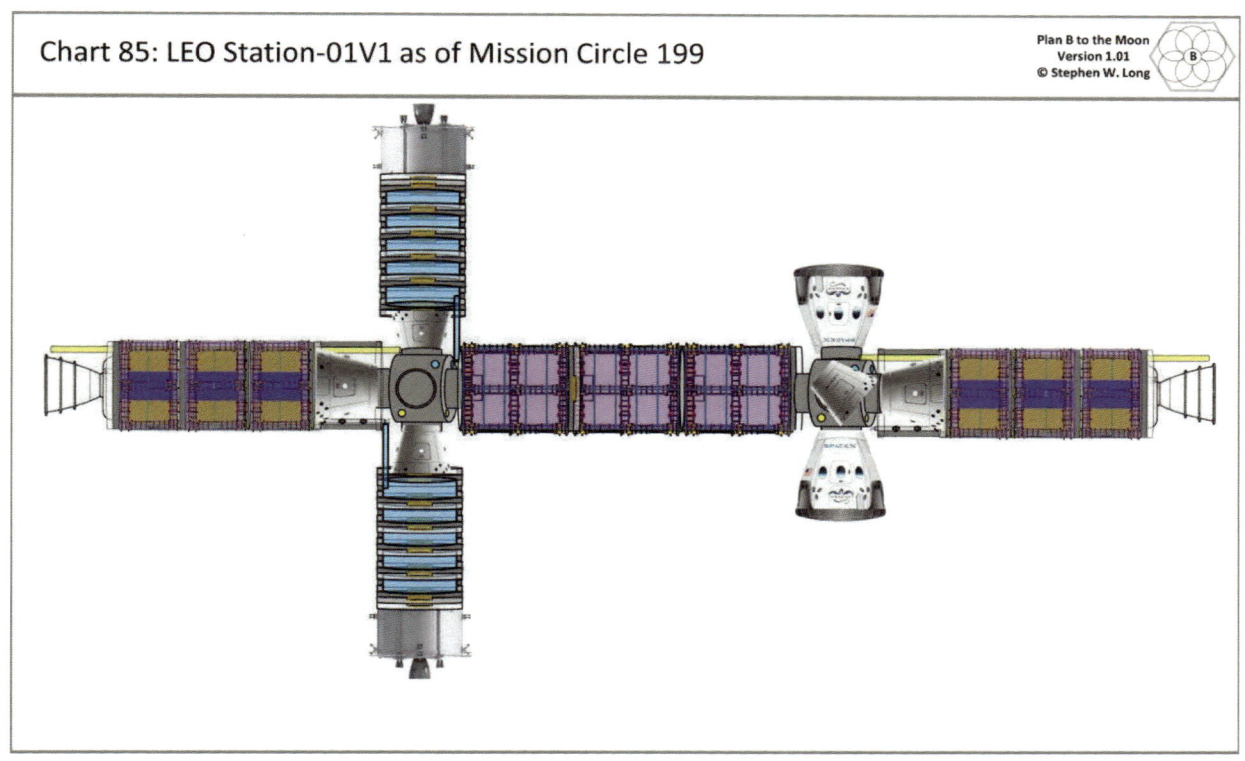

Chart 85: LEO Station-01V1 as of Mission Circle 199

Plan B to the Moon
Version 1.01
© Stephen W. Long

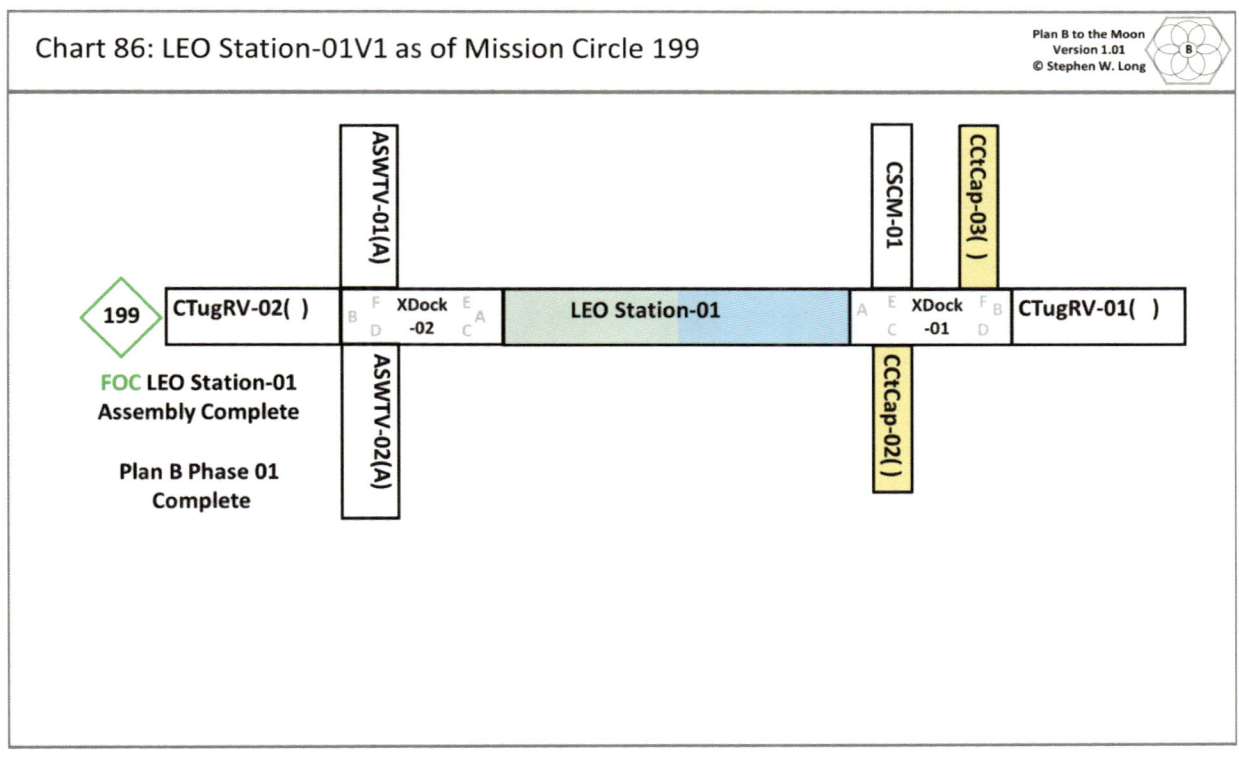

Chart 86: LEO Station-01V1 as of Mission Circle 199

Plan B to the Moon
Version 1.01
© Stephen W. Long

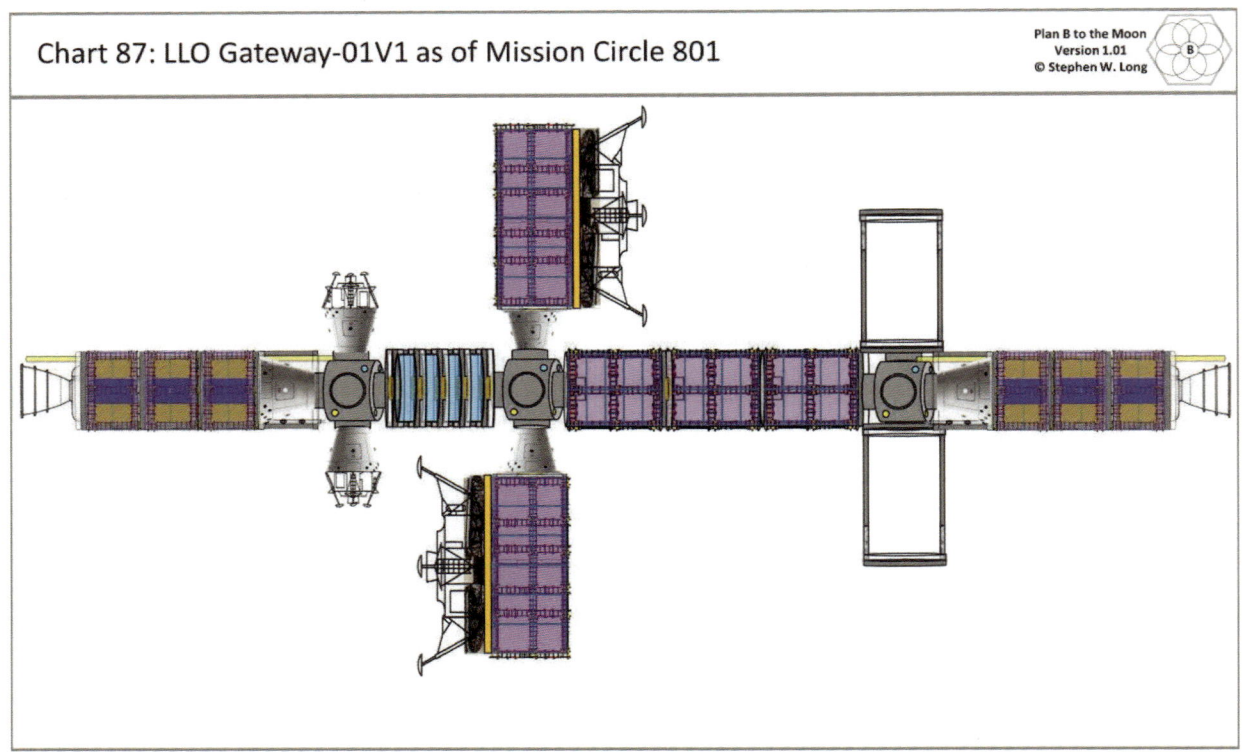

Chart 87: LLO Gateway-01V1 as of Mission Circle 801

Plan B to the Moon
Version 1.01
© Stephen W. Long

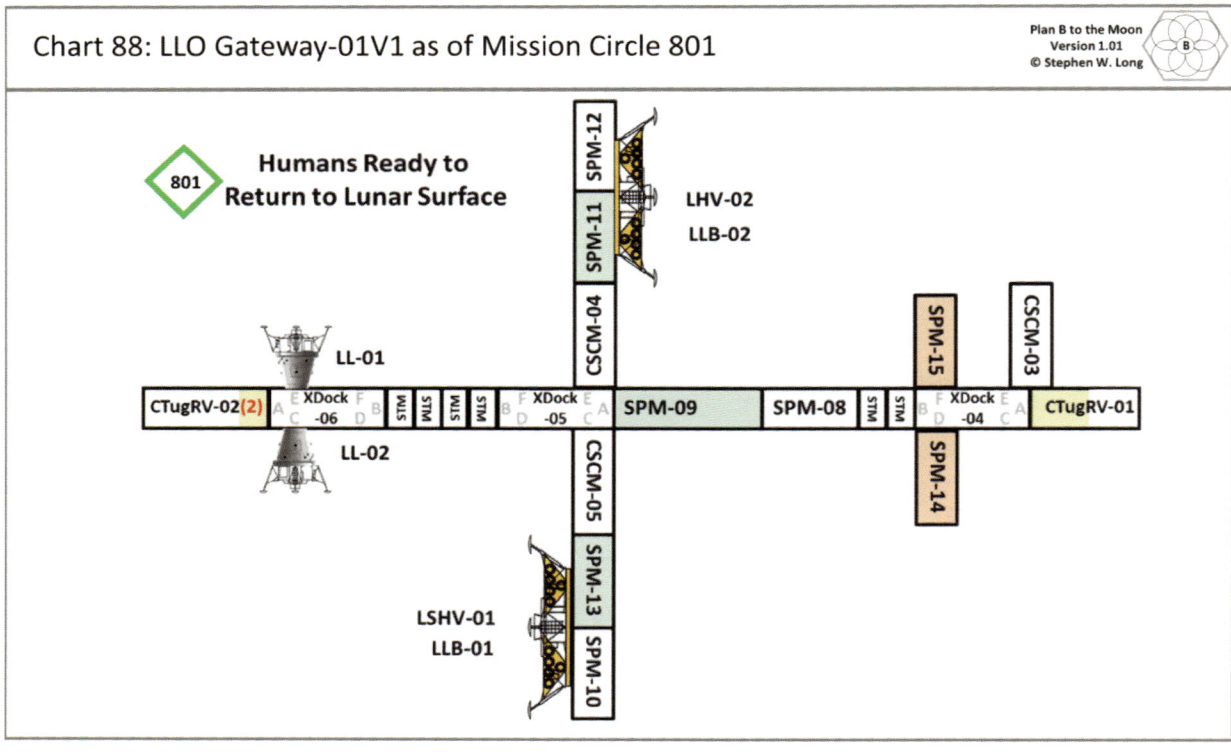

Chart 88: LLO Gateway-01V1 as of Mission Circle 801

Plan B to the Moon
Version 1.01
© Stephen W. Long

8.10 Autonomous-Space Water-Transfer-Vehicle

Chart 89 depicts the Autonomous-Space Water-Transfer-Vehicle (ASWTV). ASWTV uses the Common-Space-Command-Module (CSCM) for Common Communications, Command, Control, Computers, and Attitude control thrusters. The Common-Space-Service-Module (CSSM) provides On-Orbit Main Thrust, Vehicle Electric Power, and Attitude Control Thrusters. Plan_B mission sequences use the ASWTV as an Optionally Manned craft, with primary Unmanned operations.

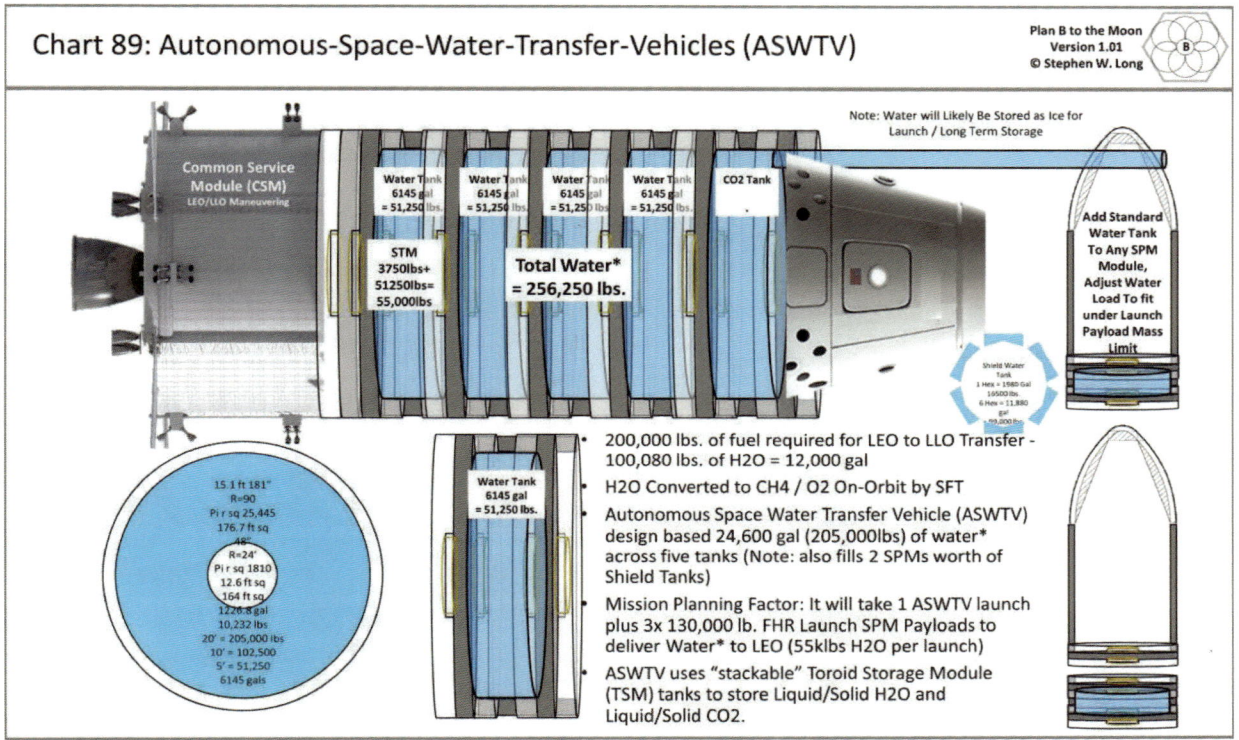

Chart 89 depicts the ASWTV using stackable Storage Toroid Module (STM) tanks to store Liquid/Solid H2O and Liquid/Solid CO2. STM tanks are also optional components for SPM cargo launches–removing the need to burden PVM and SPM designs with Liquid/Solid storage systems. The Toroid design (donut hole) allows eventual crew passage through the STM, enabling STM use with chained / coupled SPMs. Note that the ASWTVs and CTugRVs are, by choice, very similar in designs. The smaller Delta-V of the CSSM suffices to move Water* in LEO from one craft to another. The raw Water* is not a combustible propellent fuel, so it needs the CSSM to host the rocket engine and the rocket fuel storage tanks.

A standard PVM is 15.1-foot wide. Given a center hole (toroid hole) of 48 inches, five linear feet of STM length will store 6145 gallons, or 51,250 pounds of water. The empty (dry) STM design weight is 3750 pounds. The wet STM weight is 55,000 pounds. A stack of five STMs provides the most useful size for an on-orbit Water* transfer vehicle–transporting a total max on-orbit wet weight of 256,250 pounds.

The STM supports multiple spacecraft and design stack applications. It is a **Plan_B Precept** that not only are STMs combined to build ASWTVs; we add STMs to standard FH Rocket launch stacks, providing an easily adjustable ballast component to any launch manifest (Chart 90). We can add STMs and adjust the water carried by the STMs to maximize the payload delivered into LEO. Plan_B Mission Circle planning documents (see Chapter 9) often add adjustable ballast loads to maximize launch mass value.

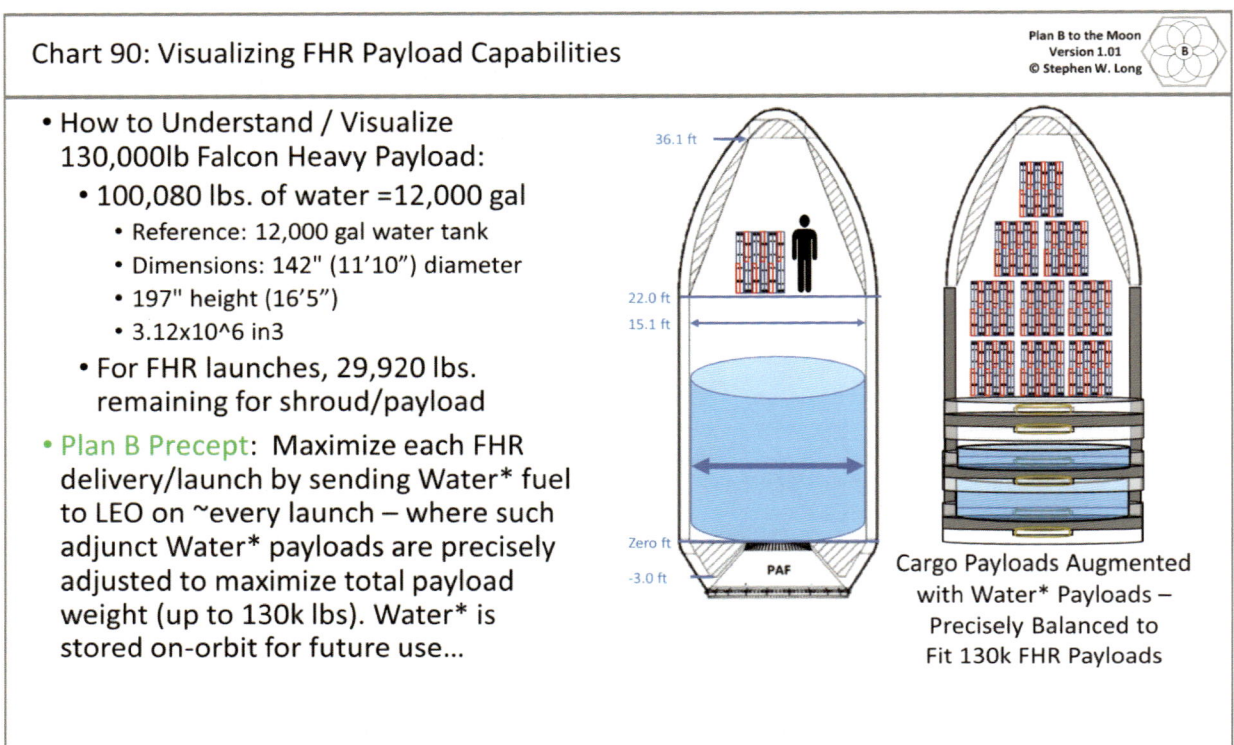

Chart 90: Visualizing FHR Payload Capabilities

Plan B to the Moon
Version 1.01
© Stephen W. Long

- How to Understand / Visualize 130,000lb Falcon Heavy Payload:
  - 100,080 lbs. of water =12,000 gal
    - Reference: 12,000 gal water tank
    - Dimensions: 142" (11'10") diameter
    - 197" height (16'5")
    - 3.12x10^6 in3
  - For FHR launches, 29,920 lbs. remaining for shroud/payload
- Plan B Precept: Maximize each FHR delivery/launch by sending Water* fuel to LEO on ~every launch – where such adjunct Water* payloads are precisely adjusted to maximize total payload weight (up to 130k lbs). Water* is stored on-orbit for future use...

Cargo Payloads Augmented with Water* Payloads – Precisely Balanced to Fit 130k FHR Payloads

8.11   Water-Transfer-Vehicle

Water-Transfer-Vehicles (WTV) are simplified ASWTVs used to ferry water from Earth to LEO (Chart 91). WTV designs maximize water payload, thus the WTV is just a stack of STMs, with no CCE or CSCM. Future analysis will determine methods for on-orbit small thruster maneuvers.

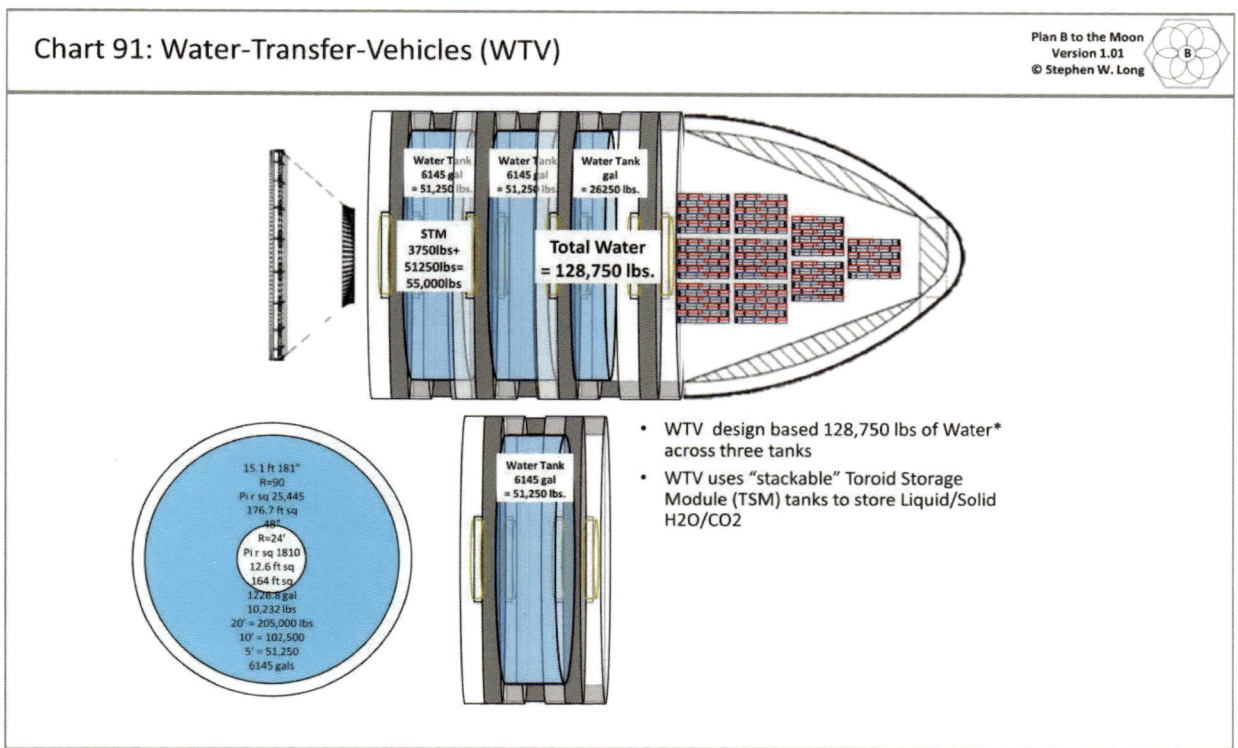

Chart 91: Water-Transfer-Vehicles (WTV)

Plan B to the Moon
Version 1.01
© Stephen W. Long

- WTV design based 128,750 lbs of Water* across three tanks
- WTV uses "stackable" Toroid Storage Module (TSM) tanks to store Liquid/Solid H2O/CO2

8.12 Common-Tug-and-Refueling-Vehicle

Chart 92 depicts the Common-Tug-and-Refueling-Vehicle (CTugRV). The Common-Space-Command-Module (CSCM) provides Common Communications, Command, Control, Computers, and Attitude control thrusters. The CTugRV uses the Common Conventional Engine (CCE). The CCE provides On-LEO refueling station Thrust, and Translunar Orbit Insertion Thrust to travel from LEO to the Moon and back to LEO. The CTugRV is an Optionally Manned craft, with primary operations as Manned, with unmanned or Remotely Piloted operations possible.

The CTugRV uses stackable Cryogenic-Storage-Module (CSM) tanks to store cryogenic liquids such as liquid CH_4 and liquid O_2 (Chart 93). By design premise the ASWTVs and CTugRVs have very similar designs. The ESA CSSM appears to have sufficient Delta-V to move Water* in LEO from one craft to another. Since raw Water* is not a usable propellent fuel, the CSSM will need to use their interior tanks for propellent fuel. By design premise, the CTugRV has sufficient Delta-V thrust to reach the Moon. To reach the Moon, you need a lot of fuel. The Plan_B Team had a late 2018 writing cycle epiphany. A vehicle with large enough cryogenic fuel tanks to reach the Moon was a vehicle with large enough fuel tanks to refuel / transport cryogenic fuel. This led to a CTugRV design refinement to support both cryogenic refueling and Trans Lunar Orbit thrust requirements. Chart 94 provides the calculation details for the storage tanks and modules. As of Plan_B Version 1.00, Cryogenic-Fuel-Storage-Nodules (CFSN) measure 29-inches radius by 120-inches long (see calculations in Chart 94). To store 200,400 pounds of cryogenic fuel (LCH4/LOX), the CTugRV needs 21 CFSNs. These 21 CFSNs will fit within the standard PVM

15.1-foot diameter shell. More refined system designs should allow for larger tanks and or better tank packing to reduce any wasted cubic footage.

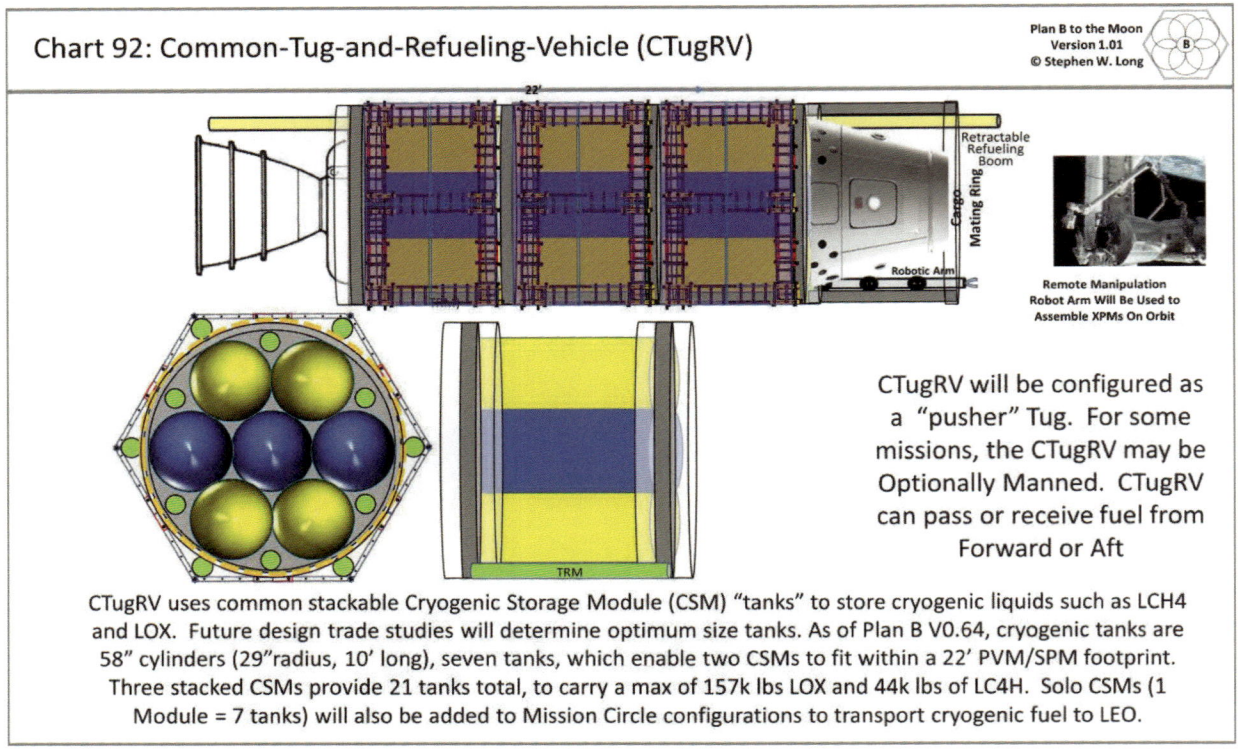

Chart 92: Common-Tug-and-Refueling-Vehicle (CTugRV)

CTugRV will be configured as a "pusher" Tug. For some missions, the CTugRV may be Optionally Manned. CTugRV can pass or receive fuel from Forward or Aft

CTugRV uses common stackable Cryogenic Storage Module (CSM) "tanks" to store cryogenic liquids such as LCH4 and LOX. Future design trade studies will determine optimum size tanks. As of Plan B V0.64, cryogenic tanks are 58" cylinders (29"radius, 10' long), seven tanks, which enable two CSMs to fit within a 22' PVM/SPM footprint. Three stacked CSMs provide 21 tanks total, to carry a max of 157k lbs LOX and 44k lbs of LC4H. Solo CSMs (1 Module = 7 tanks) will also be added to Mission Circle configurations to transport cryogenic fuel to LEO.

8.13   Cryogenic Refueling Vehicle

The Cryogenic Refueling Vehicle (CRV) is a simplified CTugRV shell used to ferry cryogenic fuel from Earth to LEO (Chart 95). The CRV design maximizes cryogenic payload; thus, the CRV is just a stack of CSMs with no CCE or CSCM. Future analysis will determine methods for on-orbit small thruster maneuvers.

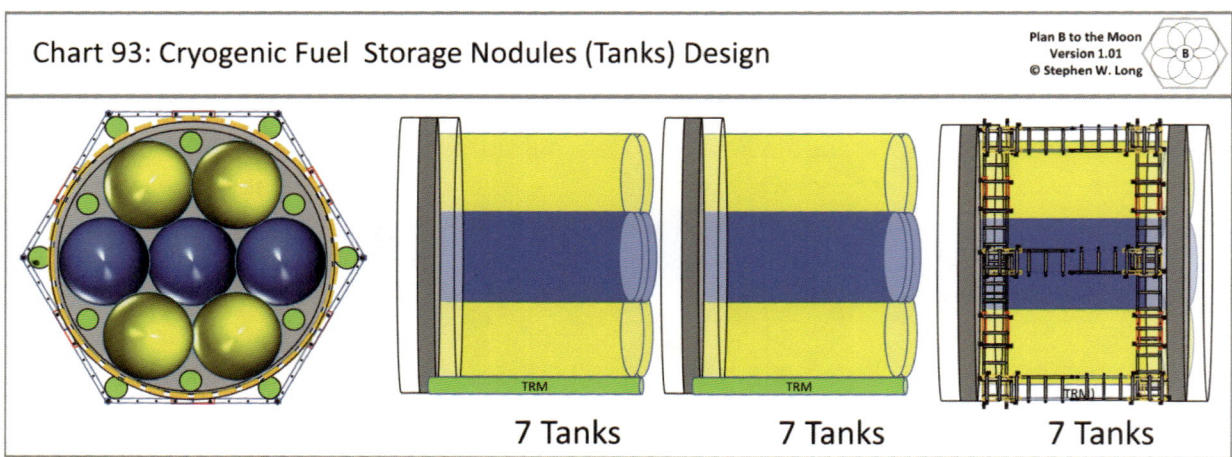

Chart 93: Cryogenic Fuel Storage Nodules (Tanks) Design

7 Tanks   7 Tanks   7 Tanks

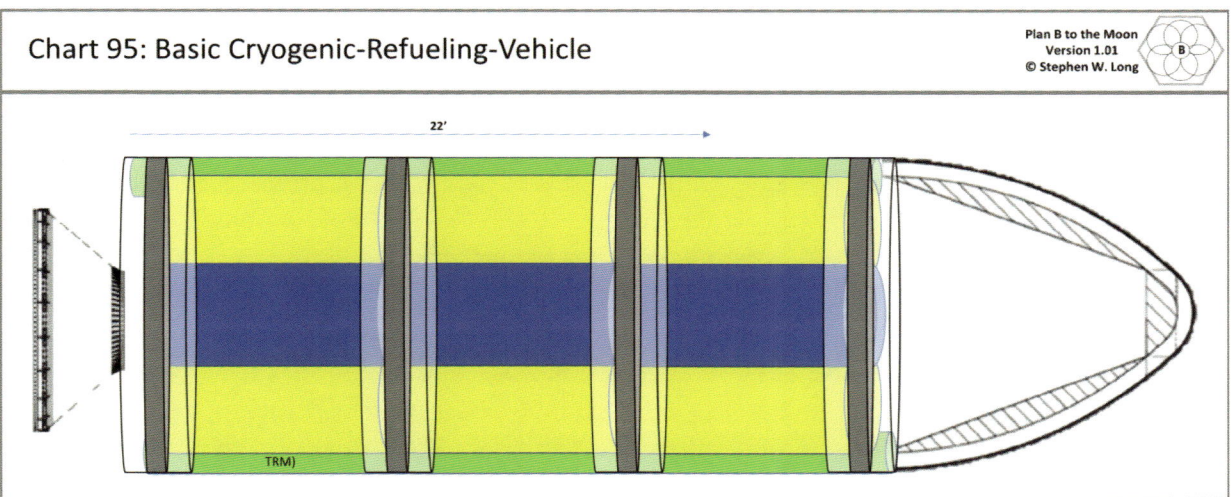

Chart 94: Cryogenic Fuel Storage Nodules (Tanks) Calculations

Plan B to the Moon
Version 1.01
© Stephen W. Long

		Ft	in	lbs per in^3	1 Nodule lbs	Component lbs.	Required Fuel		lbs	kg	w Aluminum	
Nodule Radius r in	r	24.00						Cyrogenic Nodule				
Nodule Vol in^3	V	57,905.84						Aluminum Sphere	r	29.2500	in r	0.2500	
Nodules	66	3,821,785.16						Tank Height	h in	120.00 10.00 0.00		
LC4H Nodules	14.3478 830,822.86	in^3	0.0153	887.00	12,726.5			V	322,539.46	in^3	
LOX Nodules	51.6522 2,990,962.30	in^3	0.0412	2,384.86	123,183.1			r	29.0000	in r	0.0985	540.80 Cylinder AL Tank
	66 3,821,785.16				135,909.6		Tank Height	h in	120.00 10.00 0.00		59.20 Other Weight
				CTugRV Weight	64,100.0			V	317,049.53	in^3	600.00 One Cylinder lbs
				ELTV Non-load lbs	200,009.6				5,489.9332		12600.00 All Tanks lbs

LCH4 Tank Radius r in	r in	29.00	29.00							
Tank Height	h in	120.00 10.00	0.00							
Tanks Vol in^3	V	317,049.53		0.0140	4,428.69	4,428.69				
Adjustment				0.9119						
LOX Tank Radius r in	r in	29.00	29.00							
Tank Height	h in	120.00 10.00	0.00							
Tanks Vol in^3	V	317,049.53		0.0376	11,907.33	11,907.33				
Adjustment				0.9119						
Total Diameter			116.00							
LC4H Tanks	9	2,853,445.78		in^3	0.0153	4,856.56	43,709.0	43,565.2		
Adjustment				1.0000			143.8			
LOX Tanks	12	3,804,594.37		in^3	0.0412	13,057.72	156,692.6	156,834.8		
Adjustment				1.0000			-142.1			
	21.00	6,658,040.14					200,401.6			
				CTugRV Weight lbs	64,100.0					
				LTV Non-Payload lbs	264,501.6					

Chart 95: Basic Cryogenic-Refueling-Vehicle

Plan B to the Moon
Version 1.01
© Stephen W. Long

22'

TRM)

8.14 Earth-Lunar-Transport-Vehicle

The CTugRV supports two missions–cryogenic refueling and Earth to Lunar payload transport. Payload sections loaded at the front, plus a CTugRV, form a combined vehicle called the Earth-Lunar-Transport-Vehicle (ELTV). The CTugRV is a pushing-tug, not a pulling-tug. No one has ever heard of push-boats but everyone knows about Tug Boats, so the CTugRV is a pushing-vehicle named a Tug.

As depicted in Chart 96, the ELTV has a robust push-frame that allows the CTugRV section to couple with any standard SPM section. The inherent strength of the core PVM and its mating collar from the Launch Vehicle Payload Adapter facilitates this push-frame design.

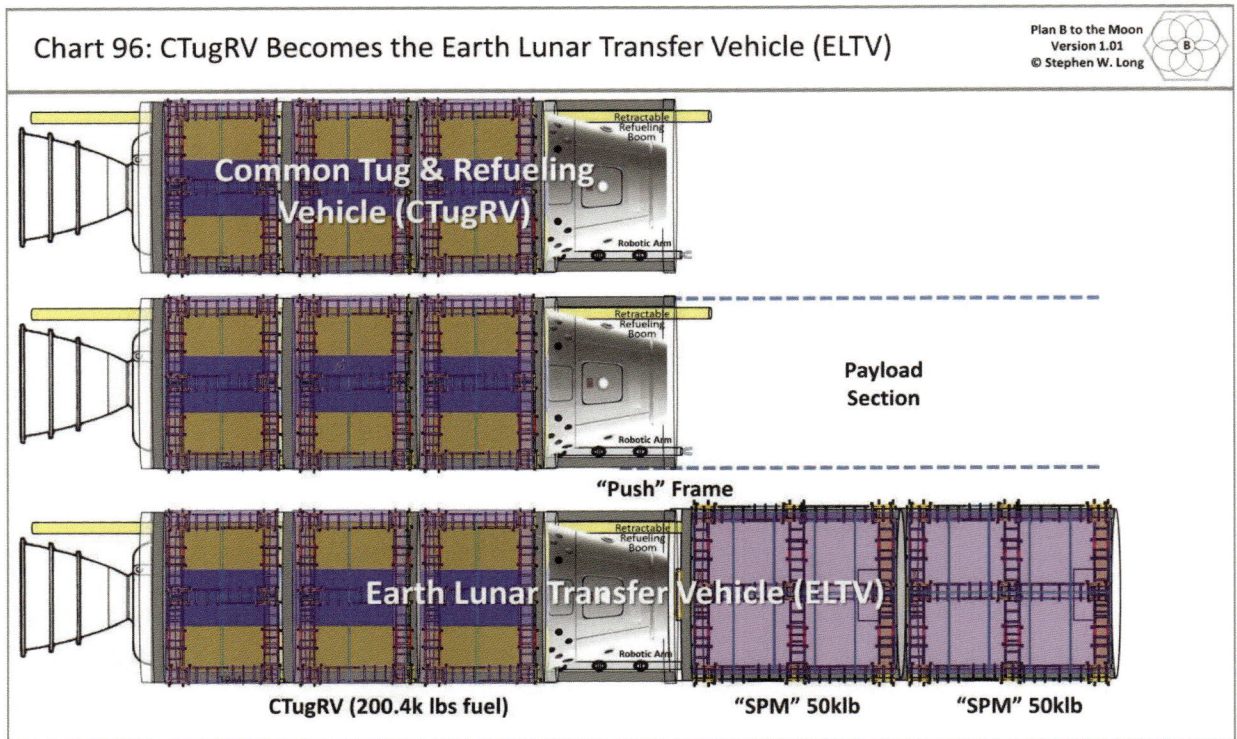

Chart 96: CTugRV Becomes the Earth Lunar Transfer Vehicle (ELTV)

For Lunar delivery missions, the Lunar SPMs are mission-manifest-limited to 100,000 pounds total (nominal 50,000 pounds each). This conforms to the **Plan_B Precept** that 100,000 pounds of Lunar payloads require 200,000 pounds of trans-lunar thrust fuel. For any given ELTV load, one SPM could exceed 50,000 pounds if the other SPM has a comparable reduction in its mass.

The CTugRV components of the ELTV need a retractable refueling boom for forward and aft refueling operations. Plan_B Team analysis determined bi-directional refueling will provide the most flexible design to maximize mission planning options. The XDock design drawings (Chart 82) also depict fuel and Water* attachment ports, to facilitate on-orbit refueling (fluid transfers) of spacecraft.

The CTugRV design includes components that enable the craft to serve as the all-purpose space construction support vehicle. To fill this role, CTugRV will need a Robotic Remote Manipulation Arm (RRMA)–similar to the Canada Arm deployed on the ISS and first used on the US Space Shuttle. Crew members in the CTugRV will use the RRMA to extract cargo modules from PVMs, Attach and extract Lunar-Landers and Lunar-Long-Boats from their PVM shells. The CTugRV will also serve as a multi-purpose construction crane / lever to move and hold FCCs and other XPMs in place. This helping hand will support space-walking astronauts and or robots performing assembly tasks.

8.15   Earth-Lunar-Transport-Vehicle Missions

As described in Plan_B Chapter 1.1, Plan_B mission designs and resultant spacecraft apportion the mass required to return to the Moon into smaller, bite-sized payloads. These smaller payloads fit within the max launch payload mass for AVAILABLE (already flying) commercial launch vehicles. This inexorably leads to the fundamental Plan_B design approach to send compact building blocks into LEO and assemble the components into much larger habitats. Once the building blocks are in LEO, Earth-Lunar-Transit Vehicle (ELTV) Missions deliver the payloads to Low Lunar Orbit. Plan_B Version 1.0 specifies four ELTV missions to deliver the mass and fuel to return to, land upon and begin colonization of the moon (through Mission Circle 899). By design, the ELTV missions are as near-to-identical as possible, meaning they become routine (a good thing). After Plan_B Version 1.0, numerous additional ELTV missions will deliver more and more XPM/SPM mass to the moon.

8.16   Lunar-Long-Boat

In the historic sailing ship era of human exploration, the Long Boat was a secondary vessel carried by the larger sailing ship. Chart 97 depicts a model of a water Long Boat. When the sailing ship set anchor off shore (not at a dock in a harbor), the Long Boat transported humans and supplies to the beach.

The sailing ship carries the Long Boat on its deck and it is lowered by ropes over the side of the ship. Once the boat was in the water, sailors scrambled down the ropes and manned the oars of the Long Boat. Some Long Boats even had a mast and simple sails to facilitate longer trips from the big ship to the shore. Long Boats moved from one island to another island while the big ship remained at anchor.

Borrowing from this historic premise, the Lunar-Long-Boat (LLB) (Chart 97 bottom right) is an exo-atmospheric vehicle. It will transport cargo / supplies / vessels from the Low Lunar Orbit Gateway to the Lunar Surface. It is a **Plan_B Precept** that the LLB is a reusable vehicle, refueled from the LLOG and the Lunar Surface, and will be used for many lunar landing missions.

To lower development costs, the LLB is an un-crewed vehicle, meaning the LLB functions as: an autonomous vehicle; a semi-autonomous vehicle; or a remotely piloted vehicle. The proposed flight control system of the LLB will:
- Have a human pilot control the departure of the LLB from the LLOG;
- The LLB auto-pilots itself to near the Lunar surface;
- Have a pilot on the surface (who landed there in the Lunar-Lander) to pilot and visually guide/fly the LLB to its landing spot on the surface.

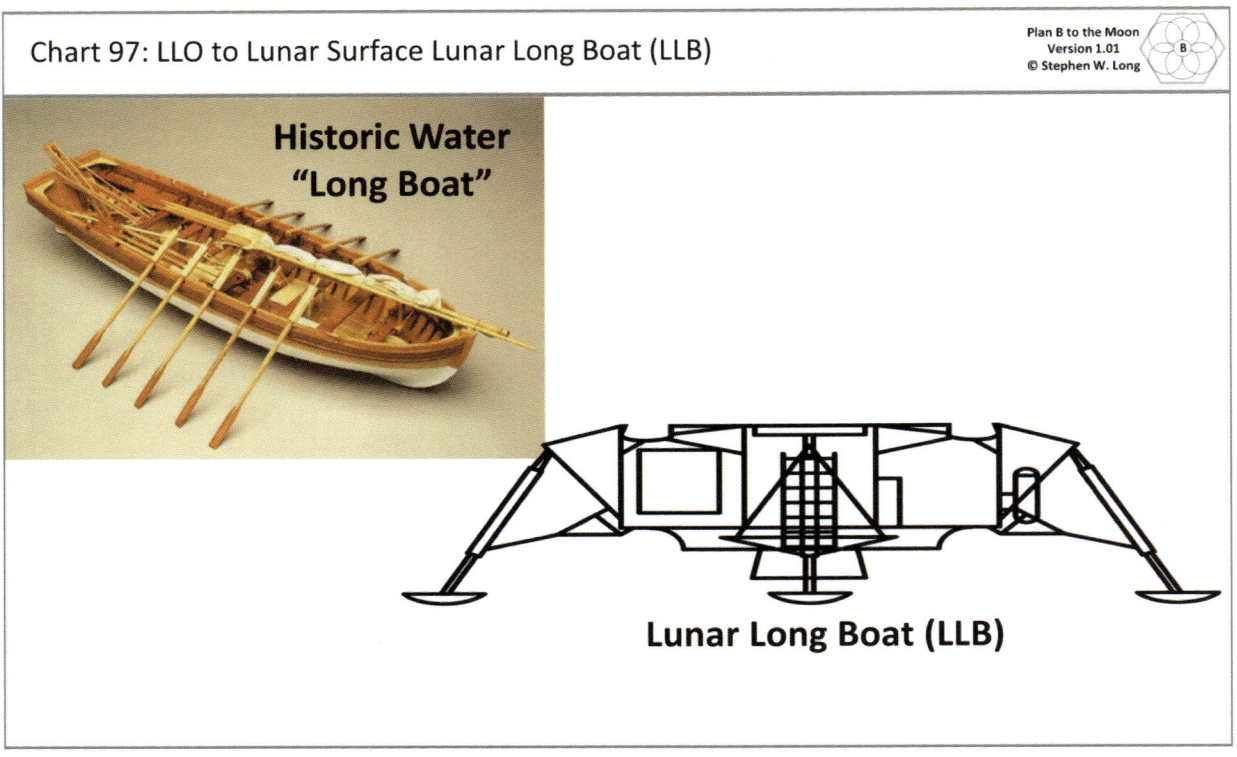

Historic Water "Long Boat"

Lunar Long Boat (LLB)

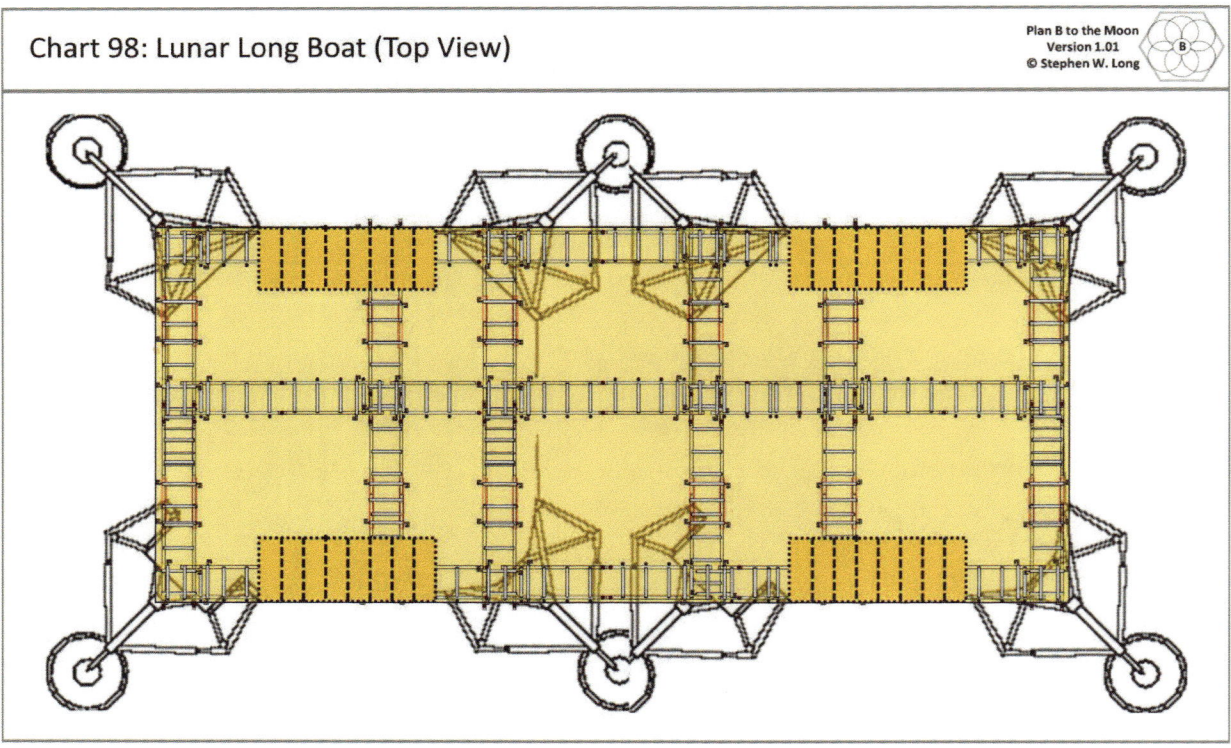

Since the LLB is exo-atmospheric, and since there is no need for human cockpits, etc. on-board, the LLB design only has core essential structural elements (Charts 97 and 98). The LLB serves as a platform to hold cargo, propulsion systems (engines and fuel/LOX tanks), and landing gear. The LLB refuels from the LLOG and from the Lunar Surface once we make fuel from Lunar ice. We found trying to build and deliver a crane lifting system to unload cargo off the LLB platform to the Lunar Surface would be logistically complex. Instead, the Program Team design requires that initial cargo loads from the LLB to the Lunar surface self-deploy (drive off) the LLB. This concept enables the easy deployment of the Lunar-Surface-Habitat-Vehicle (LSHV see Chart 104). This is very similar to the premise of USAF large cargo aircraft. Self-powered vehicles drive off the back (or front of a C-5), down a ramp, to their intended destinations.

No one would propose equipping a four-engine jetliner with different make and model engines that each use different fuels at each of the four engine nacelle mounts. Thus, too the LLB should have near identical engine components with the Lunar-Lander (LL) and use the same LCH4/LOX fuel as the CTugRV and all other Plan_B spacecraft. The Plan_B Team recommends that the propulsion systems (rocket engines) used for the LLB be compatible with if not identical to the Lunar-Lander propulsion systems. All such engines would use LCH4/LOX fuels. This reduces development costs, enables fuel sharing, and provides maximum parts flexibility for long term Lunar Surface survival. In such a concept, perhaps the LL only needs a single LCH4/LOX engine, but the LLB needs two or more LCH4/LOX engines (more mass needs more thrust).

The LLB folded-while-stowed design will allow the launch and delivery to LLO of two LLBs in one PVM shell (see Chart 99). This LLB PVM choice requires a longer than normal PVM shell.

The LLB has a design mass of 21,750 pounds and a design-to-cost of $108.75 M.

8.17   Lunar-Lander

The Lunar-Lander (LL) (Chart 100) will use a Common-Space-Command-Module as the crew compartment, coupled with a unitary–combination descent / assent engine. It is a **Plan_B Precept** that the LL is a reusable vehicle, refueled from the LLOG (and Lunar Surface), and it will support many lunar landing missions. A little-known over-build of the original Apollo Lunar Excursion Module (LEM) inspired the Plan_B LL combination descent / ascent engine. The original LEM engine design had twice the thrust required to land on the Moon. Early in the LEM development cycle, mission planning required the LEM engines to return the entire LEM mass back to Earth (direct return to Earth). The LEM mission profile changed to use Lunar Orbital Rendezvous (instead of direct-return-to-Earth). It was too late in the development schedule to replace the engine with a smaller engine. NASA changed the name of the lander from the LEM to the LM–but everyone continued to call the lander the LEM.

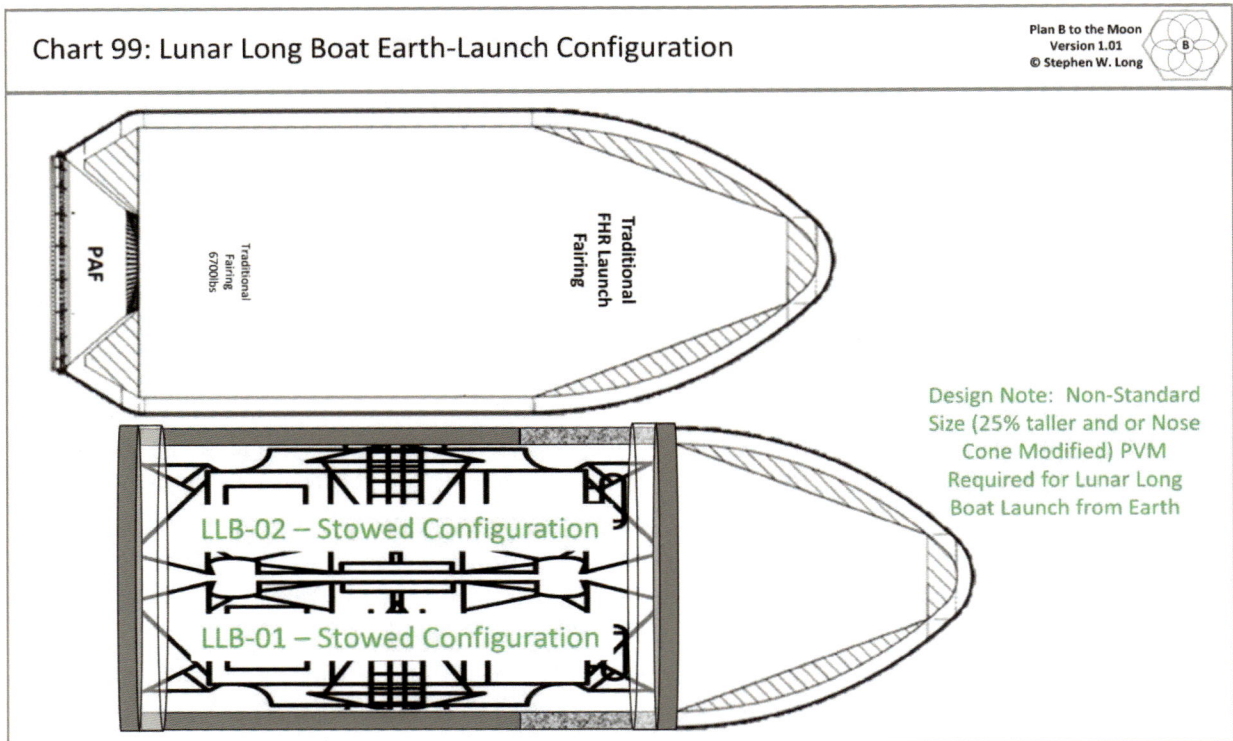

Chart 99: Lunar Long Boat Earth-Launch Configuration

Plan B to the Moon
Version 1.01
© Stephen W. Long

PAF

Traditional
Fairing
6700lbs

Traditional
FHR Launch
Fairing

LLB-02 – Stowed Configuration

LLB-01 – Stowed Configuration

Design Note: Non-Standard Size (25% taller and or Nose Cone Modified) PVM Required for Lunar Long Boat Launch from Earth

The LL folded-while-stowed design enables the delivery of two LLs to the Moon in one PVM shell for earth launch (Chart 101). This design choice requires a longer protective PVM shell to launch the LLs.

It is a **Plan_B Precept** to segregate the roles / missions of the LLB and the LL, meaning only the LL is human-rated. The LL (and only the LL) will ferry crew members from the LLOG to the Lunar Surface and back to the LLOG. The Lunar-Long-Boat will only deliver cargo (SPMs and more). Further analysis will determine if it is best to have a remote-human-pilot-in-the-loop for the Long Boat. A human pilot would control the departure of the LLB from the LLOG. The LLB will auto-pilot itself to near the Lunar surface. A human pilot that previously landed in the LL will take over control of the LLB and visually guide the LLB to its landing spot on the surface.

As an analog, consider the LL to be a small jeep used to scout out the terrain (what is the Lunar equivalent of terrain?). The LLB is the analog of the recently developed self-driving tractor-trailer that delivers pallets of cargo and goods.

Plan B to the Moon: A How-to Guide to Return to and Colonize the Lunar Surface

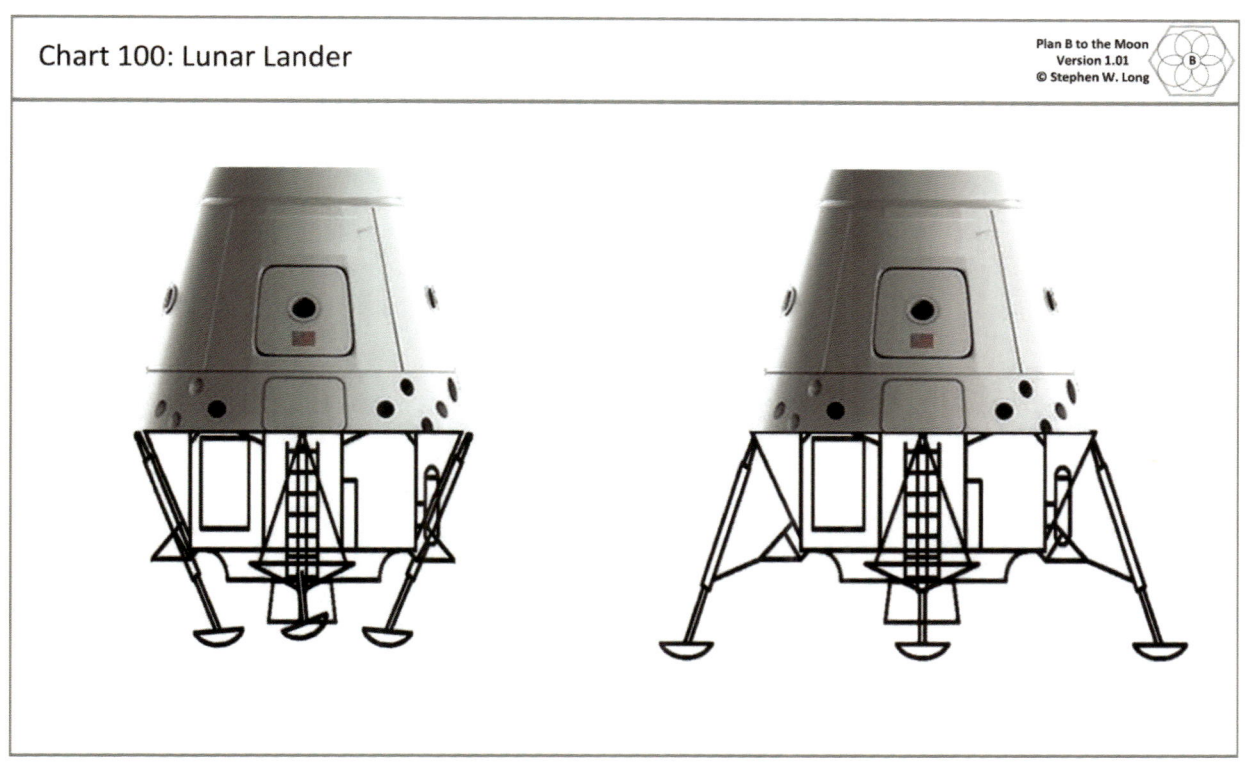

Chart 100: Lunar Lander

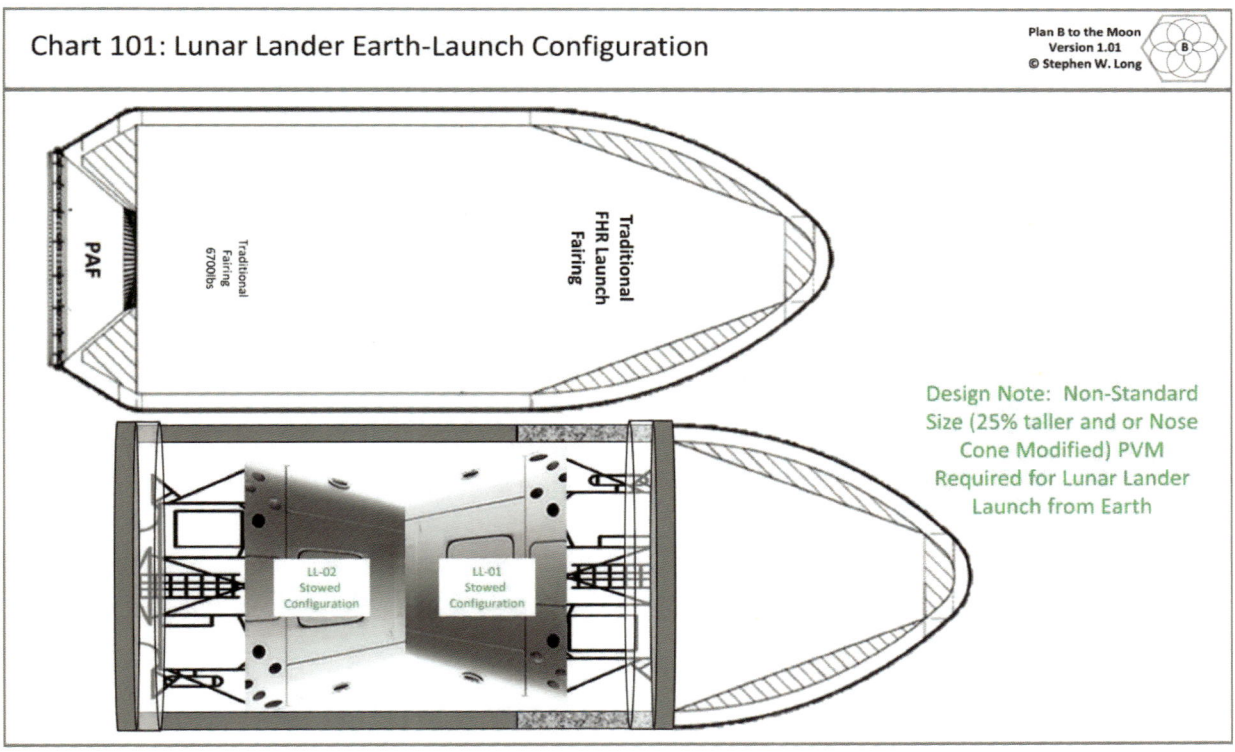

Chart 101: Lunar Lander Earth-Launch Configuration

Design Note: Non-Standard Size (25% taller and or Nose Cone Modified) PVM Required for Lunar Lander Launch from Earth

8.18   Lunar-Surface-Habitat-Vehicle

There is probably no greater exemplar of the modern American spirit of common-man exploration than owning a Recreational Vehicle (RV). To anyone that has ever owned and traveled extensively in RVs, such travel experiences are spiritually transformative. The Plan_B Program Director lived in a 30-foot Class-A RV for over a year, traveling across the USA. That trip was inspirational, wonderful and profoundly sad. The PD bought the RV, installed a lot of high technology gear and took his wife with her Stage-4-colon-cancer-diagnosis on a yearlong tour of the USA. That year in the RV was their goodbye tour. Touring the USA was something long-planned to do in retirement and after the cancer diagnosis, they knew they had about a year, "to do things together sooner than later." Living in a small but comfortable home on wheels for a year was a profound lesson in togetherness, exploration, and wonder at the beauty of America.   That journey yielded hundreds of joyful interactions with salt-of-the-earth Americans. There was the unexpected joy of living without the excess trappings of routine suburban or urban life. That trip traversed forty-two states; 65,000 road miles; and produced enough memories to fill a dozen lifetimes.

A modern Class-A RV is a sophisticated and comfortable home on wheels. Such a vehicle:
- Provides self-locomotion—the house part of the RV sits on top of a commercially derived truck chassis. In fact, many modest-sized Class-A vehicles use the identical chassis found in standard school buses.
- Provides its own electrical power, in the form of a generator. Many RVs also have solar panels on the roof and battery banks to provide quiet electric power.
- Potable water storage tanks.
- Grey and Black water storage tanks for wastewater (showers are gray water; toilets are black water).
- Provides commercial-off-the-shelf two-way satellite communications (via dishes on the roof the RV), allowing the occupants to communicate with and connect to the world-wide internet.
- Living environment—kitchen, bathrooms, bedrooms, recreation (TV, etc.), office (work surfaces and supporting computer systems), storage areas.

Chart 102 depicts a very sophisticated (and expensive) modern RV. This super-RV has sophisticated satellite communications systems and a unique a self-contained garage that internally stores a smaller vehicle (car) for excursion trips.

Living in a home you can move is the premise and inspiration behind the Plan_B Lunar-Surface-Habitat-Vehicle (LSHV). Profound flexibility derives from exploring while you live, instead of being forced to stay in a fixed location and make only short excursion trips. There is no commercial-off-the-shelf RV that is ready to explore the moon. But it is a **Plan_B Precept** that a properly equipped and capable home-on-wheels (tracks) will be the first substantive habitat for Lunar explorers.

Chart 102: Advanced Class A Recreational Vehicle

Plan B to the Moon
Version 1.01
© Stephen W. Long

Are there exemplar mobile platforms (~homes on wheels) created to survive in hostile Earth environments-that allow groups of humans to live, work, and explore together? Yes, there are such vehicles. Chart 103 depicts several such vehicles, including vehicles designed for troop / combat support in the Artic.

Chart 103: Inspiration Ideas for Lunar Colony Habitat Vehicles

Plan B to the Moon
Version 1.01
© Stephen W. Long

The Plan_B Team contacted several manufacturers of snow-tractors, and at least one vendor already has an electric-motor-tracked vehicle in development.

Chart 104 depicts a notional LSHV. Two SPMs and one CSCM mounted onto an electric-powered tracked vehicle are the core elements of the LSHV. After assembly on-orbit, the LLB delivers the LSHV to the lunar surface, and the LSHV drives off the LLB to become a functioning, transportable lunar habitat. The Lunar-Surface-Habitat-Vehicle (LSHV) has a habitable design volume of 8126 cubic feet (1 CSCM + 2 SPM). The LSHV design has two decks, each ~7 feet tall at the center. This is equivalent to a ~1015 sq foot earth surface home with 8-foot ceilings. By regulation, a large earth RV has a maximum size of 450 sq ft. The two-story LSHV design provides a living space larger than two modern maximum size earth RVs.

Why is it so important for Lunar explorers to have a home they can move? Because today we know little about what it takes to live on the Moon. Remember, we have only sent tourists to the moon, not colonists. We could spend a near infinite number of Lunar-Lander tourist trips to the Moon's surface looking for ice, or Helium 3, or other resources. The likely locations for discovering lunar ice may be unsuitable for safe LL landings. The ability to pack-up-and-move the initial human settlement on the Moon to new places will enhance the explorer's ability to find what they are looking for. Travelers seeking new lands make those folks the first colonists, not tourists.

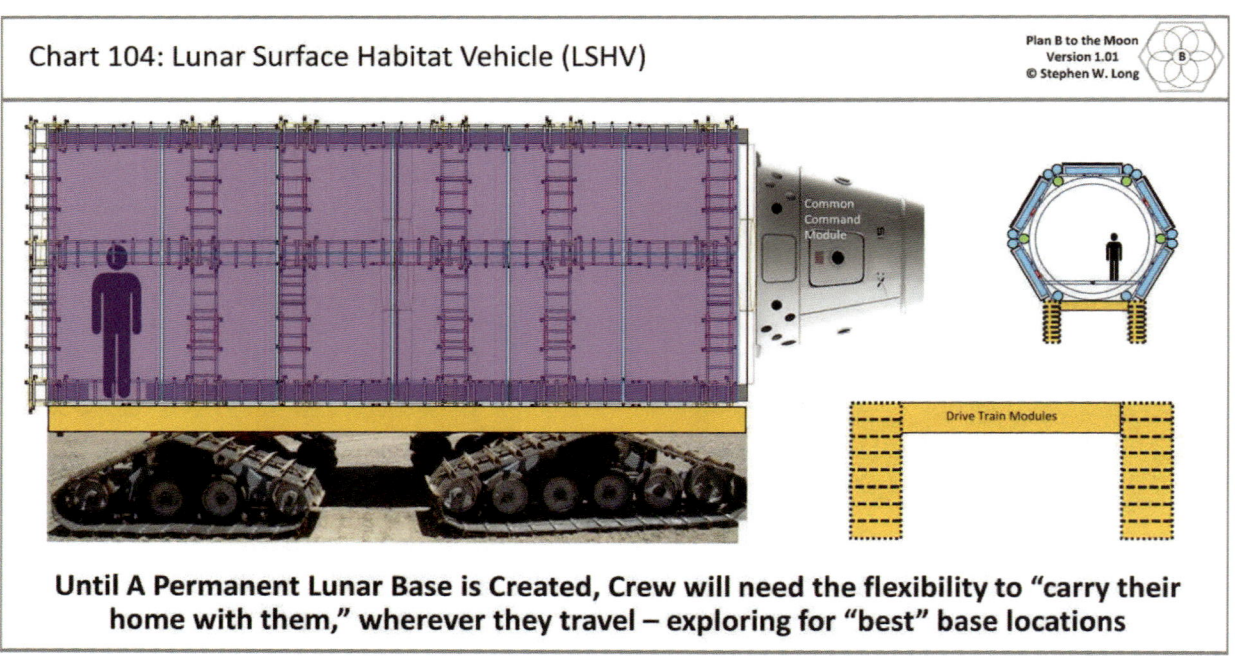

Chart 104: Lunar Surface Habitat Vehicle (LSHV)

Plan B to the Moon
Version 1.01
© Stephen W. Long

Common Command Module

Drive Train Modules

Until A Permanent Lunar Base is Created, Crew will need the flexibility to "carry their home with them," wherever they travel – exploring for "best" base locations

It is a **Plan_B Precept** that the LSHV will use a Common-Space-Command-Module for housing the computers/communications/command and control of the LSHV. The LSHV will have the CSCM as the human-rated capsule on the LSHV to serve as a lifeboat / emergency shelter within the larger LSHV design. The CSCM provides seats/drive controls, and all other C4 elements for the LSHV. The CSCM will have small windows for the crew members to observe and marvel at the landscape as they travel across the Lunar surface. They will want better views, provided by super-high-resolution forward-facing 8K HDR (8000 lines of resolution, High Dynamic Range) cameras with images sent to monitors in the cockpit to serve as large virtual windows. As a historical footnote, USAF strategic bombers and refueling tankers were (are) equipped with forward looking vision systems. These systems enable the pilots to take off and fly the aircraft with the cockpit windows covered. The vision system monitors show the taxiways and runway for takeoff. The window coverings protect the aircrews from the postulated dangerous wartime scenario of nuclear explosion bright flashes in the distance. The flash through the windows would blind the crew and prevent their ability to takeoff and fly the aircraft.

Engineering analysis and trade studies may allow a glorious glass portal for the LSHV crew to experience the Lunar surface as they explore. If not, a virtual windshield is a viable design choice.

SPM segments provide the habitat, laboratory, and exploration work areas for LSHVs (Chart 105). Electric engines / tractor treads attach under the SPMs, making the entire habitat mobile. After assembly at LLOG, the LLBs deliver the LSHVs to the LS.

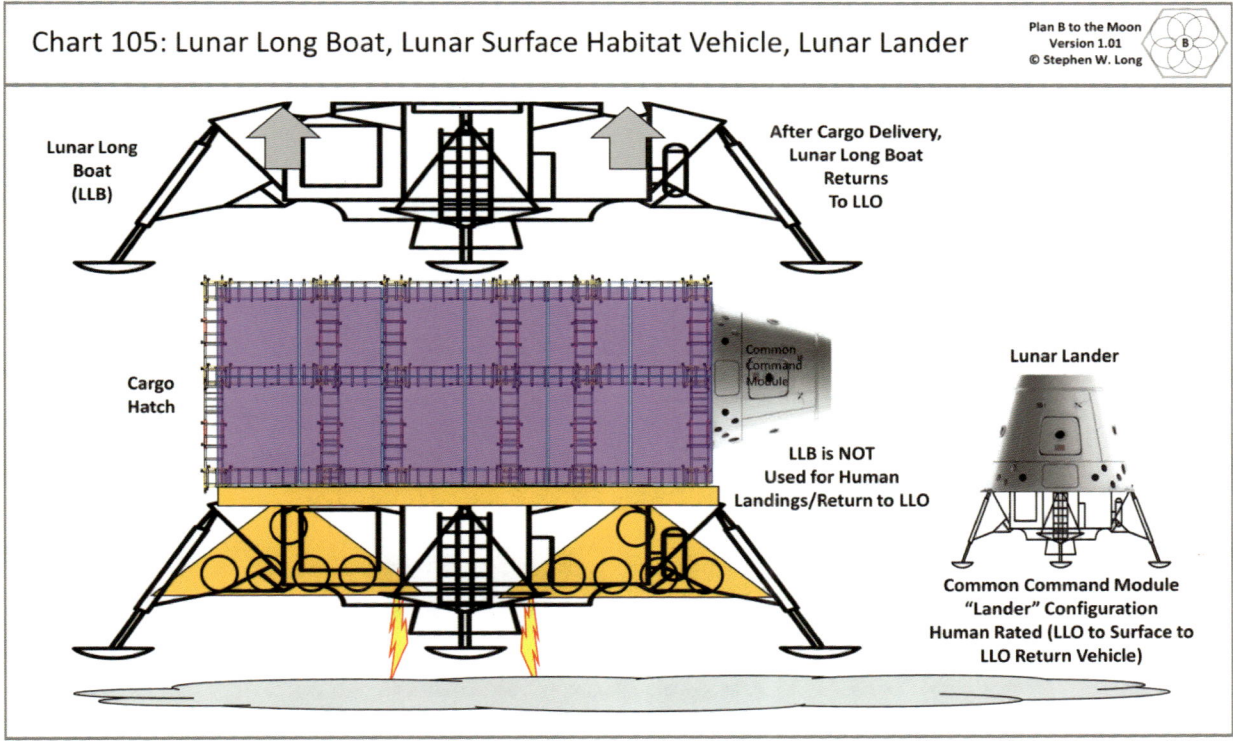

8.19 Lunar-Habitats

Plan_B Version 2.0 transports additional SPM based components to the Lunar Surface, for assembly into more permanent structures for the Lunar Surface (Chart 106). The simplest Lunar-Habitat has a habitable design volume of 15,552 cubic feet (4 SPM). The LH design has two decks each ~7 feet tall at the center. This is equivalent to a ~1900 sq foot earth surface home with 8-foot ceilings.

Chart 107 depicts a preliminary concept for a larger Lunar Base, built within easy transport distance of discovered Ice or Helium 3 deposits on the Moon. Therefore, some of the most important early LS exploration missions would be to find Lunar Ice.

The depicted ten module Lunar Base (LB) (Chart 107) has a notional habitable design volume of 38,800 cubic feet (10 SPM). Fifty percent of the LB design has two decks each ~7 feet tall at the center. This 10 SPM design would therefore be equivalent to a ~4,000 sq foot earth surface small factory/warehouse. 4000 sq foot is the size of a small strip mall shopping center with four Mom and Pop stores with 10-foot ceilings.

One concept of Lunar surface operations would be to link LLO Gateway operations and the Lunar Base (LB) operations. For instance, humans might live on the LLOG for 14 days, and return to the Lunar Base (LB) for 14 days. Duty tours are defined by solar cycles.

The LLO Gateway or a constellation of communications satellites (see Section 8.4) should be in an Earth-facing Low Lunar Polar Orbit (LLPO). Such orbits enable near continuous exposure to sunlight and enable near-continuous line-of-sight communications with the Earth. Multiple Lunar communications satellites will need to be orbiting 120 degrees offset from the LLOG. Such offsets would enable the LB and the LLOG to remain within line-of-sight communications with each other and with communications relays to planet Earth.

8.20 Mobile Colonies Follow the Sun

The Plan_B Team debated the concern of having to invent high-power storage systems (batteries or fuel cells) to power the LSHV for 14+ days. The moon has 336 hours of darkness during lunar nights. It may prove difficult for humans to inhabit any portion of the shadowed surface of the moon for extended periods (Lunar Day is 14 earth days; Lunar Night is 14 earth days). One approach may be to construct facilities near/at the Lunar pole where solar power arrays around the circumference of the pole provide continuous electric power. Since the forever-shadowed deep craters at the poles are the likely source of primordial ice deposits, the Lunar poles will be the likely, early locations for Lunar Bases.

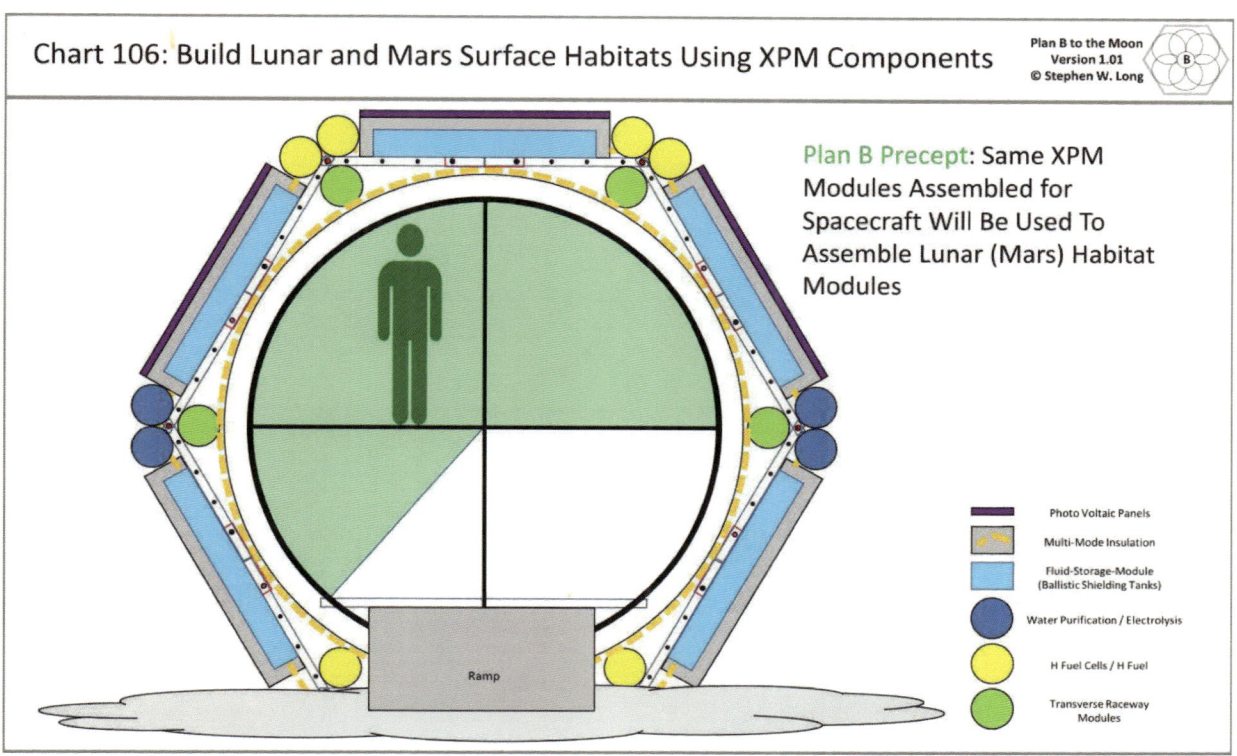

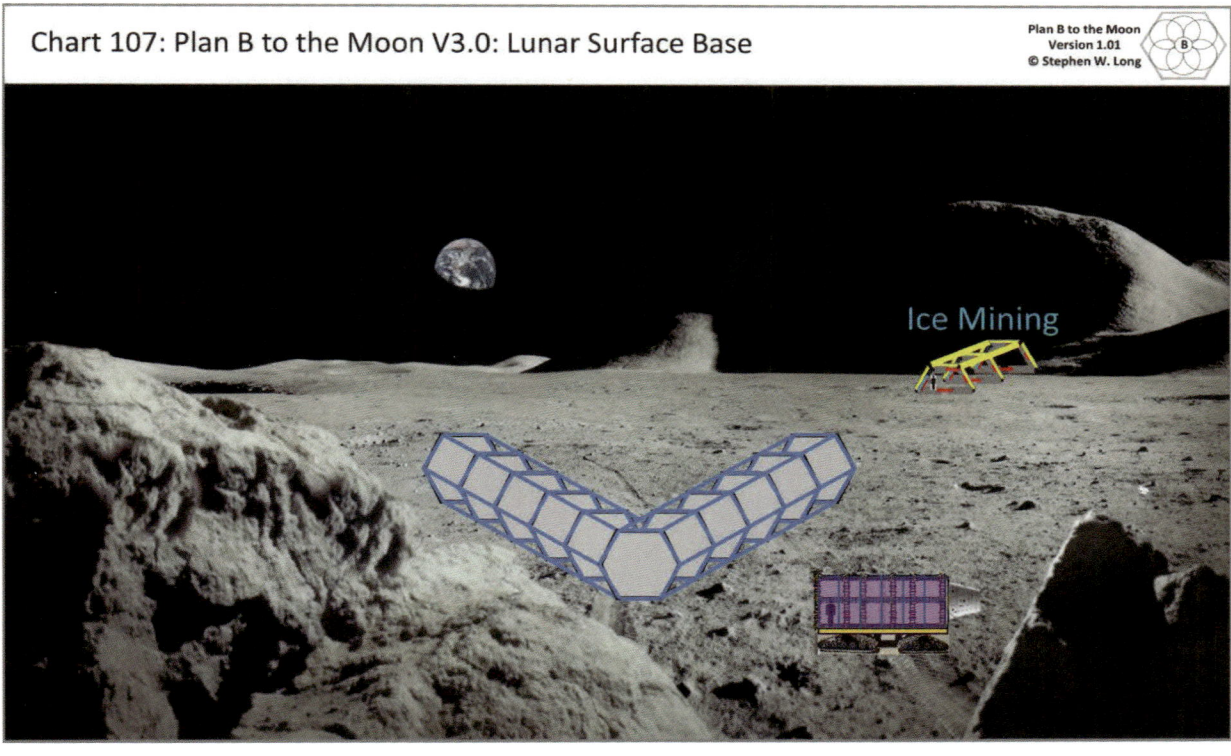

One somewhat speculative concept would be to deploy superconducting power cables operating within covered lunar trenches over very-long surface distances. This concept would transport electricity via superconductors from Lunar Surface sites A to B to C with zero electric power transmission loss. Some engineers would say it would be impossible to encircle the pole with electric cables to provide continuous electric power. But if continuous illumination is not possible, what if using long super-conducting cables were enough to add a week (from 14 days to 21 days) out of 28 days?

These dream-out-loud concepts came out of interesting physics facts learned while researching cryogenic fuel production for the SFT. In the shadows of the Moon, the low temperature is 100 Kelvin (minus 173 Celsius). Superconductivity capable Yttrium based Cuprate Perovskite material (HBCO) will function at 92 Kelvin (only 8 degrees colder than Lunar shadow free cold). The highest-temperature superconductor material to date on the Earth is a ceramic material of mercury, barium, calcium, copper and oxygen ($HgBa_2Ca_2Cu_3O_8$). These materials operate at temperatures of 133–138 Kelvin. Such materials may enable ~100 Kelvin temperature superconducting wires, constructed and laid within hundreds of miles of lunar surface trenches. When covered with a shell of Moon dust to block sunlight (heat), the native dark shade temperature of the Moon would enable a superconducting electric cable of near infinite length.

For a carefully chosen Polar Lunar Base site able to exploit grazing sun angles, the Lunar Base could have radial arms of solar power stations connected with superconducting wire. The radial arm power stations would provide the Lunar Base with near continuous power, even when the base itself is in shadow. Future studies will balance launch masses of batteries/fuel cells versus building multiple solar stations connected with superconducting cables. Fuel cells need fuel that requires mining and manufacturing.

Envision a traveling Lunar colony that moves across the Lunar Surface every ~10 days. The colonists would follow the Sun—moving from Lunar Base-Alpha (LB-A) to LB-Bravo to LB-Charlie. Expert system enabled automation would oversee the bases during dark periods. In a farther distant future, superconducting power lines would connect Lunar Base sites A, B and C, enabling the continuous inhabitation of sites A, B and C.

8.21   LEO Station-01V2, LLO Gateway-01V2

Chart 108 presents concepts for Artificial-Gravity-Modules (AGM) by Spinning SPMs. After Plan_B Version 1.0 was published, Version 2.0 addressed the need to enlarge the first Lunar Bases. Versions 3.0, 4.0, 5.0, and 6.0 started expansion of both LEO Station and LLO Gateway. Plan_B Version 6.0 addresses the need to have artificial gravity in spacecraft. We have millions of years of experience as creatures living in 1 G. We only have a few decades of experience with humans visiting micro-gravity, also known as Zero-G.

Therefore, we should patiently and methodically build LARGE artificial, low-gravity spacecraft. Chart 108 depicts a six-point star-shaped spacecraft built out of SPM shells. Spinning the shells at 0.5 RPM (30 seconds per revolution) will produce about 0.10 of artificial G at the far ends of the spinning craft points. The star points are literal Type III SPMs.

We have some, but only limited experience of growing staple crops / human consumable plants and animals (fish)–raised in Zero-G (yes fish already live in ~Zero G). Perhaps in the farther future biologists can produce GMO plants and animals that prefer Zero-G to some gravity. But until that farther future, Plan_B posits that the lack of knowledge of Zero-G plants have the risk of becoming existential threats to deep space survival. Therefore, we need to build 0.10 G Deep Space Food Modules (DSFMs) as depicted in Chart 109. We don't know if we will able to grow enough food (grains, etc.) in Zero-G to feed a medium-to-large crew–or larger groups of colonists–for years on end. We don't know how large quantities of traditional crops will survive/thrive in Zero-G.

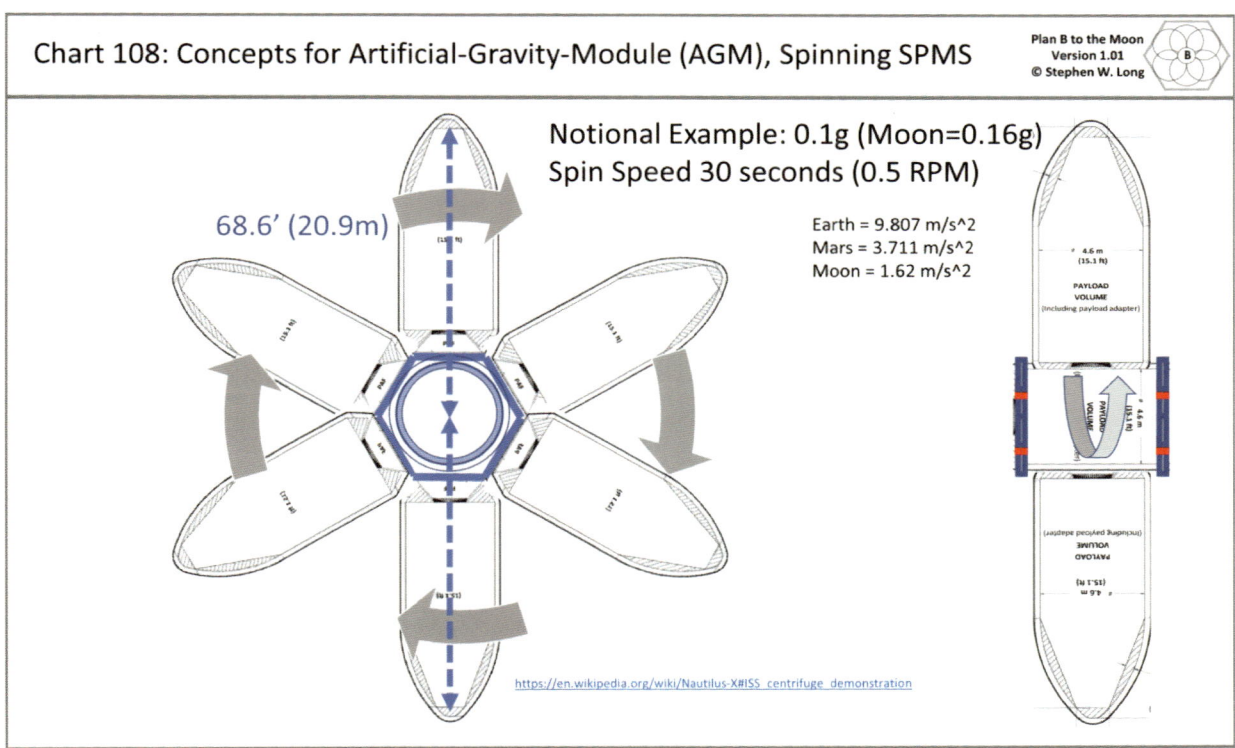

It might be even more beneficial to achieve 0.20 G–Moon equivalent gravity. The foods that grow on the Moon will be foods we have assurance will grow on long duration spaceflights. It is just an engineering decision to increase from 0.10 G to 0.20 G. You either build a larger diameter, slow-spinning craft or you spin a smaller craft faster. The human inner ear may not easily tolerate spin speeds faster than 0.5 RPM, but plants may not care one iota about the spin. Plan_B Version 2.0 assumes a 0.10 G initial baseline. Human hygiene (toilet use, bathing, etc.) and resulting long-term human comfort may significantly enjoy having even low (0.1) gravity–allowing fluids to travel in expected directions without the use of high-vacuum suction. Ponder for a moment the visual joke in the movie, 2001 A Space Odyssey. A Zero-G passenger has to read multiple pages of instructions to use a Zero-G toilet.

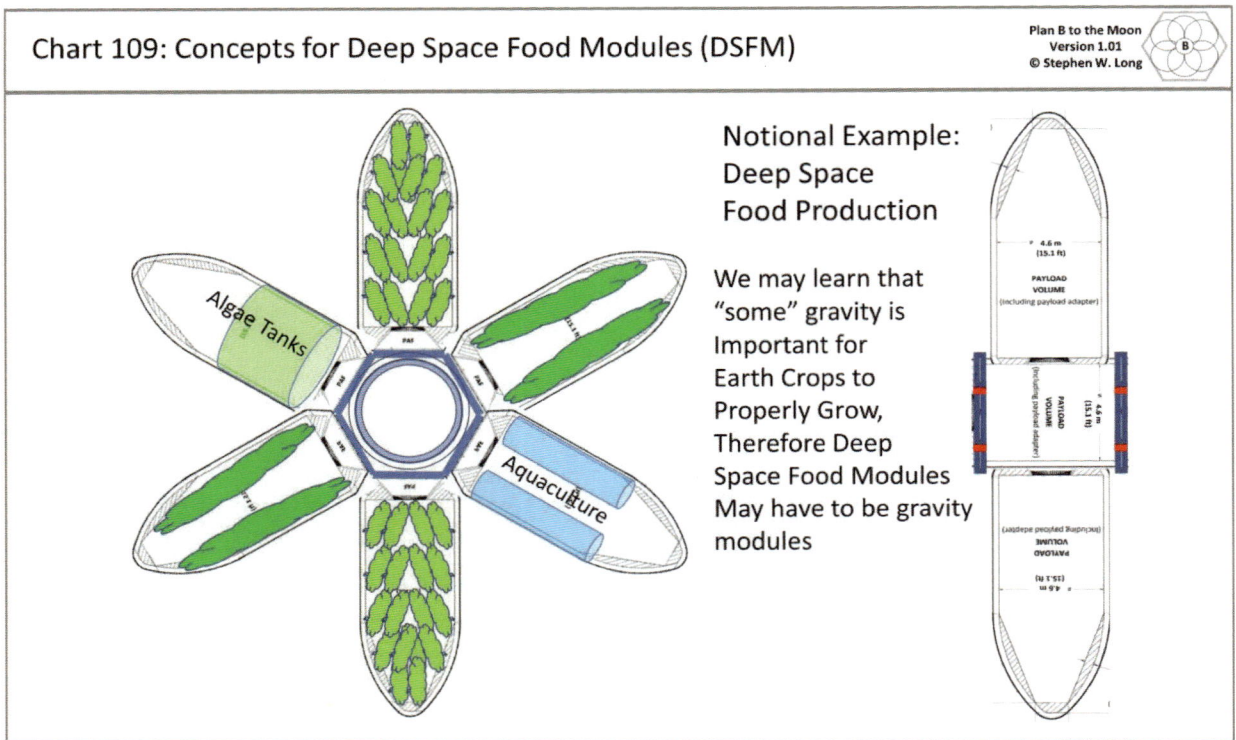

Chart 109: Concepts for Deep Space Food Modules (DSFM)

As an Action Item, the Plan_B Team calls for new research to determine the best mix of humans and machines to perform on-orbit and lunar surface operations. How do you strike the balance between having a smaller ship with few people and lots of robots compared to having a large human construction crew to build larger habitats? The deeper answers to such questions are beyond Plan_B Version 1.0, but we plan for Versions 2.0 and beyond to start such required analysis.

Chart 110 presents concepts for future LEO Stations and LLO Gateways. Artificial Gravity Modules enable very large space structures. There is basically no limit to the size of future space stations, up to the limit of the attachment structural strength of the SPM Pressure-Bulkhead-Modules and XDocks.

8.22   Linear-Accelerator-Relativistic-Ion-Thrust Engine

Plan_B Version 1.0, Chapter 4, Chart 50, briefly introduced the concept known as Linear Accelerator Relativistic Ion Thrust Engines–LARIT-E. Plan_B Version 1.0 does not use or require development of LARIT Engines. This Chapter 8.22 (see Chart 111) further explains the concepts behind such future engines. Plan_B wants readers to understand very advanced future propulsion engines are possible for deep space exploration, continuing to use the Water*-as-Fuel paradigm. Said another way, Plan_B advocates building spacecraft with Water* converted to LCH4/LOX now. As LARIT-Engine research progresses, future craft will use native Water (water molecules) instead of LCH4 and LOX fuel.

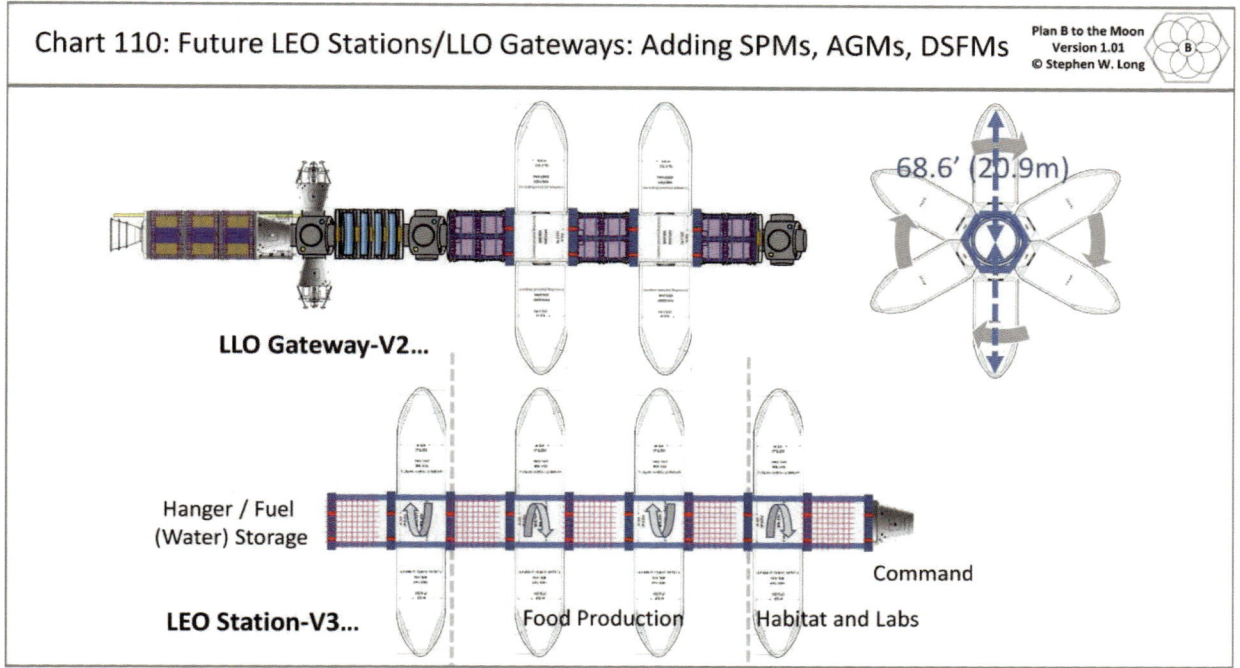

Chart 110: Future LEO Stations/LLO Gateways: Adding SPMs, AGMs, DSFMs

Plan B to the Moon
Version 1.01
© Stephen W. Long

LLO Gateway-V2...

68.6' (20.9m)

Hanger / Fuel
(Water) Storage

Command

LEO Station-V3...

Food Production

Habitat and Labs

LARIT-Engines will be an element of Plan_B Version 14.0. By including LARIT-Engine concepts this early in Plan_B Version 1.0, we hope to inspire a current young reader to make building a LARIT Engine her life's work.

The basic premises behind Ion engines are the physics formulas that Force is the rate of change of momentum over time. Kinetic energy is half of the (propellent) mass times its velocity squared. This means that as the velocity of a particle increases, its impact upon–the increase of-kinetic energy is exponential as the speed increases. As a more familiar example, consider that a car crash at 20 MPH just bends metal. But a car crash above 60 MPH causes pieces of the car to fly out in all directions. When an aircraft crashes at several hundred miles per hour, only tiny pieces of metal remain. The kinetic energy equation means just small increases in speed causes exponentially larger energy effects.

By definition, Ions can be positively or negatively charged, meaning suitable magnetic fields can manipulate or steer the Ions. The premise of a Linear-Accelerator-Relativistic-Ion-Thrust Engine is to convert available mass–such as water molecules-into ionized particles. The science principle behind LARIT is to use common water (H_2O) and ionize the water molecules using solar radiation or electric heat. One method of Ion conversion is to apply intense solar heat to convert H_2O into HO and H_3O, which is a caustic and highly ionized substance. Vacuum isolation will reduce or eliminate the negative interactions of the caustic molecules, and intense magnetic fields will guide the flow of the highly ionized particles. Each foot of magnetic field (think a very-long tube) would impart more and more kinetic energy to the Ions. After a sufficient length of acceleration, the molecules would reach a significant percentage of the speed of light. We have technology today to accelerate particles to ~99.9997 percentage of the speed-of-light. This technology comes from the linear accelerator systems used for physics research (thus the Linear

Accelerator Relativistic portion of the LARIT engine name). As captured in Einstein's ½ M V sq formula, when more and more acceleration electrical energy pushes the particles, their masses continue to increase. Because of Newton's Third Law, such high-speed particles exiting one end of a tube (a spacecraft engine nozzle) will have an equal and opposite thrust component. The common term for producing reaction mass at high velocities is a Rocket engine.

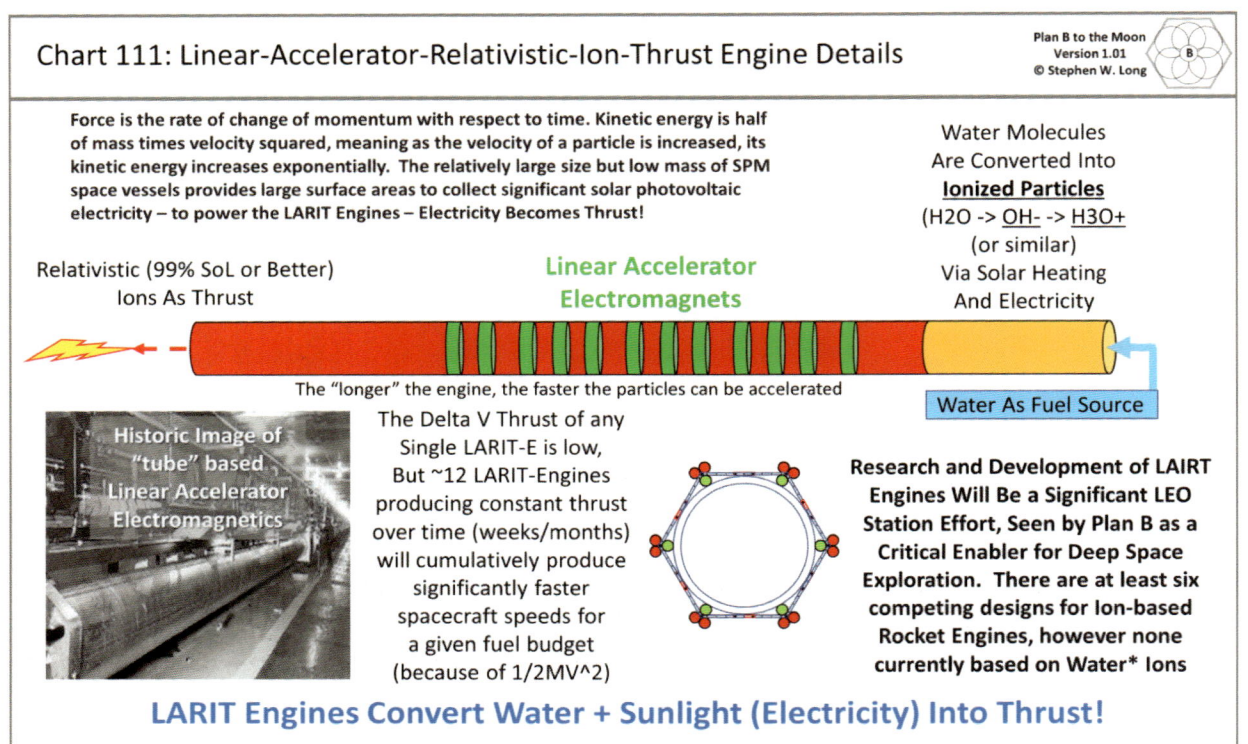

Linear acceleration magnetics would use solar energy (electric power) plus simple water to create high-speed reaction mass. The imparted total momentum energy to any single ~H_3O molecule is small. Electromagnetics, not chemistry, define the resultant energy. This means you can produce more long term thrust—a constant source of thrust for days/weeks/months/years (for interstellar travel).

Some satellites orbiting the earth today already incorporate small Ion engines to replace the conventional chemical rocket thruster engines used to date. These first-small-step Ion engines use an inert gas, typically Xenon and ionize the gas. Electric fields accelerate the Ions to a high enough velocity to provide small thruster maneuvers for spacecraft. This Ion design increases the useful lifespan of the satellite because the Ion thrusters require less fuel mass than conventional rocket fuel thrusters. We also note with a smile that the original Star Trek Chief Engineer, Mr. Scott, spoke with admiration of Ion engines on an alien spacecraft. Warp engines are way-more-cool than Ion engines. But we can forgive the 1960s era television script writers for not grasping how soon Ion rocket engines would become real things. Perhaps Plan_B Version 11.5 can incorporate Alcubierre warp engines into its Catalog of Spacecraft designs.

For the more realistic and achievable near-term future, Plan_B has allocated space for linear engine tubes within SPM designs to support future installations of LARIT Engines.

LARIT-E Linear Accelerator tubes are one form of XPM. The literal tube modules fit as a balanced configuration—three tubes spread around 360°—to enable a uniform linear accelerator thrust vector out the back of a futuristic space vessel.

8.23   Deep Space Transportation Vessels

Plan_B Phase 15 addresses building very large spacecraft to send humans into Deep Space—beyond the gravity wells of Earth and Moon (Chart 112). Building upon the artificial gravity modules defined in Phase 15, Deep Space Transportation Vessels (DSTVs) will need to have a LOT of cubic volume to grow crops. DSTVs need large outside surface areas to produce a LOT of electricity and have long-duration constant thrust engines that are significantly more fuel efficient than current chemical reaction rockets.

DSTVs need a very-long axis upon which to house the linear accelerators required for LARIT Engines. Deep Space Vessels will also need to include very-large habitat areas to carry a LOT of water (fuel). They also have to have sufficient surface area to grow food, to provide housing / labs, etc. for a medium-to-large sized crew.

The central premise of a DSTV using LARIT Engines is to use constant Ion thrust to accelerate the DSTV to some reasonable percentage of the Speed-of-Light. The Plan_B Team believes that this goal is attainable with 21st century technology, and that such a ship could reach as much as 10% of the Speed-of-Light (SoL). The DSTV would accelerate for the first 50% of the transit time to a far destination. At the mid-point the ship would flip end-over-end and decelerate back to normal velocities at the end destination. The DSTV would also use conventional LCH4/LOX engines at the beginnings and ends of the journey.

It will take a LOT of patience for the DSTV to reach 10% of SoL during the high initial mass departures from LEO / LLO. Therefore, multiple CTugRVs (see Chapter 8) will also provide the initial thrust vectors for the DSTV. After boosting the DSTV, the CTugRVs would still be close enough and have sufficient Delta V to return safely to Earth or Lunar Orbit. The DSTV would also have options to eject (discard) autonomous CTugRVs and fuel tanks, similar to ejecting fuel tanks from a fighter aircraft once empty.

The instantaneous thrust of a LARIT engine may be low—but it can run for weeks/months/years on end. Plan_B Phase 13 builds and uses at least three LARIT Engines. More Engines equals More Thrust, but it is also the case that DSTV designs need three engines to allow for periodic engine maintenance during a long space voyage. It would be most unfortunate to use the LARIT Engines to accelerate the DSTV to 10% of SoL, reach your destination and have an engine failure. Having no means to slow down the spacecraft would be terrible. Imagine spending your entire life dedicated to reaching a distant planet orbiting Proxima Centauri (Chart 113), to blow past it because you have no brakes.

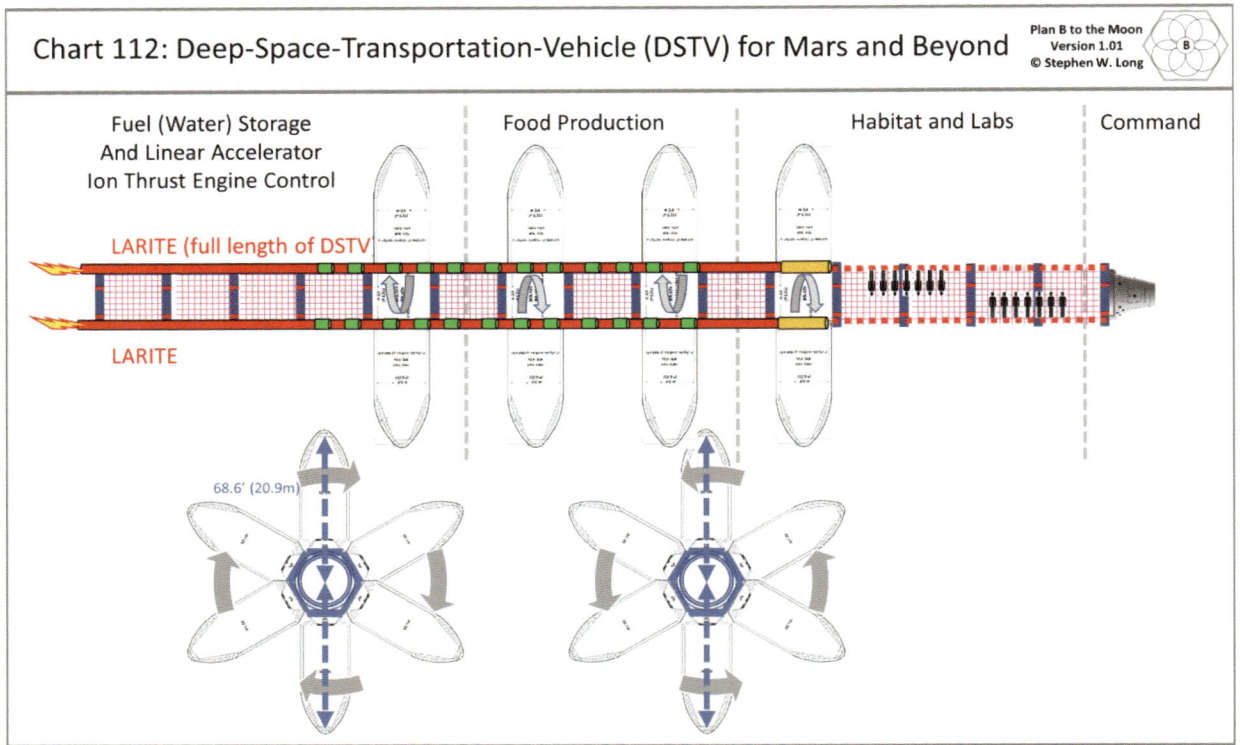

For those Readers old enough to remember the 1960s television program Lost in Space, the Robinson family got ~lost on the way to Alpha Centauri. Maybe they blew past their destination because their brakes failed. Danger Will Robinson…

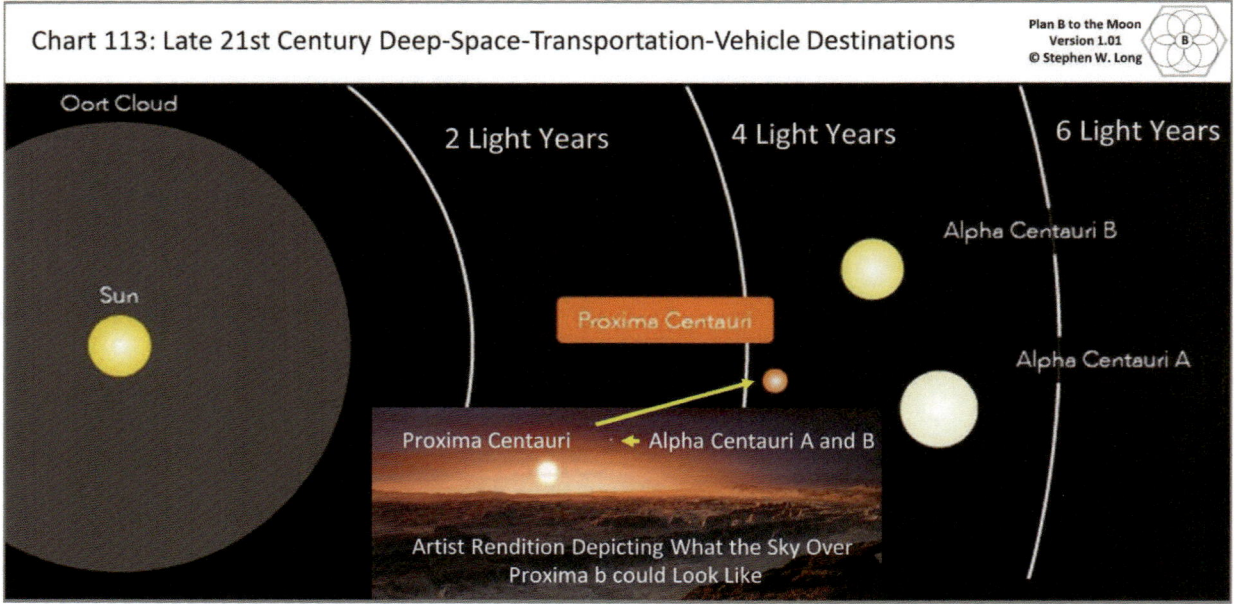

Chapter 9: Mission Plans

9.1 Mission Plans Introduction

The Six Pillars of Moon Colonization organize the **Plan_B Precepts**:
- Pillar 1: Mass into Orbit Affordably (Chapter 3)
- Pillar 2: Use X-Poly-Modules to Lower Spacecraft Costs (Chapter 4)
- Pillar 3: Leverage Existing Human Rated Spacecraft (Chapter 5)
- Pillar 4: Water* as Fuel–Common Conventional CH4/O2 (Methane/Oxygen) Rocket Engines and LARIT Engines (Chapter 6)
- Pillar 5: On-Orbit Refueling and Infrastructure (Chapter 7)
- Pillar 6: Catalog of Plan_B Spacecraft (Chapter 8)

In Chapter 9, Plan_B builds upon the Six Pillars, and defines in great detail the step-by-step activities to return to and enable the eventual colonization of the Moon. A Mission Circle defines each discrete step. As of Plan_B Version 1.00, it takes 804 Mission Circles to land on the Moon and by Mission Circle 899 Lunar Colonization begins. We note here that Plan_B Version 2.0 (first written in 2019) picks up at Mission Circle 899 and starts the first Lunar Colony (Lunar Base-01). Plan_B organizes the Mission Circles into logical overall Return-to-the-Moon phases, as depicted in Chart 114.

Chart 114: Plan B to Return to the Moon: Mission Phases	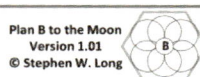

- Phase 1A: LEO Station-01 Assembly Start
- Phase 1B: LEO Station-01 Assembly Complete
- Phase 2A: SFT-01 Assembly Start
- Phase 2B: SFT-01 Assembly Complete
- Phase 3A: ELTV-01 Assembly Start (LLO Gateway-01)
- Phase 3B: ELTV-01 Arrival - Humans Return to LLO
- Phase 4A: ELTV-02 Lunar Components Delivery
- Phase 4B: ELTV-02 Arrival LLO
- Phase 5A: ELTV-03 Lunar Components Delivery
- Phase 5B: ELTV-03 Lunar Components Delivery
- Phase 6A: ELTV-04 Lunar Components Delivery
- Phase 6B: ELTV-04 Lunar Components Delivery
- Phase 7A: LLO Lunar Components Assembly
- Phase 7B: LLO Lunar Components Assembly
- Phase 8A: Humans Land on Lunar Surface / Lunar Base-01 Location Selection
- Phase 8B: Lunar Base-01 Establish

- Phase 09: Expand Lunar Base-01V2 / Produce Food
- Phase 10: Expand Lunar Base-01V3 / Lunar Mining
- Phase 11: Expand LEO Station-01V2 / Artificial Gravity Module
- Phase 12: Expand LEO Station-01V3 / Build Deep Space Food Modules
- Phase 13: Build Prototype LARIT Engine
- Phase 14: Build Deep Space Transportation Vehicle
- Phase 15: Send Humans to Mars
- Phase 16: Comet or Asteroid Rodeo
- Phase 17: Europa Exploration

In overview summary:

- Phase 1A and 1B builds the LEO Station-01 in Low Earth Orbit. LEO is the staging location for the assembly and marshaling of Plan_B SPMs for delivery to the Moon.
- Phase 2A and 2B builds the Space-Fuel-Terminal (see Chapter 7.2 for details).
- Phase 3A and 3B builds the first Earth-Lunar-Transfer-Vehicle payloads and delivers them to LLO. **Phase 3B (return to LLO) achieves Mission Parity with Plan_A EM-3.**
- Phases 4A, 4B, 5A, 5B, 6A and 6B deliver more ELTV payloads to LLO.
- Phase 7A and 7B assemble the vehicles/components for Humans to return to the Moon's surface.
- Phase 8A Humans Land on the Lunar Surface multiple times, seeking and choosing a location for the First Permanent Lunar Base (LB). In Phase 8B, the Astronaut Crews build the first LB outpost. If they volunteer to do so, 4 to 8 of those crew members transition from being rotational crew members to becoming Lunar Colonists.

Chart 115 provides the visual key to the various Mission Circle activities (numbers, shapes and colors). Mission Circles use a three-digit sequential numerical count (001–999), where the first digit defines the applicable Phase of the mission step. To facilitate understanding the function defined by the number, the numbers are given colors or shapes:

- Launch (purple circle), Rendezvous / Maneuvers (dark orange circle–since maneuvers make flames), Park (a hexagon with text Park), Dock (a hexagon with text Dock), Undock (a hexagon with text Undock), Space Autonomous Refueling or SAR (a hexagon with text SAR and sometimes a fluid direction arrow), State Change or Note (clear circle),
- Extract (a hexagon with text Extract), Attach (a hexagon with text Attach) and Un-attach (a hexagon with text Un-attach).
- Milestone (green diamond), Reserved (dark circle).

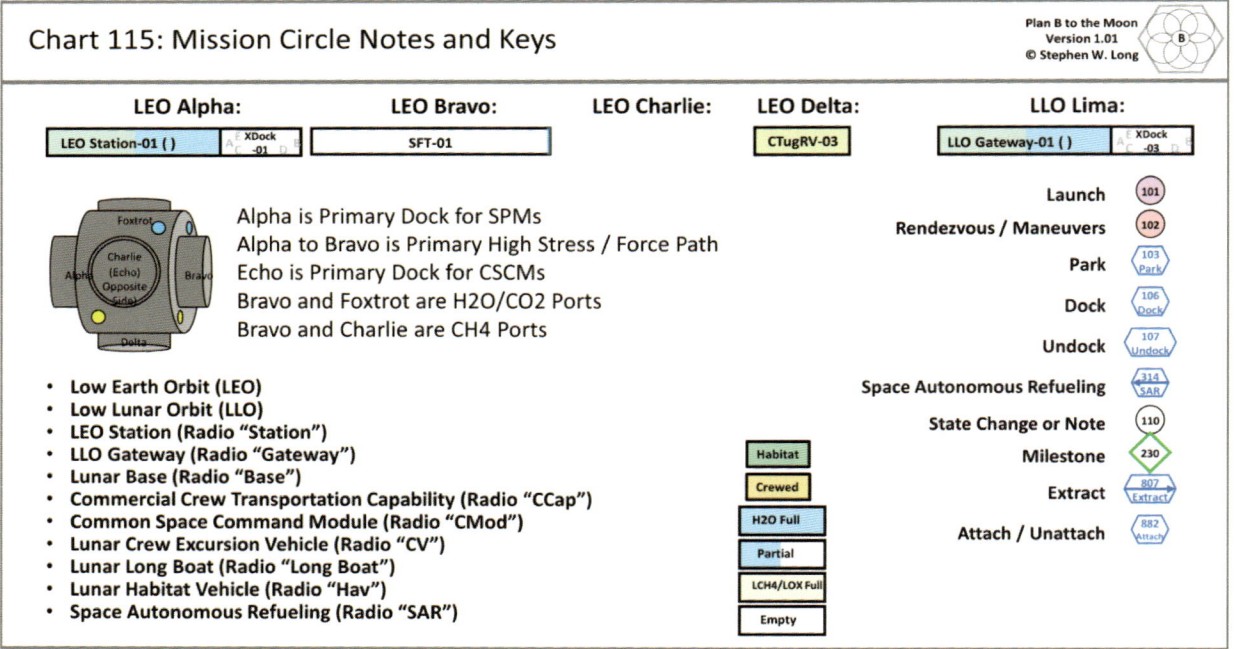

Chapter 9 depicts spacecraft as rectangles, with the name and the number of the entity (such [SPM-01]) which corresponds to the ID name used in the MISSION MATRIX. Color-codes define if a spacecraft is a Habitat (green), Crewed (yellow), and the water state (blue) from full / partial / empty. The cryogenic fuel state (yellow) depicts full / partial / empty.

Combinations of Mission Circles portray process-flow-diagrams depicting the major step-by-step tasks within a Plan_B Phase. The flow diagrams visually portray the step-by-step execution of the mission—this is the How-to element of the Plan_B title. Not all Mission Circles are equal—some circles specify the Launch of a Billion-dollar complex spacecraft; other Mission Circles report that a Phase is complete.

A significant percentage of Mission Circles describe the Rendezvous phases and exquisite 3D dances of spacecraft as they maneuver around each other. Plan_B added many such fine details over time (Plan_B Version 0.4 had less than 100 steps to land on the Moon). The Plan_B Team quickly learned it would take many (many) steps to define the sequential assembly of building blocks on-orbit to create functional spacecraft and habitats. We call these many steps the 3D-Dance-of-the-SPMs.

Every Mission Circle has a purpose. After accomplishing hundreds of Mission Circles—in the correct order—they define the literal How-to step-by-step plans to return to and colonize the Moon. Whew.

Chart 115 also defines a preliminary (nominal) radio call-sign index. The Plan_B Team adjusted the various names of spacecraft entities so that radio chatter would not lead to misunderstandings about who was talking to whom. For instance, the structure in LEO is a Station, and the structure in LLO is also a station. We changed its name to Gateway (NASA Plan_A also calls their Lunar Orbit structure a Gateway) so there would be no confusion between talking about LEO Station and LLO Gateway. Human pilots being human will always shorten a radio call sign to use the minimum words required. The full name of the structure in Earth Orbit is the *Low Earth Orbit Station Zero One*, radio calls will shorten this to, *Station*. The *Low Lunar Orbit Gateway Zero One* is *Gateway*. The *Lunar Base Zero One* is *Base*. The Commercial-Crew-Transportation-Capability (spacecraft launched from Earth ferrying astronauts to LEO and returning them to Earth) is *CCap-Zero-One* (Zero Two…). The *Lunar-Surface-Habitat-Vehicle Zero One* is *Hab Zero One*. The *Lunar-Lander Zero One* is *Lander One*. The *Lunar-Long-Boat Zero One* is *Boat One*. The Plan_B Team admits, with big smiles, that we had great fun practicing radio calls to one another. We practiced radio calls spanning hundreds of maneuvers on-orbit to assemble the various stations, gateways, and spacecraft.

9.2 Mission Phase 1A: LEO Station-01 Assembly Start

Overview: The Phase 1A Mission Objective is to Launch and Assemble in LEO SPM-01 (Habitat), CSCM-01, XDock-01, SPM-02, XDock-02, SPM-03, and four STMs.

CCtCap-01 delivers six crew members into orbit. FH rockets launch the first of several CTugRV vehicles and crew pilots fly the CTugRV as a tug boat. CTugRVs maneuver SPMs and associated modules into the LEO Station-01 configuration. The crew builds LEOS-01 by assembling and configuring PVMs, FCCs, PBMs, SSTs, TRMs, LFCs, WPEs, and PVPs.

Chart 116 (Phase 1A) defines the following Mission events:

- Mission Circle 101 Launches SPM-01+XDock-01+CSCM-01. Mission Circle 102 Maneuvers and Parks the SPM-01 stack into Orbital position LEO Alpha. This is the first autonomous delivery of components for LEO Station-01 (LEOS-01). SPM-01 is a Crew Habitat module (QRMs pre-fill the SPM-01 interior).

- Mission Circle 103 Launches CCtCap-01 (6 crew) and Rendezvous with LEOS-01. Mission Circle 104 CCtCap-01 Docks at LEOS-01 XDock-01 Foxtrot. Mission Circle 105 Launches CTugRV-01 (Autonomous) and Rendezvous with LEOS-01. Mission Circle 106 CTugRV-01 Docks at LEOS-01 XDock-01 Delta. Mission Circle 107 Crew (6) begins Assembly of LEOS-01 (adding FCCs, WSTs, etc.).

- Mission Circle 108 Starts a complex maneuver sequence to validate/check-out all XDock hatches and functions. Four Crew board CTugRV-01, Undocks from XDock-01 Delta. Mission Circle 109 CTugRV-01 Maneuvers to XDock-01 Echo and Mission Circle 110 Docks at XDock-01 Echo. Mission Circle 111 CTugRV-01 Undocks and Mission Circle 112 Maneuvers where CTugRV-01 performs Mission Circles 113, 114, and 115. CTugRV-01 repeats Docking and Undocking as necessary to cycle through all docking ports to validate dock/undock functions. Pilots also rotate seats, giving each pilot stick-time to certify all pilots as able to perform docking maneuvers. This is equivalent to multiple pilots rotating through the left and right pilot seats in military aircraft where each pilot performs multiple approaches / touch-and-goes / landings / takeoffs. Rotation sequence concludes with Mission Circles 116 and 117.

- Mission Circle 118 CTugRV-01 Maneuvers to the CSCM-01 nose hatch and Mission Circle 119 CTugRV-01 Attaches (not a full Dock) to CSCM-01. For Mission Circle 120 the combined CTugRV-01 and CSCM-01 Undocks and Mission Circle 121 Maneuvers CSCM-01 to XDock Echo. Mission Circle 122 Docks CSCM-01 to XDock-01 Echo. Mission Circle 123 CTugRV-01 Un-attaches and Mission Circle 124 Maneuvers to XDock-01 Bravo and Mission Circle 125 Docks. The purpose behind moving CSCM-01 is to prepare SPM-01 to dock with SPM-02.

- Mission Circle 126 LEO Station-01 Assembly Continues. Mission Circle 127 Reserved. Mission Circle 128 Launches SPM-02+XDock-02 and Rendezvous with LEOS-01. Mission

Circle 129 CTugRV-01 (2 Crew pilots, leaving 4 crew in the SPM) Undocks from XDock-01 Bravo. Mission Circle 130 Maneuvers to the opposite end of the LEOS-01 stack. Mission Circle 131 CTugRV-01 Docks with XDock-02 Bravo. CTugRV-01 pushes the XDock-02 plus SPM-02 stack and Mission Circle 132 Docks SPM-02 into SPM-01.

- Mission Circle 133 LEO Station-01 Assembly Continues.

- Mission Circle 134 Launches SPM-03+drySTM+drySTM+WetSTM+WetSTM. Mission Circle 135 CTugRV-01 (2 Crew) Undocks from XDock-02 Bravo and Mission Circle 136 Maneuvers to the STM end of the SPM-03 STM stack.

- Mission Circle 137 CTugRV-01 Docks with the STM SPM-03. Mission Circle 138 CTugRV-01 pushes the STM SPM-03 stack to cause SPM-03 to Dock with XDock-02 Bravo.

- Mission Circle 139 LEO Station-01 Assembly Continues. Mission Circle 140 CTugRV-01+SPM-03+STM+STM+STM+STM+XDock-02 Undocks from SPM-02.

- Mission Circle 141 CTugRV-01 flips (Pirouette Maneuver) the SPM-03 stack. Mission Circle 142 CTugRV-01 Undocks from SPM-03. Mission Circle 143 CTugRV-01 Maneuvers and Mission Circle 144 Docks with XDock-02 Bravo.

- Mission Circle 145 CTugRV-01 Maneuvers the stack to Mission Circle 146 Dock with SPM-03 to SPM-02.

- Mission Circle 147 CTugRV-01 Undocks from XDock-02 Bravo and Mission Circle 148 Maneuvers to the far side of the Station Stack. Mission Circle 149 CTugRV-01 Docks with XDock-01 Bravo. This is the desired LEOS-01 stack end state.

- Mission Circle 150 Crew performs a Water Transfer from the STMs to the SPM Shielding-Storage-Tanks (initial partial load). Mission Circle 151 LEO Station-01 Ops Normal.

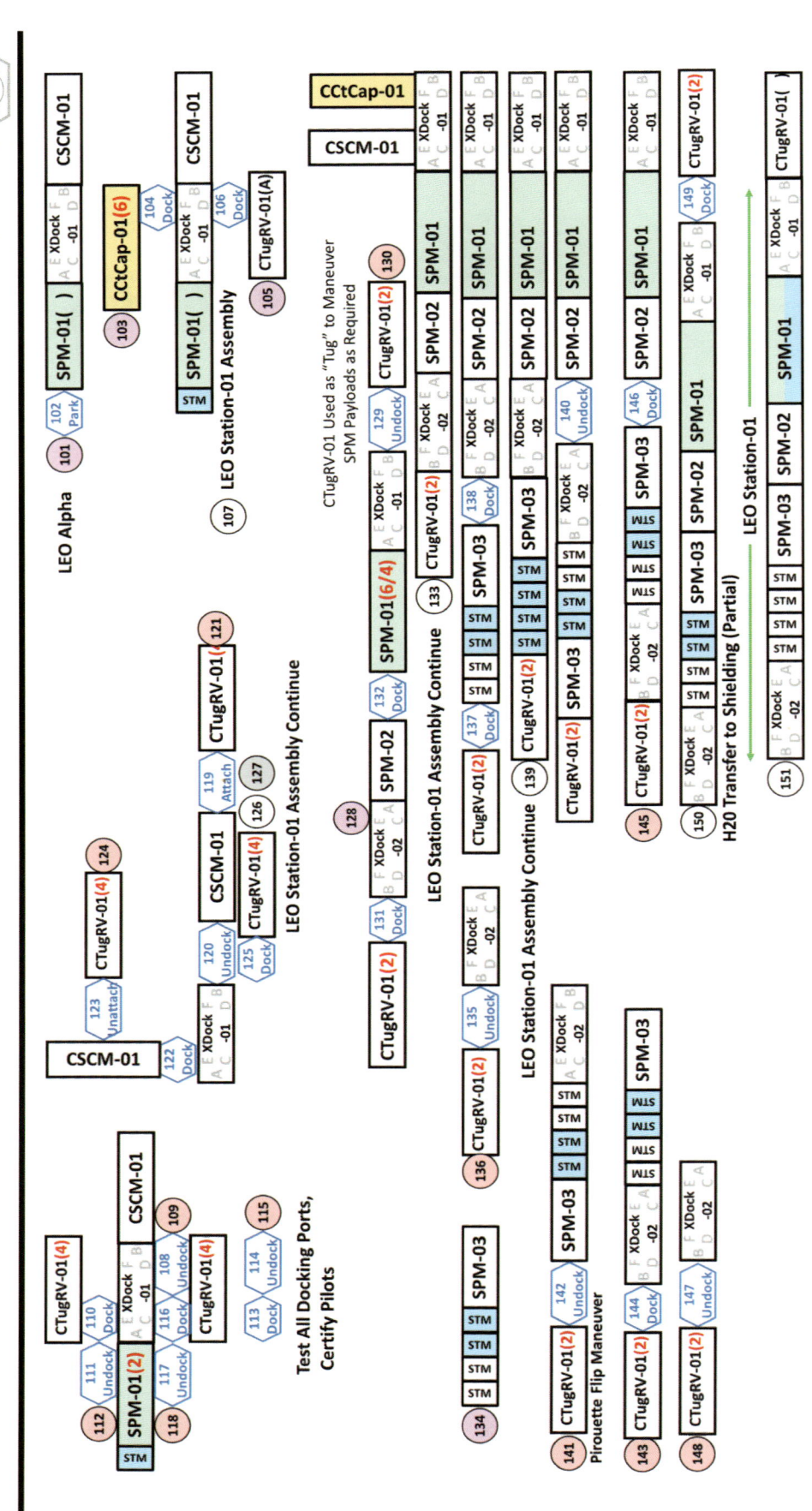

Chart 116: Phase 1A: LEO Station-01 Assembly Start

9.3 Mission Phase 1B: LEO Station-01 Assembly Complete

Overview: The Phase 1B Mission Objective is to continue and complete assembly of LEO Station-01.

Chart 117 (Phase 1B) defines the following Mission events:

- Mission Circle 151, LEOS-01 Ops Normal. It is standard practice in the Chapter 9 diagrams for the starting Mission Circle in a Phase B diagram to repeat the last Mission Circle from a Phase A diagram. This helps sequential Mission Circle tracing.

- Mission Circle 152, F9 Launch and Rendezvous of CCtCap-02 (6 crew members) with LEO Station-01. Mission Circle 153, CCtCap-02 Docks at LEOS-01 XDock-01 Delta.

- Mission Circle 154, LEOS-01 Ops Normal and Crew Rotation preparations and cross training.

- Mission Circle 155, CCtCap-01R (for Return Mode) Undocks from XDock-01 Foxtrot and Mission Circle 156 Returns to Earth with 6 Crew. Future analysis will determine if entire crews should rotate in/out, or always leave one or more experienced crew members on-board to become the Commander of the next crew. Mission Circle 157 LEOS-01 Ops Normal.

- Mission Circle 158 ASWTV-01 Launches on FHR and Rendezvous with LEOS-01.

- Mission Circle 159 ASWTV-01 performs Space Autonomous Refueling (water transfer) to STM modules on LEOS-01. Mission Circle 160 is a State Change (water transfers out) and Mission Circle 161 is a State Change (water transfers in).

- Mission Circle 162 ASWTV-02 Maneuvers and Mission Circle 163 Docks with XDock-02 Foxtrot.

- Mission Circle 164 Crew transfers Water from STMs to SPM Shielding Tanks.

- Mission Circle 165 LEOS-01 Ops Normal.

- Mission Circle 166, F9 Launch and Rendezvous of CCtCap-03 (6 crew members) and Mission Circle 167 Docks with XDock-01 Foxtrot. Foxtrot previously vacated by CCtCap-01. Mission Circle 168 LEOS-01 Ops Normal.

- Mission Circle 169 ASWTV-02 Launches on FHR and Rendezvous with LEOS-01. Mission Circle 170 ASWTV-02 performs Space Autonomous Refueling (water transfer) to STM modules on LEOS-01. Mission Circle 171 is a State Change (water transfers out) and Mission Circle 172 is a State Change (water transfers in).

Plan B to the Moon: A How-to Guide to Return to and Colonize the Lunar Surface

- Mission Circle 173 ASWTV-03 Maneuvers and Mission Circle 174 Docks with XDock-02 Delta.

- Mission Circle 175 Crew transfers Water from STMs to SPM Shielding Tanks. Mission Circle 176 LEOS-01 Ops Normal.

- Mission Circle 177 ASWTV-03 Launches on FHR and Rendezvous with LEOS-01. Mission Circle 178 ASWTV-04 performs Space Autonomous Refueling (water transfer) to STM modules on LEOS-01. Mission Circle 179 is a State Change (water transfers out) and Mission Circle 180 is a State Change (water transfers in).

- Mission Circle 181 ASWTV-03 Maneuvers and Parks at LEO Delta, becoming an on-orbit spare spacecraft. This may be a candidate for the lower-cost WTV vehicle.

- Mission Circle 182 Crew transfers Water from STMs to SPM Shielding Tanks. Mission Circle 183 Water transfers to Shielding Tanks Complete. SPM-01+SPM-02+SPM-03 shielding tanks are full and enable nominal shielding of crew habitable spaces. Mission Circle 184 LEOS-01 Ops Normal.

- Mission Circle 185 Launches CTugRV-02 (Autonomous) and Rendezvous with LEOS-01. Mission Circle 186 CTugRV-03 Docks at LEOS-02 XDock-02 Bravo.

- Mission Circle 187 Launches CTugRV-03 (Autonomous) and Rendezvous with CTugRV-01 (docked at LEOS-01 XDock-01 Bravo).

- Mission Circle 188 performs Cryogenic Space Autonomous Refueling (CSAR) and transfers cryogenic fuel to CTugRV-01. Mission Circle 189 is a State Change (fuel transfers out) and Mission Circle 190 is a State Change (fuel transfers in).

- Mission Circle 191 CTugRV-03 Maneuvers to CTugRV-02 and Mission Circle 192 performs CSAR and transfers cryogenic fuel to CTugRV-02. Mission Circle 193 is a State Change (fuel transfers out) and Mission Circle 194 is a State Change (fuel transfers in).

- Mission Circle 195 CTugRV-03 Maneuvers and Mission Circle 196 Docks to XDock-02 Charlie. Mission Circle 197 LEOS-01 Ops Normal. Mission Circle 198 Reserved.

- Mission Circle 199 MILESTONE. LEOS-01 Assembly Complete and LEO Station-01 is Fully Operational Capable (FOC), Plan_B Phase 01A&B Complete.

Plan B to the Moon
Version 1.01
© Stephen W. Long

Chart 117: Phase 1B: LEO Station-01 Assembly Complete

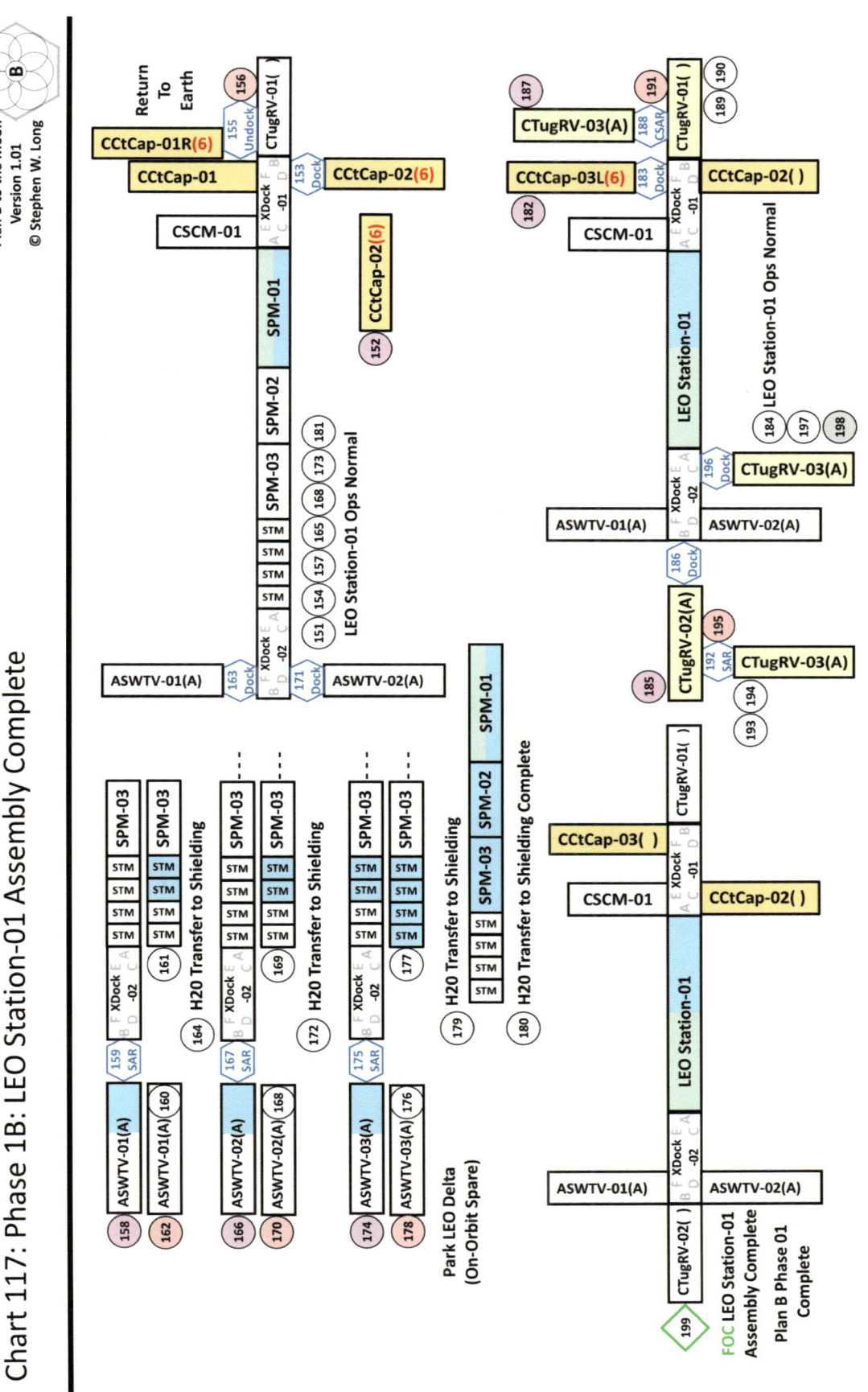

9.4 Mission Phase 2A: Space-Fuel-Terminal-01 Assembly Start

Overview: The Phase 2A Mission Objective is to build Space-Fuel-Terminal Zero One (SFT-01). Plan_B FHR Launches deliver payloads into LEO, and Crew on-orbit use the CTugRVs as tug boats to maneuver SPM and associated modules into the SFT-01 configuration.

Chart 118 (Phase 2A) defines the following Mission events:

- Mission Circle 201 Launches SPM-04+CSCM-02+WetSTM+WetSTM. Mission Circle 201 Reserved.

- Mission Circle 203 CTugRV-02 (2 crew) Undocks from XDock-02 Bravo and Mission Circle 204 Maneuvers to the SPM-04 stack. Mission Circle 205 CTugRV-02 Docks with the STM+STM+SPM-04+CSCM-02 stack.

- Mission Circle 206 CTugRV-02 Maneuvers the stack and Mission Circle 207 Docks with XDock-02 Echo. This is a 90-degree offset.

- Mission Circle 208 CTugRV-02+STM+STM+SPM-04 Un-attaches from CSCM-02.

- Mission Circle 209 CTugRV-02+STM+STM+SPM-04 Maneuvers and Mission Circle 210 Docks with XDock-02 Bravo.

- Mission Circle 211 CTugRV-02 Undocks from the STM stack and Mission Circle 212 Maneuvers and Mission Circle 213 Docks with XDock-01 Delta. Mission Circle 214 SFT-01 Assembly Starts.

- Mission Circle 215 Launches SPM-05+DrySTM+WetSTM+WetSTM. Mission Circle 216 Reserved.

- Mission Circle 217 CTugRV-02 Undocks from XDock-01 Alpha and Mission Circle 218 Maneuvers to and Mission Circle 219 Docks with SPM-05+STM+wetSTM+wetSTM.

- Mission Circle 220 CTugRV-02 stack Maneuvers and Mission Circle 221 Docks with the STM+STM+SPM-04 stack.

- Mission Circle 222 the CTugRV-02 stack Undocks from XDock-02 Bravo and CTugRV-02+SPM-05 stack Mission Circle 223 Undocks. Mission Circle 224 CTugRV-02+SPM-05 Maneuvers and Mission Circle 225 Docks SPM-04 to SPM-05.

- Mission Circle 226 CTugRV-02 performs a Pirouette Flip Maneuver with SPM-05+SPM-04+STM+STM+STM+STM+STM stack. CTugRV-02 Maneuvers and pushes the stack to Mission Circle 227 Dock with XDock-02 Bravo.

- Mission Circle 228 CTugRV-02 Undocks from SPM-05 and Mission Circle 229 Maneuvers and Mission Circle 230 Docks with XDock-01 Delta. Crew Rest.

- Mission Circle 231 SFT-01 Assembly Continues.

- Mission Circle 232 Launches SPM-06+WetCSM+WetCSM. Mission Circle 233 Reserved.

- Mission Circle 234 CTugRV-01 Undocks from XDock-01 Bravo and Mission Circle 235 Maneuvers and Mission Circle 236 Docks with CSM+CSM+SPM-05 stack.

- Mission Circle 237 CTugRV-01+CSM+CSM+SPM-06 stack Docks with SPM-05 and Mission Circle 238 CTugRV-01 Undocks from the CSM.

- Mission Circle 239 CTugRV-01 Maneuvers and Mission Circle 240 Docks with XDock-01 Bravo. Mission Circle 241 Reserved.

- Mission Circle 242 Launches XDock03+SPM-07+CSM+CSM+CSM stack. Mission Circle 243 Reserved.

- Mission Circle 244 CTugRV-01 Undocks from XDock-01 Bravo and Mission Circle 245 Maneuvers and Mission Circle 246 Docks with XDock-03 Bravo.

- Mission Circle 247 CTugRV-01+XDock-03 stack Maneuvers to Mission Circle 248 Docks to the CSM+CSM+SPM-06+SPM-05+SPM-04+STM+STM+STM+STM+STM+XDock-02 stack.

- Mission Circle 249 CTugRV-01 Undocks XDock03+SPM-07+CSM+CSM+CSM stack from the CSM+CSM+SPM-06 stack. Mission Circle 250 CTugRV-01 Undocks XDock-03 Alpha from SPM-07.

- Mission Circle 251 CTugRV-01 performs a Pirouette Flip Maneuver.

- Mission Circle 252 CTugRV-01 Docks XDock-03 Alpha with other end CSM… SPM-07 stack and Mission Circle 253 CTugRV-01… SPM-07 stack Docks with SPM-06.

- Mission Circle 254 CTugRV-01 Undocks from XDock-03 Bravo and Mission Circle 255 Maneuvers and Mission Circle 256 Docks with XDock-01 Bravo. Mission Circle 257 SFT-01 Assembly Continue.

Plan B to the Moon
Version 1.01
© Stephen W. Long

Chart 118: Phase 2A: Space Fuel Terminal-01 Assembly Start

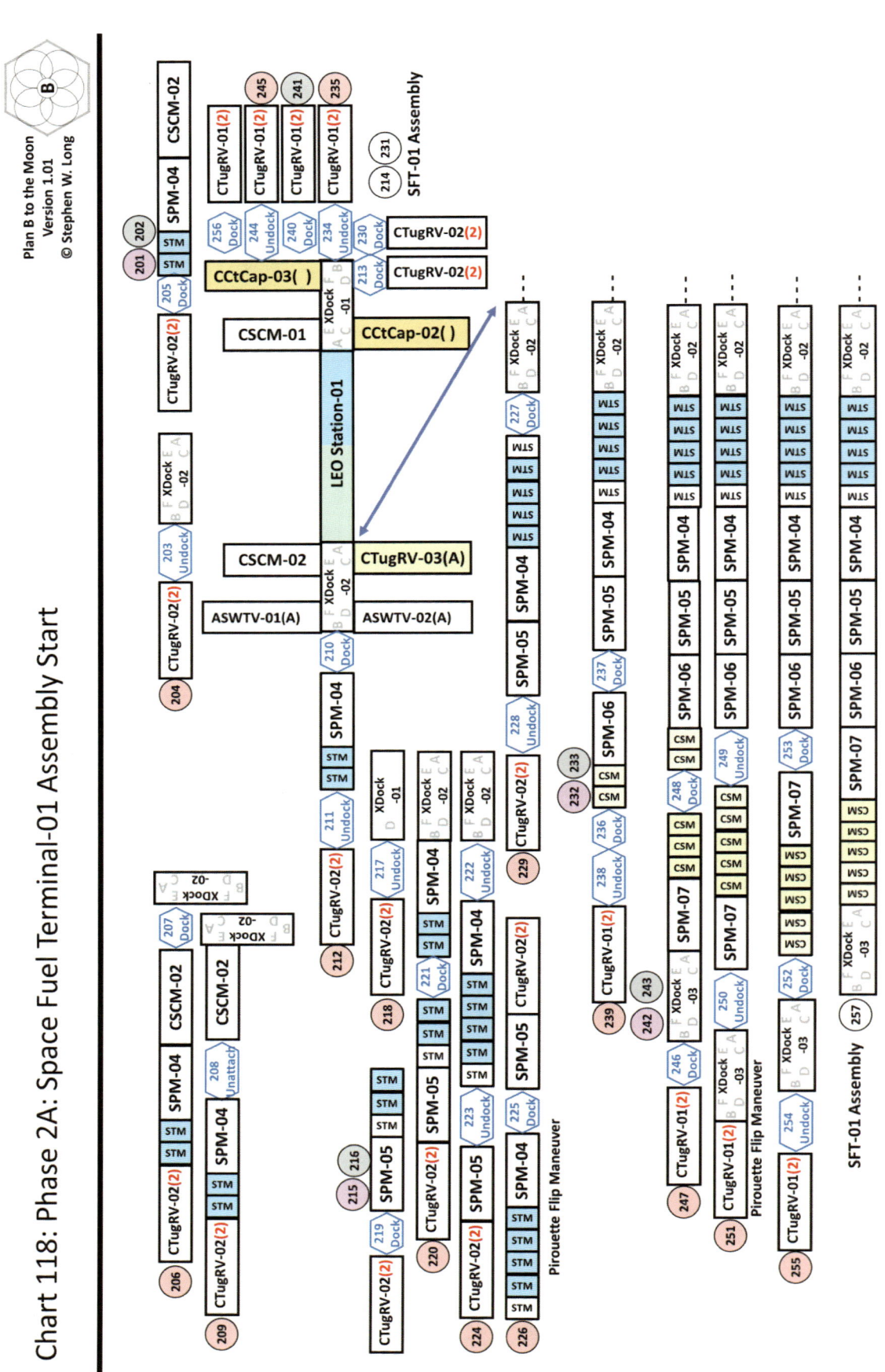

9.5 Mission Phase 2B: Space-Fuel-Terminal-01 Assembly Complete

Overview: The Phase 2B Mission Objective is to finish assembly and operationalize SFT-01.

Chart 119 (Phase 2B) defines the following Mission events:

- Mission Circle 258 Crew Assembles the SFT-01 Power Array. The SFT-01 Power Array will be the biggest power production solar array ever constructed on-orbit. Plan_B has designed CTugRV-01 and CTugRV-02 to operate as assembly support platforms during construction of the Array. SFT-01 is first assembled as one structure and separated into two elements (SFT-01A and SFT-01B). This reduces the mass to move A and B to the SFT-01 LEO parking slot LEO Bravo.

- Mission Circle 259 CTugRV-02 (4 crew members) Undocks from XDock-01 Delta and Mission Circle 260 Maneuvers and Mission Circle 261 Docks with XDock-03 Bravo. The next Maneuvers will move large spacecraft masses.

- Mission Circle 262 CTugRV-01 remains attached to the XDock-01B LEOS-01 stack and Undocks LEOS-01 from XDock-02 Alpha. Mission Circle 263 CTugRV-01 (4 crew members) Undocks from XDock-01 Bravo and Mission Circle 264 Maneuvers and Mission Circle 265 Docks with XDock-02 Alpha.

- Mission Circle 266 involves CTugRV-01 and CTugRV-02 pulling apart and Undocking the SPM-04 stack from the STM... XDock-02 stack.

- Mission Circle 267 CTugRV-02... SPM-04 Maneuvers and moves the SFT-01A structure to LEO Bravo. Mission Circle 268 CTugRV-01... STM Maneuvers and moves the SFT-01B structure to LEO Bravo and Rendezvous with SFT-01A structure.

- Mission Circle 269 has CTugRV-01 and CTugRV-02 guiding SFT-01A and SFT-01B to Dock. This is likely the most difficult docking maneuver in all of Plan_B.

- Mission Circle 270 has the both -01 and -02 Crews checkout and make SFT-01 operational. When SFT-01 System Checkout is complete, Mission Circle 271 CTugRV-02 Undocks from XDock-03 Bravo. Mission Circle 272 Maneuvers and Mission Circle 273 Docks with LEOS-01 XDock-01 Delta. Mission Circle 274 CTugRV-01 Undocks from XDock-02 Alpha and Mission Circle 275 Maneuvers and Mission Circle 276 Docks with XDock-01 Delta. Mission Circles 283 to 298 Reserved.

- Mission Circle 299 MILESTONE. SFT-01 Assembly Complete and SFT-01 is Making Fuel On-Orbit (IOC). Plan_B Phase 02 A and B Complete.

Plan B to the Moon
Version 1.01
© Stephen W. Long

Chart 119: Phase 2B: Space Fuel Terminal-01 Assembly Complete

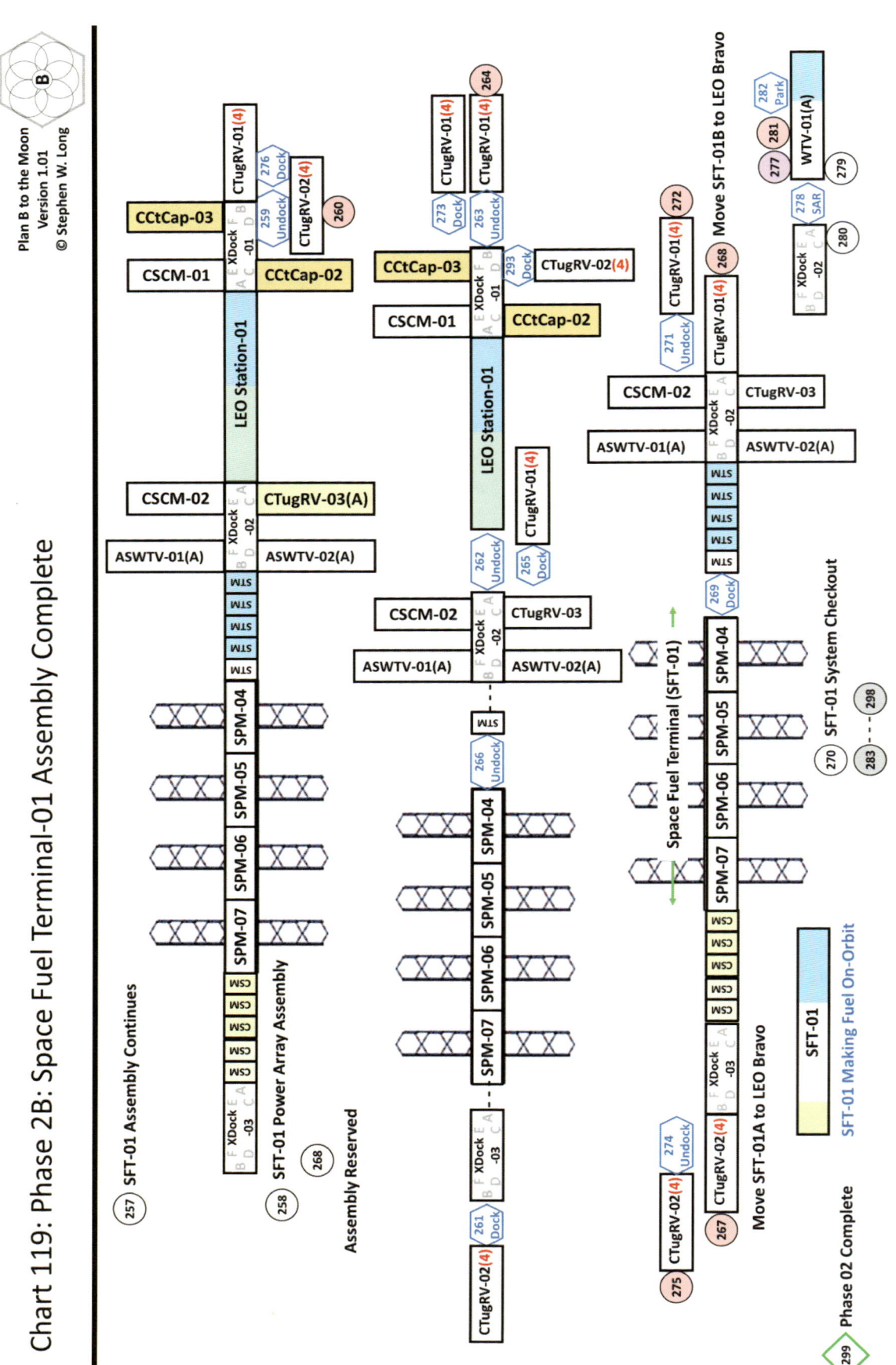

9.6 Mission Phase 3A: ELTV-01 Assembly Start

Overview: The Phase 3A Mission Objective is to Launch and Assemble in LEO SPM-08, SPM-09 (Habitat), CSCM-03, XDock-04. Modules are sent to LLO using Earth Lunar Transit Vehicle (ELTV) Zero One (ELTV-01).

Chart 120 (Phase 3A) defines the following Mission events:
- Mission Circle 301 Launches SPM-08+CSCM-03+XDock-04+WetSTM and Rendezvous with LEOS-01.

- Mission Circle 302 CTugRV-02 (2 crew) Undocks from XDock-01 Delta and Mission Circle 303 Maneuvers to the SPM-08 stack. Mission Circle 304 CTugRV-02 Docks with the STM+SPM-08+CSCM-03+XDock-04 stack.

- Mission Circle 305 CTugRV-02 Maneuvers the stack and Mission Circle 306 Docks with LEOS-01 (original SPM-03). Mission Circle 307 CTugRV-02+ STM+SPM-08+CSCM-03 Undocks from XDock-04 Bravo.

- Mission Circle 308 CTugRV-02 stack Maneuvers and Mission Circle 309 Docks with XDock-04 Echo (Note this is 90 Degree Offset from LEOS-01). Mission Circle 310 CTugRV-01+STM+SPM-08 Un-attaches from CSCM-03 and Mission Circle 311 Maneuvers and Mission Circle 312 CTugRV-02 Undocks from STM+SPM-08. Mission Circle 313 CTugRV-02 Maneuvers and Mission Circle 314 Docks with XDock-01 Delta.

- Mission Circle 315 ASWTV-02(A) Undocks from SFT-01 XDock-02 Delta and Mission Circle 316 Maneuvers and Rendezvous with LEOS-01.

- Mission Circle 317 ASWTV-02 performs Space Autonomous Refueling (water transfer) from STM module to ASWTV-02. Mission Circle 318 is a State Change (water transfers out) and Mission Circle 319 is a State Change (water transfers in).

- Mission Circle 320 ASWTV-02 Maneuvers and Parks on-orbit on standby for later Mission Circle 328 (Location TBD).

- Mission Circle 321 LEOS-01 Crew starts ELTV-01 Assembly. Mission Circle 322 LEOS-01 Ops Normal.

- Mission Circle 323 Launches STM+SPM-09 (Habitat) and Rendezvous with LEOS-01.

- Mission Circle 324 CTugRV-02 Undocks from XDock-01 Delta and Mission Circle 325 Maneuvers and Mission Circle 326 Docks with SPM-09. Mission Circle 327 CTugRV-02 Maneuvers / Station Keeps SPM-09 Stack steady.

- Mission Circle 328 ASWTV-02 Maneuvers and Rendezvous with the SPM-09 stack. Mission Circle 329 ASWTV-02 performs Space Autonomous Refueling (water transfer) from STM module to ASWTV-02. Mission Circle 330 is a State Change (water transfers out) and Mission Circle 331 is a State Change (water transfers in).

- Mission Circle 332 ASWTV-02 Maneuvers and Rendezvous with SFT-01. Mission Circle 333 ASWTV-02 performs Space Autonomous Refueling (water transfer) to SFT-01. Mission Circle 334 is a State Change (water transfers out) and Mission Circle 335 is a State Change (water transfers in).

- Mission Circle 336 ASWTV-02 Maneuvers and Mission Circle 337 Docks with XDock-02 Delta.

- Mission Circle 338 CTugRV-02+SPM-09+STM Maneuvers and Mission Circle 339 Docks with STM+SPM-08+XDock-04 stack.

- Mission Circle 340 CTugRV-02+SPM-09+STM+STM+SPM-08 stack Undocks from XDock-04 Bravo. Mission Circle 341 CTugRV-02+SPM-09 Undocks from SPM-08. Mission Circle 342 Maneuvers CTugRV-02+SPM-09 to Mission Circle 343 Dock with SPM-08 and Mission Circle 344 Dock with XDock-04 Bravo.

- Mission Circle 345 LEOS-01 Crew continues ELTV-01 Assembly.

- Mission Circle 346 MILESTONE ELTV-01 Assembly Complete.

- Sequence 347 through 379 Reserved for Contingency Fuel Delivery Missions if SFT-01 does not produce fuel and are therefore subject to revision. Duplicated Mission Circle numbers show a potential contingency branch point.

- Mission Circle 347 CRV-04(A) Launch. CTugRV-01 with two crew Mission Circle 348 Undocks from XDock-01 Bravo and Mission Circle 349 Maneuvers to join CRV-04A.

- Mission Circle 350 CRV-04(A) performs Cryogenic Space Autonomous Refueling (CSAR) with CTugRV-01. Mission Circle 351 is a State Change (Cryogenic fuel transfers out) and Mission Circle 352 is a State Change (Cryogenic fuel transfers in).

- Mission Circle 353 CRV-04(A) Maneuvers and Mission Circle 354 Docks with XDock-03F.

- Mission Circle 355 CRV-04(A) Launch. CTugRV-01 with two crew Mission Circle 356 Undocks from XDock-01 Bravo and Mission Circle 357 Maneuvers to join CRV-05A.

- Mission Circle 358 CRV-04(A) performs CSAR with CTugRV-01. Mission Circle 359 is a State Change (Cryogenic fuel transfers out) and Mission Circle 360 is a State Change (Cryogenic fuel transfers in).

- Mission Circle 363 SFT-01 Fuel is Ready.

- Mission Circle 364 CTugRV-03(A) Undocks from XDock-03 Delta and Mission Circle 365 Maneuvers.

- Mission Circle 366 CTugRV-03(A) performs Cryogenic Space Autonomous Refueling (CSAR) with SFT-01. Mission Circle 367 is a State Change (Cryogenic fuel transfers out) and Mission Circle 368 is a State Change (Cryogenic fuel transfers in).

- Mission Circle 369 CTugRV-03(A) Maneuvers and Mission Circle Rendezvous with LEOS-01.

- Mission Circle 370 CTugRV-02 (Crew 2) Undocks ELTV-01 from LEOS-01.

- Mission Circle 371 CTugRV-03(A) Maneuvers to CTugRV-02.

- Mission Circle 372 CTugRV-03(A) performs CSAR with SFT-01. Mission Circle 373 is a State Change (Cryogenic fuel transfers out) and Mission Circle 374 is a State Change (Cryogenic fuel transfers in).

- Mission Circle 375 CTugRV-03(A) Maneuvers to LEO Bravo and Mission Circle 376 Docks with XDock-03 Delta.

- Mission Circle 376 is an SFT-01 MILESTONE—Fully Operational Capable (FOC): Water* delivered, Cryogenic Fuel Created, and Refueling Completed to the Lunar Destination vehicle.

- Mission Circle 377 LEO Station Ops Normal.

- Mission Circle 378 ELTV-01 with Six Crew members Depart for LLO—on the way to Returning Humans to the Lunar Orbit. Mission Circle 379 CTugRV-01 (Crew 2) Departs for LLO. This could occur at Circle 357. Mission Circle 380 MILESTONE Phase 03A Complete.

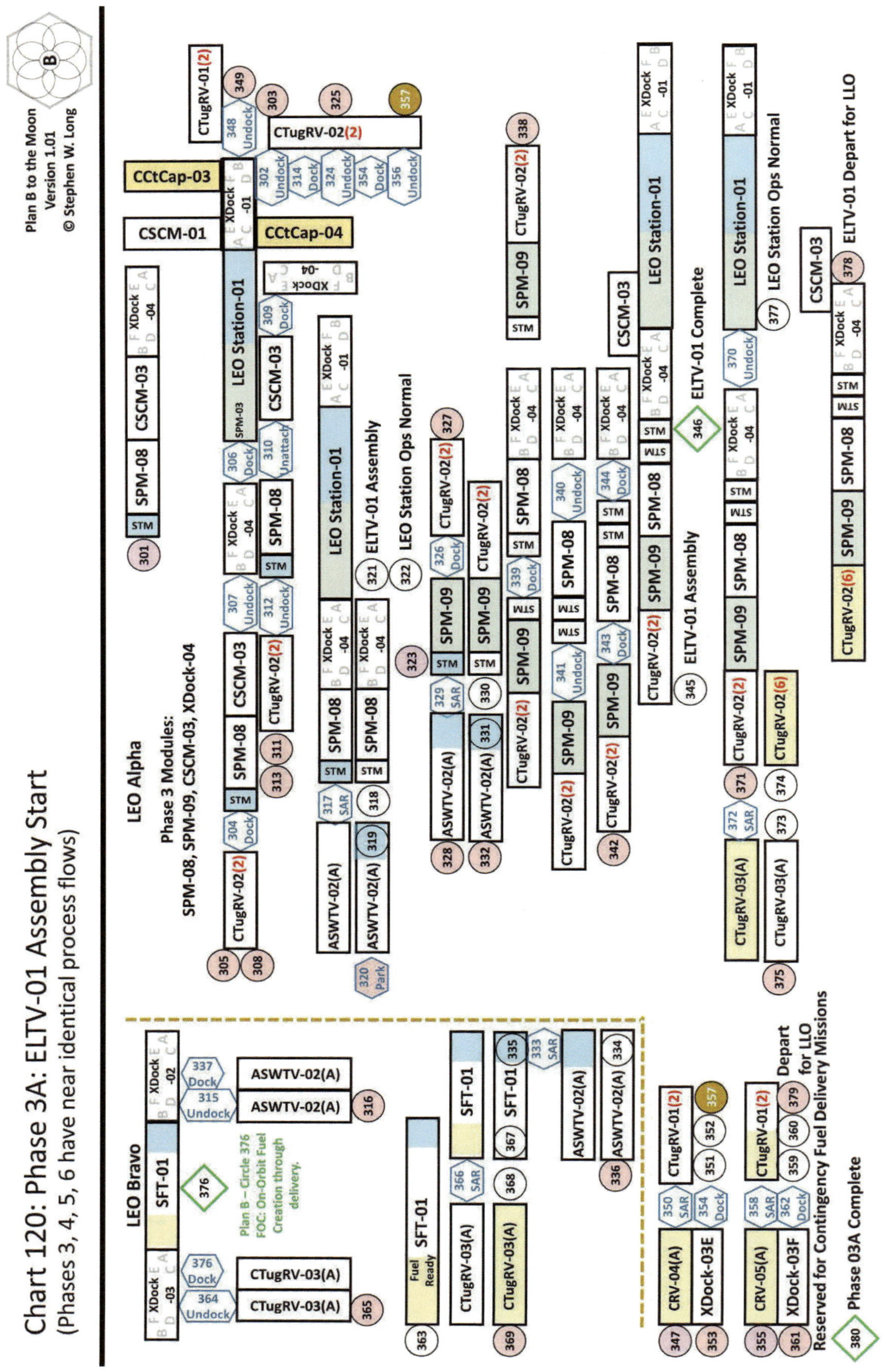

Plan B to the Moon
Version 1.01
© Stephen W. Long

Chart 120: Phase 3A: ELTV-01 Assembly Start
(Phases 3, 4, 5, 6 have near identical process flows)

LEO Alpha

Phase 3 Modules:
SPM-08, SPM-09, CSCM-03, XDock-04

LEO Station-01

LEO Station-01 Assembly

LEO Station Ops Normal

ELTV-01 Assembly

ELTV-01 Complete

LEO Station Ops Normal

CSCM-03

ELTV-01 Depart for LLO

LEO Bravo

SFT-01

Plan B – Circle 376
FOC: On-Orbit Fuel
Creation through
delivery.

Fuel Ready SFT-01

Reserved for Contingency Fuel Delivery Missions

Depart for LLO

Phase 03A Complete

CCtCap-03   CSCM-01   CCtCap-04

CTugRV-01(2)   CTugRV-02(2)   CTugRV-02(2)

SPM-08   CSCM-03   SPM-08   SPM-09

CTugRV-02(2)   CTugRV-02(2)   CTugRV-02(6)   CTugRV-03(A)

ASWTV-02(A)   CRV-04(A)   XDock-03E   CRV-05(A)   XDock-03F

9.7 Mission Phase 3B: ELTV-01 Arrives in LLO-Humans Return to LLO

Overview: The Phase 3B Mission Objective is for Humans to Return to Lunar Orbit and begin assembly of Low Lunar Orbit Gateway Zero One (LLOG-01).

Chart 121 (Phase 3B) defines the following Mission events:
* Mission Circle 381 MILESTONE when Humans Return to Lunar Orbit. ELTV-01 arrives and parks in LLO Lima orbital slot. Mission Circle 382 CTugRV-01 Arrives in LLO. CTugRV-01 is delivering the Sustainment Fuel for Lunar Spacecraft Operations. Mission Circle 383 Reserved.

* Mission Circle 384 CTugRV-01 Maneuvers and Mission Circle 385 Docks with LLOG-01A XDock-04 Alpha. Mission Circle 386, Crew Starts LLOG-01A Assembly on LLO.

* Mission Circle 387 CTugRV-02 Undocks and Mission Circle 388 Maneuvers/Returns to LEOS-01.

* Mission Circle 389, Crew on LLOG-01 performs LLO Polar Reconnaissance searching (observing) for Lunar Polar-Landing Sites.

* Mission Circle 390 CTugRV-02 Arrives in LEO and Mission Circle 391 Rendezvous with LEOS-01. Mission Circle 392 CTugRV-02 Docks with LEOS-01 XDock-01 Delta.

* Mission Circle 393 LEOS-01 Ops Normal.

* Mission Circle 394 CCtCap-04 (Six Crew) Launches and Rendezvous with LEOS-01. Mission Circle 395 CCtCap-04 Docks with LEOS-01 XDock-01 Bravo. Mission Circle 396 LEO Station Ops Normal and Crew Training and Handover.

* Mission Circle 397 CCtCap-02R (Six Crew) Undocks and Mission Circle 398 Maneuvers to Return to Earth.

* Mission Circle 399 MILESTONE Plan_B Phase 03 Complete—Humans Begin Continuous Occupation of Lunar Orbit.

* Occupancy in Lunar Orbit! LLOG-01 IOC.

Chart 121: Phase 3B: ELTV-01 Arrival – Humans Return to LLO

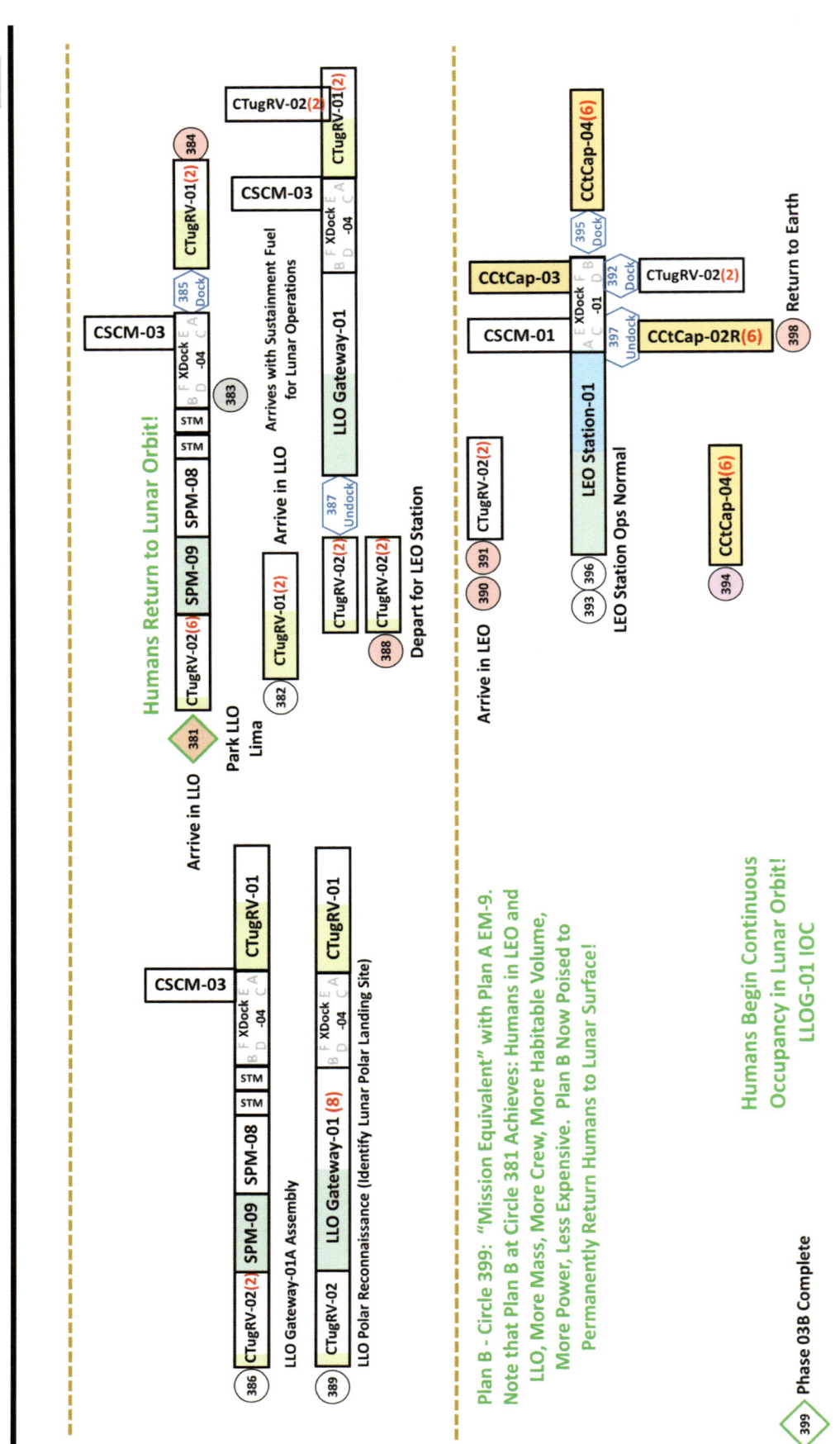

Plan B to the Moon: A How-to Guide to Return to and Colonize the Lunar Surface

Plan B to the Moon
Version 1.01
© Stephen W. Long

Humans Return to Lunar Orbit!

Arrive in LLO

Park LLO
Lima

CSCM-03

CTugRV-02(6) | SPM-09 | SPM-08 | STM | STM | CTugRV-02(2) | F XDock E / B D -04 C A | Dock 385 | CTugRV-01(2) | 384

CTugRV-02(2)

381

382 CTugRV-01(2)

383

Arrive in LLO

Arrives with Sustainment Fuel
for Lunar Operations

CSCM-03

CTugRV-02(2) | B F XDock E / D -04 C A | LLO Gateway-01

CTugRV-02(2)

387 Undock

CTugRV-02(2) 388

Depart for LEO Station

CTugRV-02(2) | SPM-09 | SPM-08 | STM | STM | CSCM-03 | B F XDock E / D -04 C A | CTugRV-01

386

LLO Gateway-01A Assembly

CTugRV-02 | LLO Gateway-01 (8) | B F XDock E / D -04 C A | CTugRV-01

389

LLO Polar Reconnaissance (Identify Lunar Polar Landing Site)

**Plan B - Circle 399: "Mission Equivalent" with Plan A EM-9.
Note that Plan B at Circle 381 Achieves: Humans in LEO and
LLO, More Mass, More Crew, More Habitable Volume,
More Power, Less Expensive. Plan B Now Poised to
Permanently Return Humans to Lunar Surface!**

Arrive in LEO

390 391 CTugRV-02(2)

CCtCap-03 | A E XDock F / B C -01 D B | Dock 395 | CCtCap-04(6)

392 Dock | CTugRV-02(2)

CSCM-01 | LEO Station-01 | 397 Undock | CCtCap-02R(6) | 398

393 396

LEO Station Ops Normal

394 CCtCap-04(6)

Return to Earth

**Humans Begin Continuous
Occupancy in Lunar Orbit!
LLOG-01 IOC**

399 Phase 03B Complete

9.8 Mission Phase 4A: ELTV-02 Assembly Start

Overview: The Phase 4A Mission Objective is to Launch and Assemble in LEO SPM-10, SPM-11 (Habitat), CSCM-04, XDock-05. From LEO, the modules are sent to LLO using Earth Lunar Transit Vehicle (ELTV) Zero Two (ELTV-02).

Chart 122 (Phase 4A) defines the following Mission events:

- Mission Circle 401 Launches SPM-10+CSCM-04+XDock-05+WetSTM and Rendezvous with LEOS-01.

- Mission Circle 402 CTugRV-02 (2 crew) Undocks from XDock-01 Delta and Mission Circle 403 Maneuvers to the SPM-10 stack. Mission Circle 404 CTugRV-02 Docks with the STM+SPM-10+CSCM-04+XDock-05 stack.

- Mission Circle 405 CTugRV-02 Maneuvers the stack and Mission Circle 406 Docks with LEOS-01 (original SPM-03). Mission Circle 407 CTugRV-02+ STM+SPM-10+CSCM-04 Undocks from XDock-05 Bravo.

- Mission Circle 408 CTugRV-02 stack Maneuvers and Mission Circle 409 Docks with XDock-05 Echo (Note this is 90 Degree Offset from LEOS-01). Mission Circle 410 CTugRV-01+STM+SPM-10 Un-attaches from CSCM-04 and Mission Circle 411 Maneuvers and Mission Circle 412 CTugRV-02 Undocks from STM+SPM-10. Mission Circle 413 CTugRV-02 Maneuvers and Mission Circle 414 Docks with XDock-01 Delta.

- Mission Circle 415 ASWTV-02(A) Undocks from SFT-01 XDock-02 Delta and Mission Circle 416 Maneuvers and Rendezvous with LEOS-01.

- Mission Circle 417 ASWTV-02 performs Space Autonomous Refueling (water transfer) from STM module to ASWTV-02. Mission Circle 418 is a State Change (water transfers out) and Mission Circle 419 is a State Change (water transfers in).

- Mission Circle 420 ASWTV-02 Maneuvers and Parks on-orbit on standby for later Mission Circle 428 (Location TBD).

- Mission Circle 421 LEOS-01 Crew starts ELTV-02 Assembly. Mission Circle 422 LEOS-01 Ops Normal.

- Mission Circle 423 Launches STM+SPM-11 (Habitat) and Rendezvous with LEOS-01.

- Mission Circle 424 CTugRV-02 Undocks from XDock-01 Delta and Mission Circle 425 Maneuvers and Mission Circle 426 Docks with SPM-09. Mission Circle 427 CTugRV-02 Maneuvers / Station Keeps SPM-09 Stack steady.

- Mission Circle 428 ASWTV-02 Maneuvers and Rendezvous with the SPM-09 stack. Mission Circle 429 ASWTV-02 performs Space Autonomous Refueling (water transfer) from STM module to ASWTV-02. Mission Circle 430 is a State Change (water transfers out) and Mission Circle 431 is a State Change (water transfers in).

- Mission Circle 432 ASWTV-02 Maneuvers and Rendezvous with SFT-01. Mission Circle 433 ASWTV-02 performs Space Autonomous Refueling (water transfer) to SFT-01. Mission Circle 434 is a State Change (water transfers out) and Mission Circle 435 is a State Change (water transfers in).

- Mission Circle 436 ASWTV-02 Maneuvers and Mission Circle 437 Docks with XDock-02 Delta.

- Mission Circle 438 CTugRV-02+SPM-11+STM Maneuvers and Mission Circle 439 Docks with STM+SPM-10+XDock-05 stack.

- Mission Circle 440 CTugRV-02+SPM-09+STM+STM+SPM-10 stack Undocks from XDock-05 Bravo. Mission Circle 441 CTugRV-02+SPM-09 Undocks from SPM-10. Mission Circle 442 Maneuvers CTugRV-02+SPM-09 to Mission Circle 443 Dock with SPM-10 and Mission Circle 444 Docks with XDock-05 Bravo.

- Mission Circle 445 LEOS-01 Crew continues ELTV-02 Assembly.

- Sequence 446 through 470 Reserved for Contingency Fuel Delivery Missions if SFT-01 does not produce fuel and are therefore subject to revision. Some highlighted Mission Circle numbers show a potential contingency branch point.

- Mission Circle 446 CRV-06(A) Launches and Rendezvous at SFT-01. CTugRV-03 (A) Mission Circle 447 Undocks from XDock-03 Delta and Mission Circle 448 Maneuvers to join CRV-06A.

- Mission Circle 449 CRV-06(A) performs Cryogenic Space Autonomous Refueling (CSAR) with CTugRV-03(A). Mission Circle 450 is a State Change (Cryogenic fuel transfers out) and Mission Circle 451 is a State Change (Cryogenic fuel transfers in).

- Mission Circle 452 CRV-06(A) Maneuvers and Parks at LEO Bravo Location TBD (because XDock-03 docks full or reserved).

- Mission Circle 453 CTugRV-03 Maneuvers and Mission Circle 454 Docks with XDock-03 Delta.

- Mission Circle 455 CRV-07(A) Launch. CTugRV-03 (A) Mission Circle 456 Undocks from XDock-03 Delta and Mission Circle 457 Maneuvers to join CRV-07(A).

- Mission Circle 458 CRV-07(A) performs Cryogenic Space Autonomous Refueling (CSAR) with CTugRV-01. Mission Circle 459 is a State Change (Cryogenic fuel transfers out) and Mission Circle 460 is a State Change (Cryogenic fuel transfers in).

- Mission Circle 461 CRV-07(A) Maneuvers and Parks at LEO Bravo Location TBD (because XDock-03 docks full or reserved).

- Mission Circle 462 CTugRV-03 Maneuvers and Mission Circle 463 Docks with XDock-03 Delta.

- Mission Circle 464 SFT-01 Fuel is Ready.

- Mission Circle 465 CTugRV-03(A) Undocks from XDock-03 Delta and Mission Circle 466 Maneuvers.

- Mission Circle 467 CTugRV-03(A) performs CSAR with SFT-01. Mission Circle 468 is a State Change (Cryogenic fuel transfers out) and Mission Circle 469 is a State Change (Cryogenic fuel transfers in).

- Mission Circle 470 CTugRV-03(A) Maneuvers and Rendezvous with CTugRV-02 (ELTV-02).

- Mission Circle 471 CTugRV-02 (Crew 2) Undocks ELTV-02 from LEOS-01.

- Mission Circle 472 CTugRV-02 Maneuvers to join CTugRV-03(A).

- Mission Circle 473 CTugRV-03(A) performs CSAR with CTugRV-02. Mission Circle 474 is a State Change (Cryogenic fuel transfers out) and Mission Circle 475 is a State Change (Cryogenic fuel transfers in).

- Mission Circle 476 CTugRV-03(A) Maneuvers to LEO Bravo and Mission Circle 477 Docks with XDock-03 Delta.

- Mission Circle 478 LEO Station Ops Normal.

- Mission Circle 479 ELTV-02 with Six Crew Departs for LLO.

- Mission Circle 480 MILESTONE Phase 04A Complete.

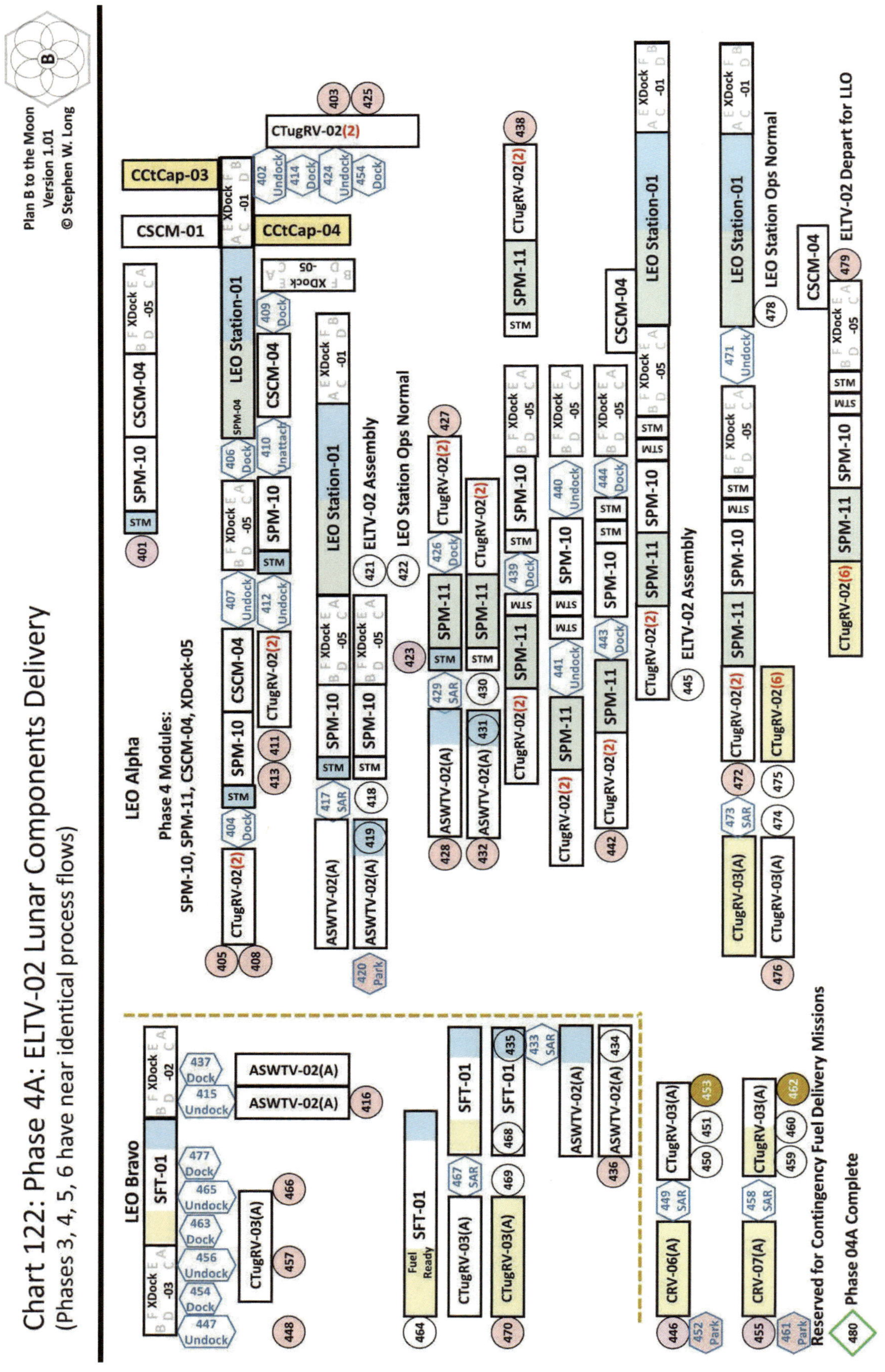

Chart 122: Phase 4A: ELTV-02 Lunar Components Delivery
(Phases 3, 4, 5, 6 have near identical process flows)
Plan B to the Moon
Version 1.01
© Stephen W. Long
LEO Alpha
Phase 4 Modules:
SPM-10, SPM-11, CSCM-04, XDock-05
LEO Bravo
CCtCap-03
CSCM-01
CCtCap-04
CTugRV-02(2)
LEO Station-01
SPM-10
SPM-11
CSCM-04
XDock -05
XDock -01
XDock -02
XDock -03
STM
SFT-01
ASWTV-02(A)
CTugRV-03(A)
CTugRV-02(6)
CRV-06(A)
CRV-07(A)
CSCM-04
ELTV-02 Assembly
LEO Station Ops Normal
ELTV-02 Depart for LLO
Fuel Ready
Dock
Undock
SAR
Park
Unattach
Reserved for Contingency Fuel Delivery Missions
Phase 04A Complete

9.9 Mission Phase 4B: ELTV-02 Arrives in LLO, Assembly Continues

Overview: The Phase 4B Mission Objective is to integrate ELTV-02 into LLOG-01.

Chart 123 (Phase 4B) defines the following Mission events:
- Mission Circle 481 ELTV-02 Arrives in LLO.

- Mission Circle 482 ELTV-02 (CTugRV-02) stack Mission Circle 482 Undocks SPM-11 from SPM-10 and Mission Circle 483 Maneuvers to the other end of LLOG-01.

- Mission Circle 484 CTugRV-02+SPM-11 Docks with XDock-05A.

- Mission Circle 485 CTugRV-02 stack Maneuvers and Mission Circle 486 Docks SPM-10 with SPM-09. Mission Circle 487 Crew Continues LLOG-01 Assembly.

- Mission Circle 488 CTugRV-02 (Six Crew) Undocks from LLOG-01A and Mission Circle 489 Maneuvers and Departs LLO to return to LEOS-01.

- Mission Circle 490 LLOG-01 Ops Normal.

- Mission Circle 491 CTugRV-02 Arrives in LEO and Mission Circle 492 Rendezvous with LEOS-01.

- Mission Circle 493 CTugRV-02 Docks with LEOS-01 XDock-01 Delta.

- Mission Circle 494 CCtCap-05 (Six Crew) Launches and Rendezvous with LEOS-01 and Mission Circle 495 Docks with XDock-01 Bravo.

- Mission Circle 496 LEO Station Ops Normal, Crew Swaps.

- Mission Circle 497 CCtCap-03R (Four Crew) Undocks and Mission Circle 498 Returns to Earth.

- Mission Circle 499 MILESTONE Plan_B Phase 4B Complete.

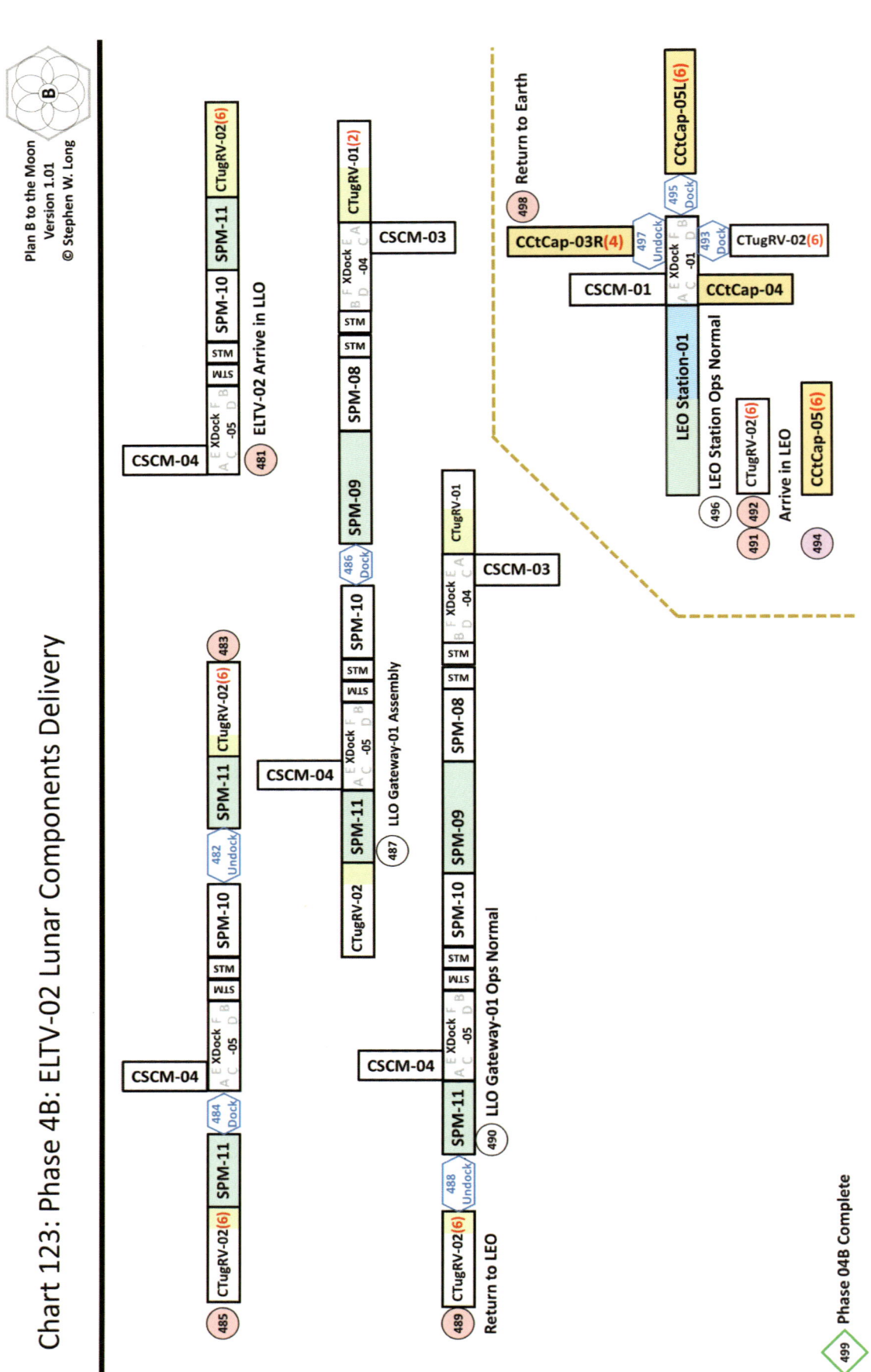

Chart 123: Phase 4B: ELTV-02 Lunar Components Delivery

Plan B to the Moon
Version 1.01
© Stephen W. Long

9.10   Mission Phase 5A: ELTV-03 Assembly Start

Overview: The Phase 5A Mission Objective is to Launch and Assemble in LEO SPM-12, SPM-13 (Habitat), CSCM-05, XDock-06. From LEO, the modules are sent to LLO using Earth Lunar Transit Vehicle (ELTV) Zero Three (ELTV-03).

Chart 124 (Phase 5A) defines the following Mission events:

- Mission Circle 501 Launches SPM-12+CSCM-04+XDock-05+WetSTM and Rendezvous with LEOS-01. Mission Circle 502 CTugRV-02 (2 crew) Undocks from XDock-01 Delta and Mission Circle 503 Maneuvers to the SPM-12 stack. Mission Circle 504 CTugRV-02 Docks with the STM+SPM-12+CSCM-04+XDock-05 stack. Mission Circle 505 CTugRV-02 Maneuvers the stack and Mission Circle 506 Docks with LEOS-01 (original SPM-03). Mission Circle 507 CTugRV-02+ STM+SPM-12+CSCM-04 Undocks from XDock-05 Bravo.

- Mission Circle 508 CTugRV-02 stack Maneuvers and Mission Circle 509 Docks with XDock-05 Echo (Note this is 90 Degree Offset from LEOS-01). Mission Circle 510 CTugRV-01+STM+SPM-12 Un-attaches from CSCM-04 and Mission Circle 511 Maneuvers and Mission Circle 512 CTugRV-02 Undocks from STM+SPM-12. Mission Circle 513 CTugRV-02 Maneuvers and Mission Circle 514 Docks with XDock-01 Delta.

- Mission Circle 515 ASWTV-02(A) Undocks from SFT-01 XDock-02 Delta and Mission Circle 516 Maneuvers and Rendezvous with LEOS-01. Mission Circle 517 ASWTV-02 performs Space Autonomous Refueling (water transfer) from STM module to ASWTV-02. Mission Circle 518 is a State Change (water transfers out) and Mission Circle 519 is a State Change (water transfers in). Mission Circle 520 ASWTV-02 Maneuvers and Parks on-orbit on standby for later Mission Circle 528 (Location TBD). Mission Circle 521 LEOS-01 Crew starts ELTV-03 Assembly. Mission Circle 522 LEOS-01 Ops Normal.

- Mission Circle 523 Launches STM+SPM-13 (Habitat) and Rendezvous with LEOS-01. Mission Circle 524 CTugRV-02 Undocks from XDock-01 Delta and Mission Circle 525 Maneuvers and Mission Circle 526 Docks with SPM-13. Mission Circle 527 CTugRV-02 Maneuvers / Station Keeps the SPM-13 Stack steady. Mission Circle 528 ASWTV-02 Maneuvers and Rendezvous with the SPM-13 stack. Mission Circle 529 ASWTV-02 performs Space Autonomous Refueling (water transfer) from STM module to ASWTV-02. Mission Circle 530 is a State Change (water transfers out) and Mission Circle 531 is a State Change (water transfers in).

- Mission Circle 532 ASWTV-02 Maneuvers and Rendezvous with SFT-01. Mission Circle 533 ASWTV-02 performs Space Autonomous Refueling (water transfer) to SFT-01. Mission Circle 534 is a State Change (water transfers out) and Mission Circle 535 is a State Change (water transfers in). Mission Circle 536 ASWTV-02 Maneuvers and Mission Circle 537 Docks with XDock-02 Delta. Mission Circle 538 CTugRV-02+SPM-13+STM Maneuvers and Mission Circle 539 Docks with STM+SPM-12+XDock-06 stack. Mission Circle 540 CTugRV-02+SPM-13+STM+STM+SPM-12 stack Undocks from XDock-06 Bravo. Mission Circle 541 CTugRV-

02+SPM-013 Undocks from SPM-12. Mission Circle 542 Maneuvers CTugRV-02+SPM-13 to Mission Circle 543 Dock with SPM-12 and Mission Circle 544 Docks with XDock-06 Bravo. Mission Circle 545 LEOS-01 Crew continues ELTV-03 Assembly.

- Sequence 546 through 570 Reserved for Contingency Fuel Delivery Missions and are subject to revision. Some Mission Circle numbers are Gold highlighted to show a potential contingency branch point. Mission Circle 546 CRV-06(A) Launches and Rendezvous at SFT-01. CTugRV-03 (A) Mission Circle 547 Undocks from XDock-03 Delta and Mission Circle 548 Maneuvers to join CRV-08A.

- Mission Circle 549 CRV-08(A) performs Cryogenic Space Autonomous Refueling (CSAR) with CTugRV-03(A). Mission Circle 550 is a State Change (Cryogenic fuel transfers out) and Mission Circle 551 is a State Change (Cryogenic fuel transfers in). Mission Circle 552 CRV-08(A) Maneuvers and Parks at LEO Bravo Location TBD (because XDock-03 docks full or reserved). Mission Circle 553 CTugRV-03 Maneuvers and Mission Circle 554 Docks with XDock-03 Delta.

- Mission Circle 555 CRV-09(A) Launch. CTugRV-03 (A) Mission Circle 556 Undocks from XDock-03 Delta and Mission Circle 557 Maneuvers to join CRV-09(A). Mission Circle 558 CRV-09(A) performs CSAR with CTugRV-01. Mission Circle 559 is a State Change (Cryogenic fuel transfers out) and Mission Circle 560 is a State Change (Cryogenic fuel transfers in).

- Mission Circle 561 CRV-09(A) Maneuvers and Parks at LEO Bravo Location TBD (because XDock-03 docks full or reserved). Mission Circle 562 CTugRV-03 Maneuvers and Mission Circle 563 Docks with XDock-03 Delta. Mission Circle 564 SFT-01 Fuel is Ready. Mission Circle 565 CTugRV-03(A) Undocks from XDock-03 Delta and Mission Circle 566 Maneuvers. Mission Circle 567 CTugRV-03(A) performs CSAR with SFT-01. Mission Circle 568 is a State Change (Cryogenic fuel transfers out) and Mission Circle 569 is a State Change (Cryogenic fuel transfers in).

- Mission Circle 570 CTugRV-03(A) Maneuvers and Rendezvous with CTugRV-02 (ELTV-03). Mission Circle 571 CTugRV-02 (Crew 2) Undocks ELTV-03 from LEOS-01. Mission Circle 572 CTugRV-02 Maneuvers to join CTugRV-03(A). Mission Circle 573 CTugRV-03(A) performs CSAR with CTugRV-02. Mission Circle 574 is a State Change (Cryogenic fuel transfers out) and Mission Circle 575 is a State Change (Cryogenic fuel transfers in).

- Mission Circle 576 CTugRV-03(A) Maneuvers to LEO Bravo and Mission Circle 577 Docks with XDock-03 Delta. Mission Circle 578 LEO Station Ops Normal. Mission Circle 579 ELTV-03 with Six Crew Departs for LLO. Mission Circle 580 MILESTONE Phase 05A Complete.

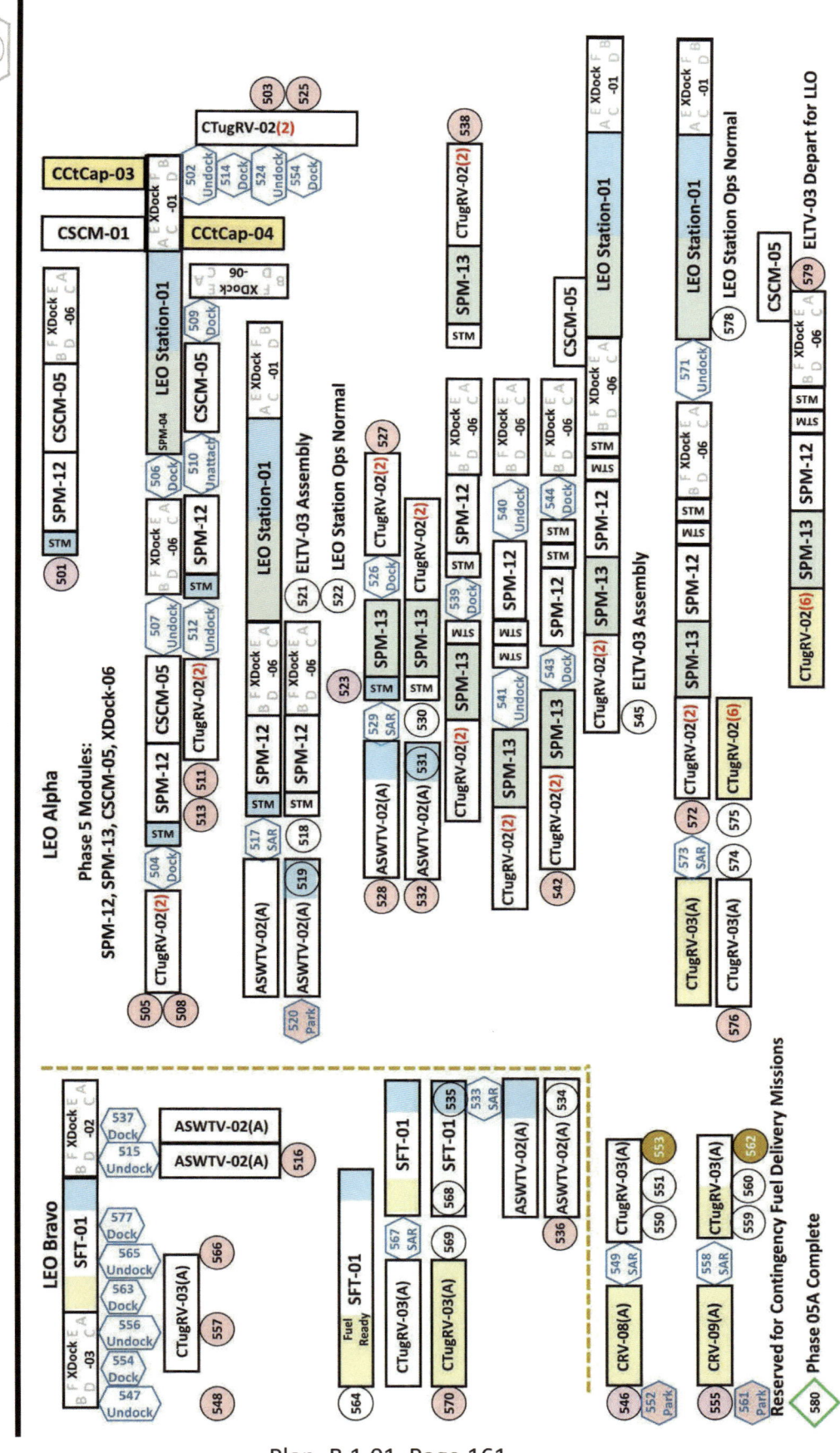

Chart 124: Phase 5A: ELTV-03 Lunar Component Delivery
(Phases 3, 4, 5, 6 have near identical process flows)

9.11 Mission Phase 5B: ELTV-03 Arrives in LLO, Assembly Continues

Overview: The Phase 5B Mission Objective is to integrate ELTV-03 into LLOG-01.

Chart 125 (Phase 5B) defines the following Mission events:

- Mission Circle 581 ELTV-03 Arrives in LLO.

- Mission Circle 582 ELTV-03 (CTugRV-02) stack Mission Circle 582 Undocks SPM-13 from SPM-12 and Mission Circle 583 Maneuvers to the other end of LLOG-01.

- Mission Circle 584 CTugRV-02+SPM-13 Docks with XDock-06A.

- Mission Circle 585 CTugRV-02 stack Maneuvers and Mission Circle 586 Docks with SPM-11.

- Mission Circle 587, Crew Continues LLOG-01 Assembly.

- Mission Circle 588 CTugRV-02 (Six Crew) Undocks from LLOG-01 and Mission Circle 589 Maneuvers and Departs LLO to return to LEOS-01.

- Mission Circle 590 LLOG-01 Ops Normal.

- Mission Circle 591 CTugRV-02 Arrives in LEO and Mission Circle 592 Rendezvous with LEOS-01.

- Mission Circle 593 CTugRV-02 Docks with LEOS-01 XDock-01 Delta.

- Mission Circle 594 CCtCap-05 (Six Crew) Launches and Rendezvous with LEOS-01 and Mission Circle 595 Docks with XDock-01 Bravo.

- Mission Circle 596 LEO Station Ops Normal, Crew Swaps.

- Mission Circle 597 CCtCap-03R (Four Crew) Undocks and Mission Circle 598 Returns to Earth.

- Mission Circle 599 MILESTONE Plan_B Phase 5B Complete.

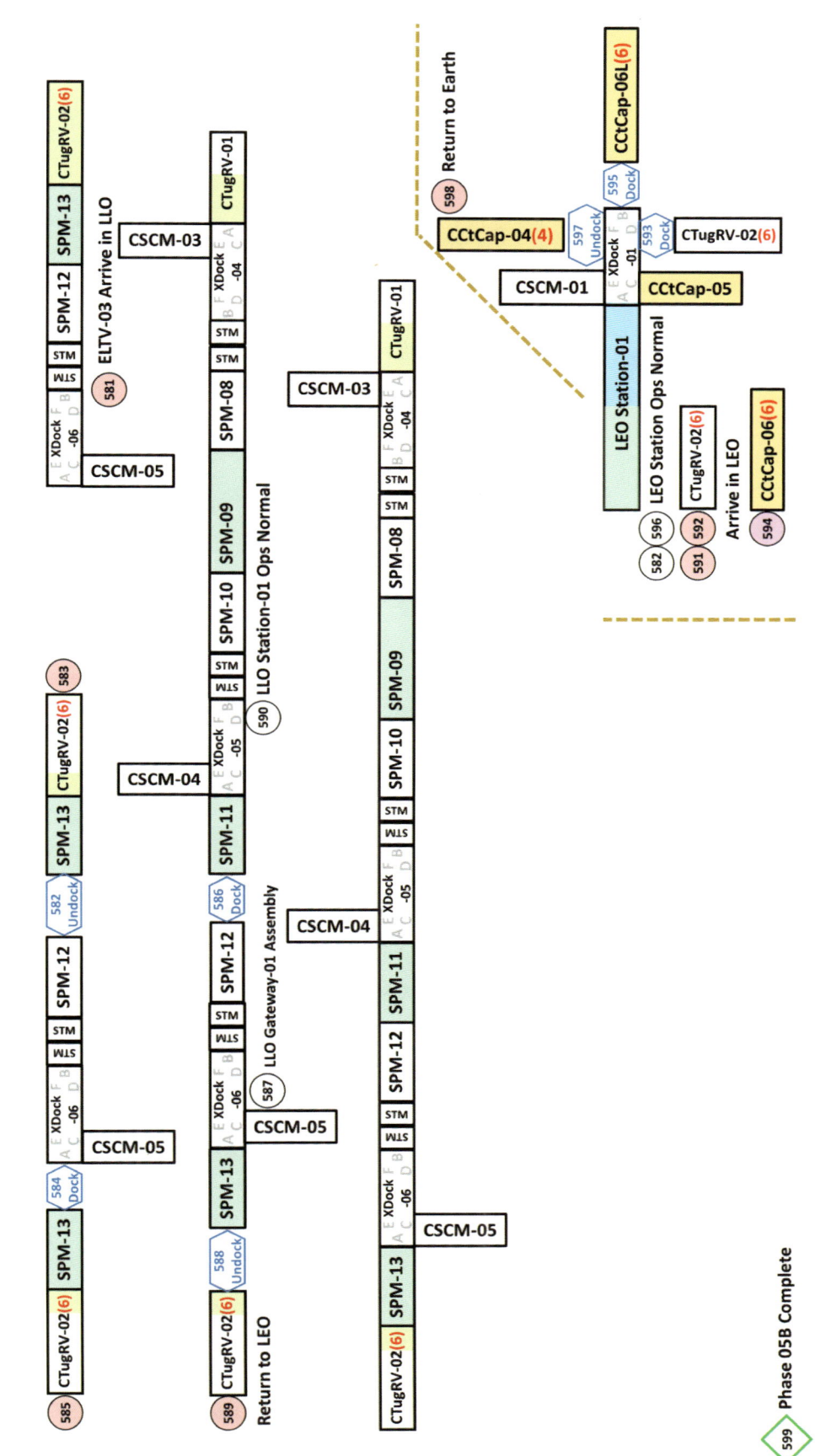

Chart 125: Phase 05B: ELTV-03 Lunar Components Delivery

9.12   Mission Phase 6A: ELTV-04 Assembly Start

Overview: The Phase 6A Mission Objective is to Launch and Assemble SPM-14 and SPM-15 in LEO. From LEO the modules are sent to LLO using Earth Lunar Transit Vehicle (ELTV) Zero-Four (ELTV-04).

Chart 126 (Phase 6A) defines the following Mission events:
- Mission Circle 601 Launches SPM-14 and Rendezvous with LEOS-01. Mission Circles 602-621 Reserved.

- Mission Circle 623 Launches SPM-15 and Rendezvous with LEOS-01.

- Mission Circle 624 CTugRV-02 Undocks from XDock-01 Delta and Mission Circle 625 Maneuvers and Mission Circle 626 Docks with SPM-14. Mission Circle 627 CTugRV-02 Maneuvers and Mission Circle 628 Docks SPM-14 to SPM-15.

- Mission Circle 629 CTugRV-02+SPM-14+SPM-15 Maneuvers and Mission Circle 630 Docks the stack to LEOS-01.

- Mission Circle 631 LEOS-01 Ops Normal. Mission Circles 632-663 Reserved.

- Mission Circle 664 SFT-01 Fuel is Ready. Mission Circle 665 CTugRV-03(A) Undocks from XDock-03 Delta and Mission Circle 666 Maneuvers.

- Mission Circle 667 CTugRV-03(A) performs Cryogenic Space Autonomous Refueling (CSAR) with SFT-01. Mission Circle 668 is a State Change (Cryogenic fuel transfers out) and Mission Circle 669 is a State Change (Cryogenic fuel transfers in).

- Mission Circle 670 CTugRV-03(A) Maneuvers and Rendezvous with CTugRV-02 (ELTV-04). Mission Circle 671 CTugRV-02 (Crew 2) Undocks ELTV-04 from LEOS-01. Mission Circle 672 CTugRV-02 Maneuvers to join CTugRV-03(A).

- Mission Circle 673 CTugRV-03(A) performs CSAR with CTugRV-02. Mission Circle 674 is a State Change (Cryogenic fuel transfers out) and Mission Circle 675 is a State Change (Cryogenic fuel transfers in). Mission Circle 676 CTugRV-03(A) Maneuvers to LEO Bravo and Mission Circle 677 Docks with XDock-03 Delta.

- Mission Circle 678 LEO Station Ops Normal.

- Mission Circle 679 ELTV-04 with Six Crew Departs for LLO. Mission Circle 680 MILESTONE Phase 06A Complete.

Plan B to the Moon
Version 1.01
© Stephen W. Long

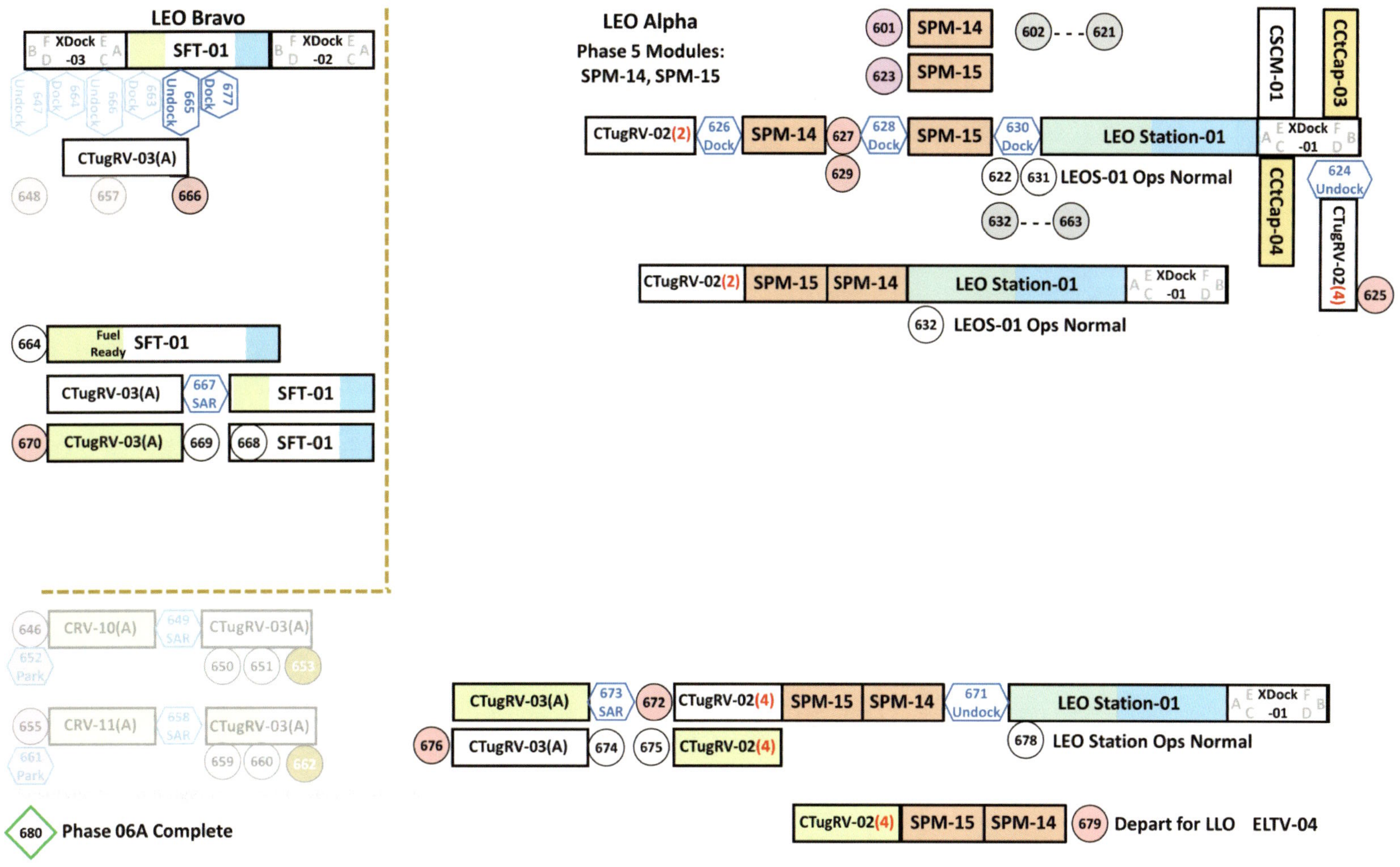

Chart 126: Phase 6A: ELTV-04 Lunar Components Delivery
(Phases 3, 4, 5, 6 have near identical process flows)

9.13   Mission Phase 6B: ELTV-04 Arrives in LLO, Assembly Continues

Overview: The Phase 6B Mission Objective is to integrate ELTV-04 into LLOG-01.

Chart 127 (Phase 6B) defines the following Mission events:

- Mission Circle 681 ELTV-04 Arrives in LLO (delivers SPM-14 and SPM-15). Mission Circle 682 CTugRV-02+SPM-15+SPM-14 Docks with XDock-04D. Mission Circle 683 CTugRV-02+SPM-15 Undocks from SPM-14. Mission Circles 684 and 685 Reserved. Mission Circle 686 CTugRV-02+SPM-15 Maneuvers and Mission Circle 687 Docks SPM-15 to XDock-04 Foxtrot. Mission Circle 688 CTugRV-02 Undocks from SPM-15 and Mission Circle 689 Maneuvers and Circle 690 Docks with XDock-05 Delta. Circle 691 LLOG-01 Ops Normal. Circles 692-698 Reserved. Mission Circle 699 MILESTONE Plan_B Phase 6B Complete.

9.14   Mission Phase 7A: Lunar Components Assemble in LLO

Overview: The Phase 7A Mission Objective is to assemble Lunar Components in LLO. Phase 7A moves large spacecraft masses to arrange the Lunar Components into the correct positions for eventual undocking and landings on the Lunar Surface.

Chart 128 (Phase 7A) defines the following Mission events:

- Mission Circle 701 CTugRV-02 (four crew) Undocks from XDock-05 Delta then Mission Circle 702 Maneuvers to SPM-13. Mission Circle 703 CTugRV-02 Docks with SPM-13. Mission Circle 704 CTugRV-02 stack Undocks SPM-10 from SPM-09 and Mission Circle 705 Undocks SPM-11 from XDock-05A. Mission Circle 706 Undocks from SPM-12 and Mission Circle 707 Undocks from XDock-06A. Mission Circle 708 CTugRV-02+SPM-13 Maneuvers and Mission Circle 709 Docks SPM-13 to CSCM-05. Mission Circle 710 CTugRV-02+SPM-13+CSCM-05 Undocks from XDock-06C and Mission Circle 711 Maneuvers and Mission Circle 712 Docks CSCM-05 to XDock-05C. Mission Circle 713 CTugRV-02 Undocks from SPM-13 and Mission Circle 714 Maneuvers and Mission Circle 715 Docks with SPM-12. Mission Circle 716 CTugRV-02 Maneuvers SPM-12+SPM-11 to Mission Circle 717 Dock with CSCM-04. Mission Circle 718 CTugRV-02 Undocks from SPM-12 and Mission Circle 719 Maneuvers and Circle 720 Docks with SPM-10.

- Mission Circle 721 CTugRV-02 pushes the large stack and Mission Circle 722 Docks with SPM-09. This is a very large mass maneuver. Mission Circle 723 CTugRV-02+SPM-10 Undocks SPM-10 from the STM+... stack. Mission Circle 724 CTugRV-02+SPM-10 Maneuvers and Mission Circle 725 Docks SPM-10 to SPM-13. Mission Circle 726 CTugRV-02 Undocks from SPM-10 and Mission Circle 727 Maneuvers to XDock-06A. Phase 7A is complete and 7B immediately starts at Mission Circle 728.

Plan B to the Moon
Version 1.01
© Stephen W. Long

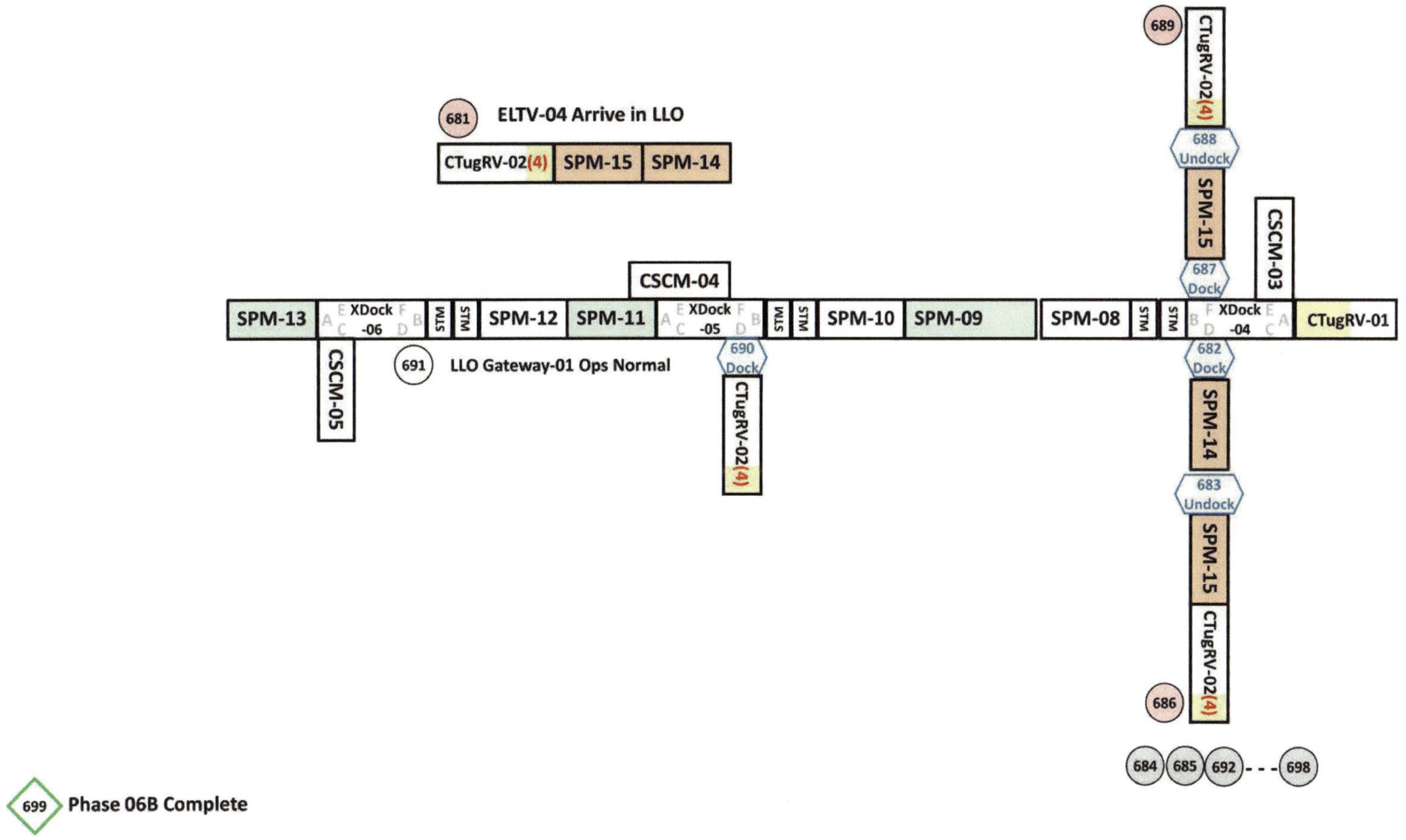

Chart 127: Phase 6B: ELTV-04 Lunar Components Delivery

CTugRV-01

CSCM-03

CTugRV-02(4) — 688 Undock — SPM-15 — 687 Dock — B F XDock E A / C D -04 C — 682 Dock — SPM-14 — 683 Undock — SPM-15 — CTugRV-02(4)

689

STM

STM

SPM-08

SPM-09

SPM-10

STM

STM

CSCM-04

A E XDock D B / C C -05 F D — 690 Dock — CTugRV-02(4)

684 685 692 - - - 698

686

681 ELTV-04 Arrive in LLO

CTugRV-02(4) — SPM-15 — SPM-14

SPM-12 — SPM-11

STM

LLO Gateway-01 Ops Normal

691

A E XDock F B / C C -06 D

SPM-13

STM

STM

CSCM-05

699 Phase 06B Complete

Chart 128: Phase 7A: LLO Lunar Components Assembly

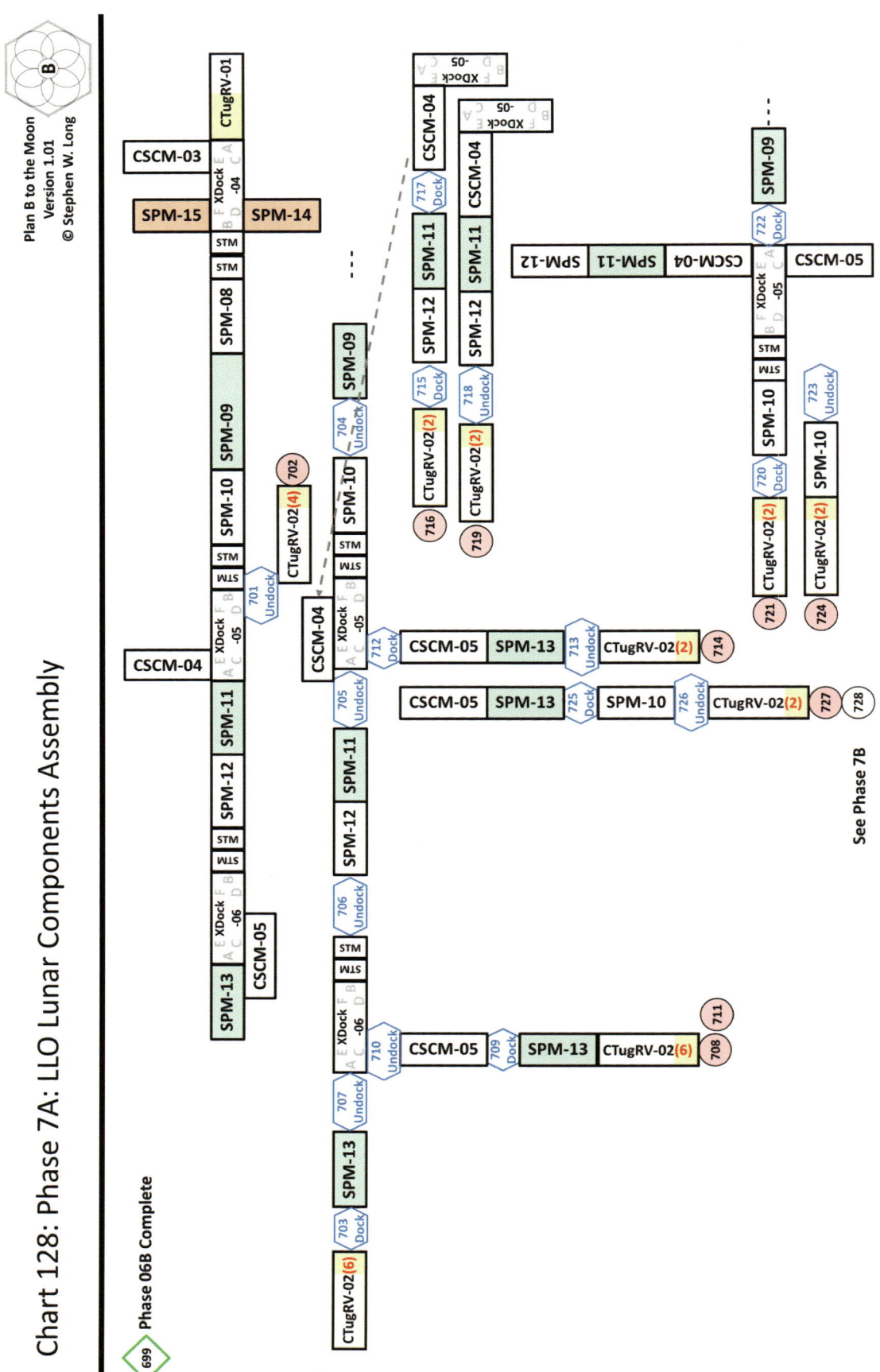

Plan B to the Moon: A How-to Guide to Return to and Colonize the Lunar Surface

9.15   Mission Phase 7B: Lunar Components Assemble in LLO

Overview: The Phase 7B Mission Objective is to continue assembly of Lunar Components in LLO. This is a large mass maneuver to arrange the Lunar Components into position for eventual undocking and landings on the Lunar Surface.

Chart 129 (Phase 7B) defines the following Mission events:

- Mission Circle 728 CTugRV-02 Docks with XDock-06A. Mission Circle 729 CTugRV-02 Maneuvers XDock-06 and Mission Circle 730 Docks XDock-06+STM+STM to the STM+STM+XDock-05 stack. Mission Circle 731 CTugRV-02 Undocks from XDock-06A and Mission Circle 732 Maneuvers to SPM-14, docked at XDock-04D.

- Mission Circle 733 CTugRV-02 uses its Robotic Remote Manipulation Arm (RRMA) to Attach to LL-02 and Mission Circle 734 Extracts LL-02 from the SPM-14 shell. Mission Circle 735 CTugRV-02+LL-02 Maneuvers the LL to XDock-06 and Mission Circle 736 Docks LL-02 to XDock-06C. Mission Circle 737 Detaches from LL-02 and Circle 738 Maneuvers to SPM-14.

- Mission Circle 739 CTugRV-02 uses its RRMA to Attach to LL-01 and Mission Circle 740 Extracts LL-01 from the SPM-14 shell. Mission Circle 741 CTugRV-02 Detaches from LL-01 and Mission Circle 742 Maneuvers to the other side of LL-01 and Mission Circle 743 Attached to LL-01. Mission Circle 744 CTugRV-02+LL-01 Maneuvers the LL to XDock-06 and Mission Circle 745 Docks LL-01 to XDock-06E. Mission Circle 746 Detaches from LL-02 and Mission Circle 747 Maneuvers and Mission Circle 748 Docks with XDock-06A.

- Mission Circle 749 LLOG-01 Ops Normal. Mission Circle 750 CTugRV-02 Undocks from XDock-06A and Mission Circle 751 Maneuvers to SPM-15. Mission Circle 752 CTugRV-02 uses its RRMA to Attach to LLB-01 and Mission Circle 753 Extracts LLB-01 from the SPM-15 shell. Mission Circle 754 CTugRV-02+LLB-01 Maneuvers and Mission Circle 755 Attaches LLB-01 to the bottom of SPM-13+SPM-10. This is the LSHV-01 Configuration.

- Mission Circle 756 CTugRV-02 Detaches from LLB-01 and Mission Circle 757 Maneuvers to SPM-15. Mission Circle 758 CTugRV-02 uses its RRMA to Attach to LLB-02 and Mission Circle 759 Extracts LLB-02 from the SPM-15 shell. Mission Circle 760 CTugRV-02+LLB-02 Maneuvers and Mission Circle 761 Attaches LLB-02 to the bottom of SPM-11+SPM-12. This is the LSHV-02 Configuration. Mission Circle 762 CTugRV-02 Detaches from LLB-02 and Mission Circle 763 Maneuvers and Mission Circle 764 Docks XDock-06A. Mission Circle 765 LLOG-01 Ops Normal. Mission Circles 766 to 798 Reserved.

- Mission Circle 799 MILESTONE Plan_B Phase 7B Complete. The Lunar Components are now in place for Humans to Return to the Lunar Surface.

Chart 129: Phase 7B: LLO Lunar Components Assembly

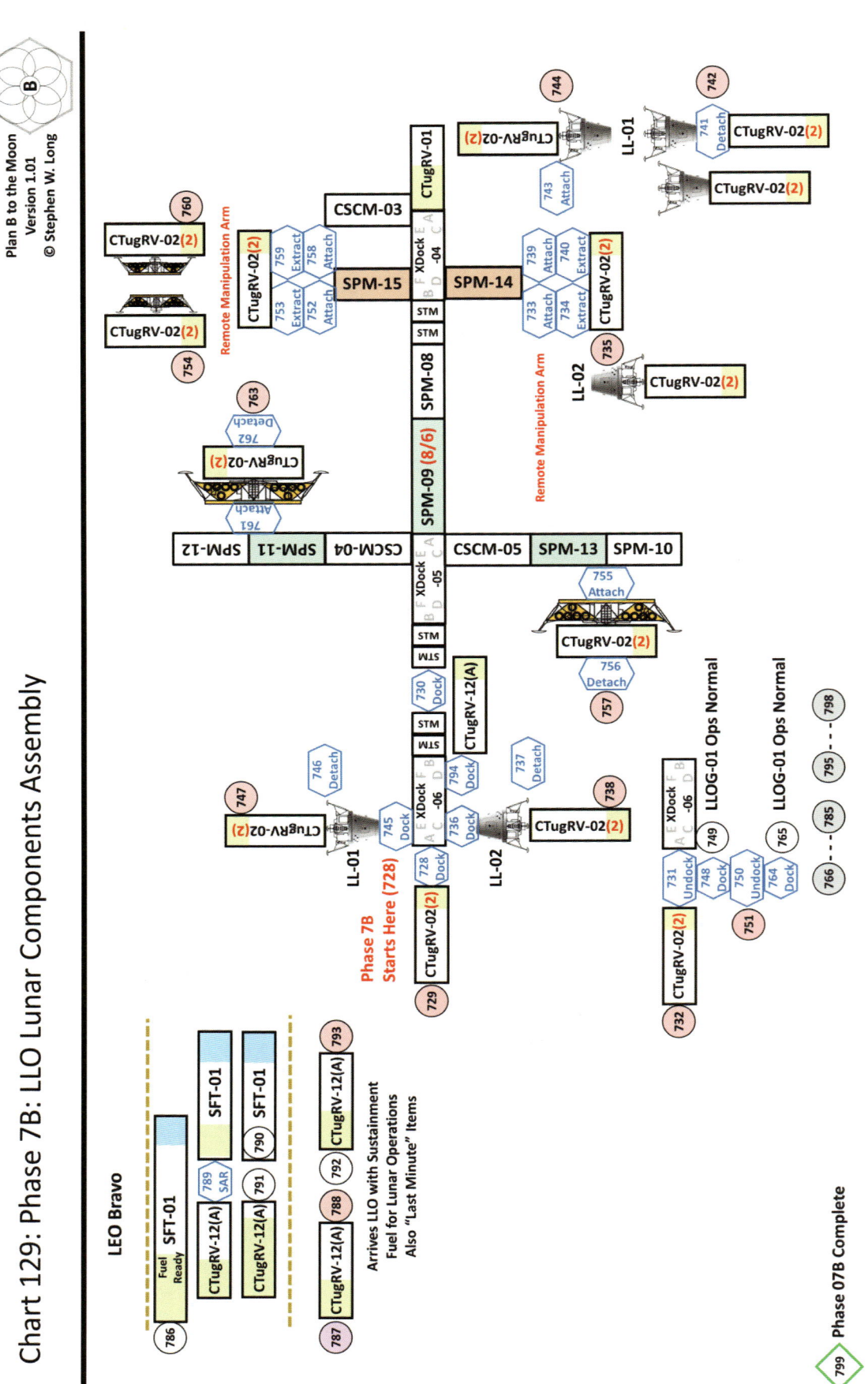

Plan B to the Moon
Version 1.01
© Stephen W. Long

9.16   Mission Phase 8A: Humans Land on Lunar Surface

Overview: The Phase 8A Mission Objective is for Humans to Return to the Lunar Surface (Land).

Chart 130 (Phase 8A) defines the following Mission events:

- Mission Circle 801 is a Strategic Pause-to-Prepare and make ready all systems to Return Humans to the Lunar Surface.

- Mission Circle 802 Lunar-Lander-01 with four Crew Undocks from XDock-06E and Mission Circle 803 Descends to the Lunar Surface.

- Mission Circle 804 LL-01 Lands on Lunar Surface. MILESTONE in Human History.

- Mission Circle 805 LL-01 Crew begins Prospecting Mission-01.

- Mission Circle 806 LL-01 Launches from Lunar Surface and Mission Circle 807 Rendezvous with LLOG-01 and Mission Circle 808 Docks at XDock-06E.

- Mission Circle 809 LL-01 Refuels at LLOG-01. Mission Circle 810 LLOG-01 Ops Normal. Mission Circle 811 Reserved.

- Mission Circle 812 Lunar-Lander-02 with four Crew Undocks from XDock-06C and Mission Circle 813 Descends to the Lunar Surface.

- Mission Circle 814 LL-02 Lands on Lunar Surface. Mission Circle 815 LL-02 Crew begins Prospecting Mission-02.

- Mission Circle 816 LL-02 Launches from Lunar Surface and Mission Circle 807 Rendezvous with LLOG-01 and Mission Circle 818 Docks at XDock-06C.

- Mission Circle 819 LL-01 Refuels at LLOG-01. Mission Circle 820 LLOG-01 Ops Normal. Mission Circle 821 Reserved.

- Mission Circles 822-831 complete Prospecting Mission-03.

- Mission Circles 832-841 complete Prospecting Mission-04.

- Mission Circles 842-851 complete Prospecting Mission-05. Plan_B V1.0 programs Five Prospecting Missions to locate the best (most optimum) sites for Lunar Base-01 to include finding desired resources (Lunar Ice and Helium 3).

Chart 130: Phase 8A: Humans Land on Lunar Surface/LB-01 Location Selection

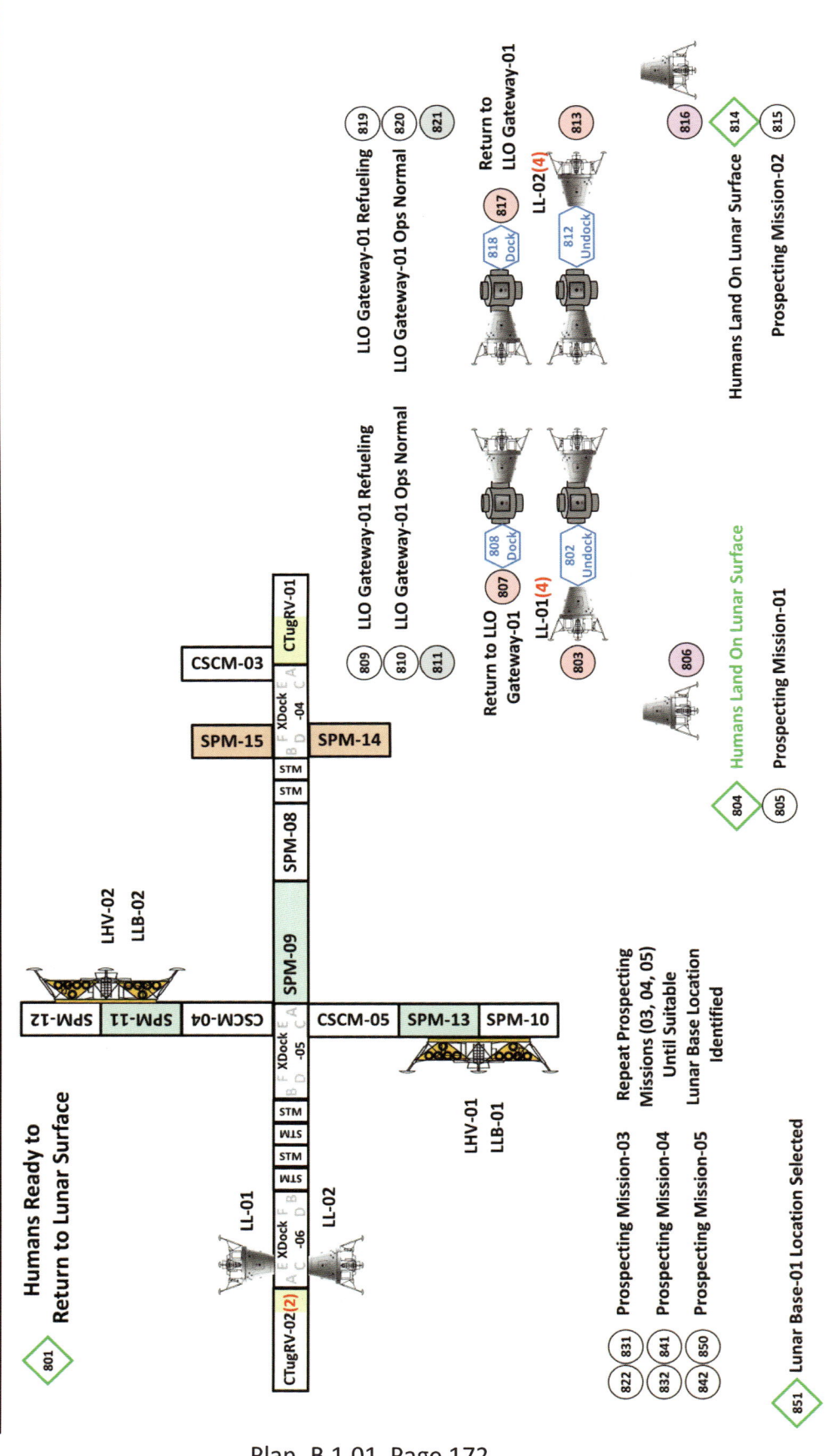

9.17   Mission Phase 8B: Lunar Base-01 Establish

Overview: The Phase 8B Mission Objective is to establish Lunar Base-01. Chart 131 (Phase 08B) defines the following Mission events:

- Mission Circle 851 MILESTONE Lunar Base-01 Location Selected. Mission Circle 852 Lunar-Lander-01 (LL-01) with Four Crew Undocks from XDock-06E. Mission Circle 853 Descends to Lunar Surface and Mission Circle 854 Lands at Lunar Base-01 (LB-01). Mission Circle 855 Lunar Habitat Vehicle-01 (LHV-01) sitting on Long Boat-01 (LLB-01) Undocks from XDock-05C and Mission Circle 856 Descends to Lunar Surface. Mission Circle 857 Lands at LB-01. The LL-01 Crew remotely pilots the LLB-01 landing. Mission Circle 858 LHV-01 Deploys (drives off LLB-01). Mission Circle 859 LHV-01 Performs Lunar Base-01 Mission Alpha 1. Mission Circle 860 LHV-01 Performs Lunar Base-01 Mission Alpha 2. Mission Circle 861 LHV-01 Mission Transitions to Colony Mission. Lunar Base-01 Ops Normal. Mission Circle 862 LLB-01 Launches from Lunar Surface and Mission Circle 863 Rendezvous with LLOG-01 and Mission Circle 864 Mates with XDock-05D. Mission Circle 865 LLB-01 Refuels from LLOG-01. Mission Circle 866 LLOG-01 Ops Normal. Mission Circle 867 LL-01 Launches from Lunar Surface and Mission Circle 868 Rendezvous with LLOG-01 and Mission Circle 869 Docks with XDock-06E. Mission Circle 870 LL-01 Refuels from LLOG-01. Mission Circle 871 LLOG-01 Ops Normal.

- Mission Circle 872 Lunar-Lander-02 (LL-02) with Four Crew Undocks from XDock-06E and Mission Circle 873 Descends to Lunar Surface. Mission Circle 874 Land at Lunar Base-01 (LB-01). Mission Circle 875 Lunar Habitat Vehicle-02 (LHV-02) sitting on Long Boat-02 (LLB-02) Undocks from XDock-05C and Mission Circle 876 Descends to Lunar Surface. Mission Circle 877 Lands at LB-01. The LL-02 Crew remotely pilots the LLB-02 landing. Mission Circle 878 LHV-02 Deploys (drives off LLB-02). Mission Circle 879 LHV-02 Performs Lunar Base-01 Mission Bravo 1. Mission Circle 880 LHV-02 Performs Lunar Base-01 Mission Bravo 2. Mission Circle 881 LHV-02 Mission Transitions to Colony Mission. Lunar Base-01 Ops Normal. Mission Circle 882 LLB-02 Launches from Lunar Surface and Mission Circle 883 Rendezvous with LLOG-01 and Mission Circle 864 Mates with XDock-05D. Mission Circle 885 LLB-02 Refuels from LLOG-01. Mission Circle 886 LLOG-01 Ops Normal. Mission Circle 887 LL-02 Launches from Lunar Surface and Mission Circle 888 Rendezvous with LLOG-01 and Mission Circle 889 Docks with XDock-06E. Mission Circle 890 LL-02 Refuels from LLOG-01.

- Mission Circle 891 LLOG-01 Ops Normal. Mission Circles 892 to 898 Reserved. Mission Circle 899 MILESTONE Phase 8 Complete. Lunar Base-01V1 Initial Operating Capability (IOC). This completes the Plan_B Version 1.0 Mission Plan to Return Humans to the Lunar Surface and begin Lunar Colonization. Plan_B Version 2.0 picks up at Mission Circle 900.

Chart 131: Phase 8B: Lunar Base-01 Establish – LB-01V1 IOC

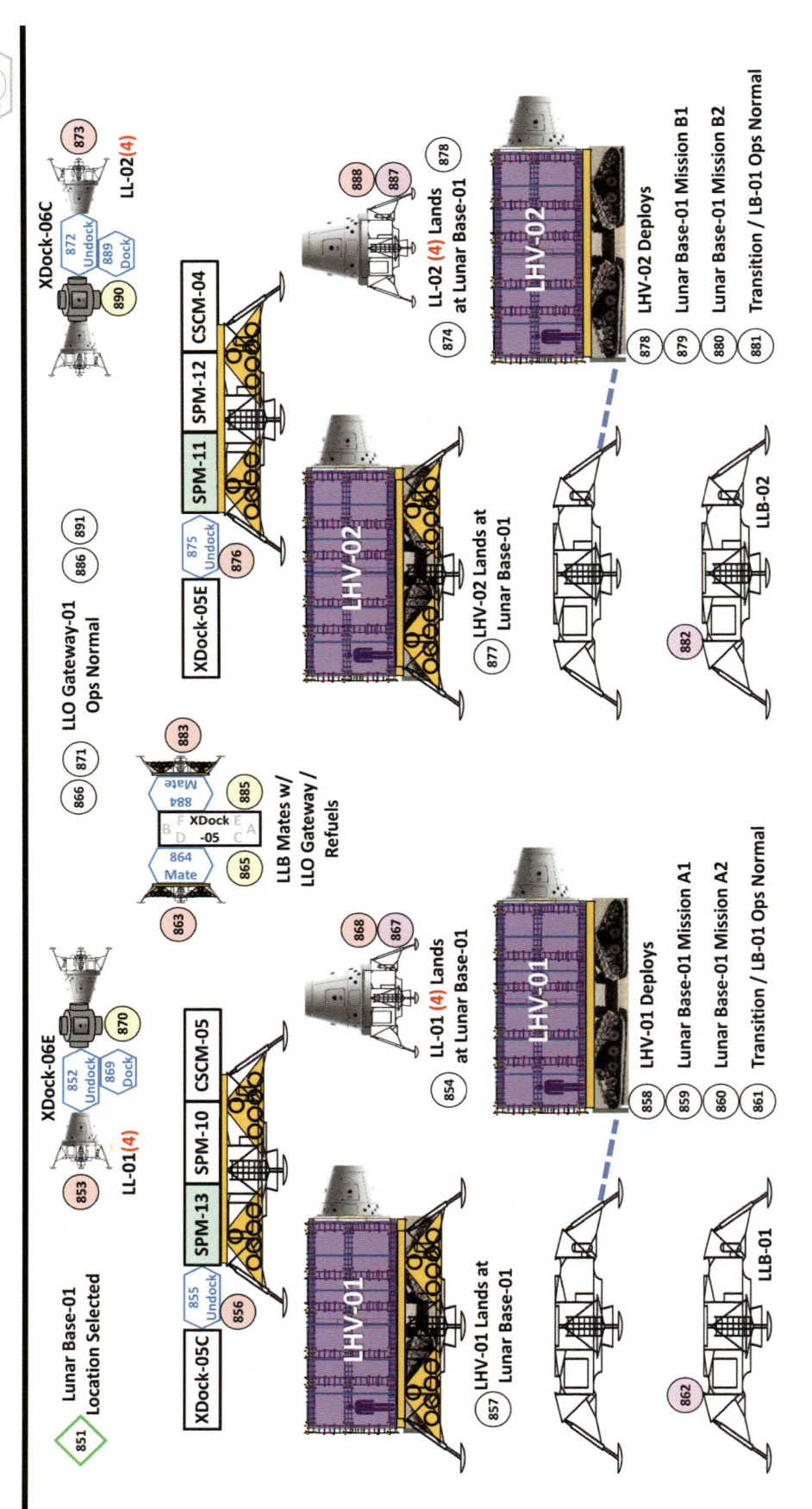

Chapter 10: Mission Matrix

10.1 Introduction

Chapter 9 is all about the Mission Circle graphics and the process flow steps between them. Chapter 10 is all about the numerical details in matrix form (X by Y cells of data). Each Mission Circle graphic number has a corresponding row of data in the MISSION MATRIX. The MISSION MATRIX row numbers match the Mission Circle ID numbers (data sheet row 804 conveys data for Mission Circle 804). As of Plan_B V1.00 every data row has 79 intersecting columns of data, where the intersection of the row and column define a data cell. Any data cell may contain a simple single number but note that many of these data cells provide linked calculation values (numerical answers), derived after analysis.

It is important to note that MISSION MATRIX calculations rely upon specified initial values, called Plan_B Defined Variables. Chapter 10.2: Plan_B Defined Variables Used in Calculations provides these numeric values. These Defined Variables are not random numbers (most of the time). For instance, mass values reflect published data (such as the mass of the Common-Space-Service-Module). Or masses reflect a design-to-weight value based on the XPM/SPM components used in the designs given in Chapter 8.

Starting with Chapter 10.3, we provide lengthy interpretation or tutorial explanations and Charts. This material explains how to understand, use and interpret the Plan_B calculation details found in the hyper-detailed Mission Matrix. While the numeric details of any give cell / row of the tutorial Charts are accurate (identical) to the formal Mission Matrix, the Tutorial removes unnecessary rows and columns.

10.2 Plan_B Defined Variables Used in Calculations

Chart 132 depicts the Defined Variables used in Plan_B calculations:
- CCtCap has a published mass of 21,300 pounds (confirmed published value).
- The Pressure-Vessel Module (PVM) has a pressure shell design weight of 10,000 pounds. Based on the size of the PVM, it has a calculated weight of 455 pounds per linear foot. The PVM LLB and LL shells have a calculated weight of 12,500 pounds.
- The Common-Space-Command-Module (CSCM) has a design weight of 10,650 pounds, designed as half of the published weight of the CCtCap. The CSCM will have two docking hatches (fore and aft), and a human access side-hatch. The CSCM will have a robust C4 infrastructure, suitable for use as the C4 components for any SPM derived vehicle.
- The Common-Space-Service-Module (CSSM) has a planned mass of 30,086 pounds, which is 4000 pounds less than the published ESA Service Module mass.
- The XDock has a design mass of 10,650 pounds.
- The Common Conventional Engine (CCE) has a design mass (dry) of 29,700 pounds, which does NOT include the mass of the LCH4 and LOX fuel tanks.
- The CTugRV design maximum fuel load (propellent—wet) mass is 200,400 pounds.

- The CTugRV vehicle dry mass is 54,600 pounds, which is the additive mass of the CSCM+CCE+CSMs*3.
- The CTugRV wet max mass is 255,000 pounds (CTugRV dry plus max fuel).
- An SPM Habitat has a QRM mass budget of 43,700 pounds which is the published weight of the Orion Capsule. The published specifications for the Orion spacecraft include requirements for long-duration living space for crews. The SPM Habitats are given Orion Capsule mass equivalence since the living space components should have comparable functions.
- Orion Capsule plus the ESA CSSM mass is 56,985 pounds (published).
- Storage-Toroidal-Module (STM) has a design dry mass of 3,750 pounds.
- STM wet payload is 51,250 pounds.
- Cryogenic-Storage-Module (CSM) has a design dry mass of 4,750 pounds (see the Fuel Tank design worksheet).
- CSM wet payload is 66,800 pounds.
- Autonomous-Space Water-Transfer-Vehicle (ASWTV) has a design dry mass of 69,486 which includes the CSCM+CSSM+PVM+TSM*5.
- ASWTV wet payload is 325,736 pounds.
- Lunar-Long-Boat (LLB) has a design dry mass of 21,750 pounds.
- Lunar-Lander (LL) has a design dry mass of 21,750 pounds, including the CSCM mass.
- CCtCap Habitable volume is 350 cubic feet (published).
- SPM (PVM empty) has a design volume of 3,888 cubic feet.
- CSCM has a habitable design volume of 350 cubic feet.
- CTugRV has a habitable design volume of 350 cubic feet (because of CSCM).
- Lunar-Surface-Habitat-Vehicle (LSHV) has a habitable design volume of 8126 cubic feet (1 CSCM + 2 SPM). The LSHV design has two decks each ~7 feet tall at the center. This is equivalent to a ~1015 sq foot earth surface home with 8-foot ceilings. By regulation, a large earth RV has a maximum size of 450 sq feet. The two-floor LSHV design provides about the living space of two maximum size earth RVs.
- The SPM solar photo-voltaic power design is 21 kW. This assumes illumination of two of the six Hex PVMs. Six Hex surfaces support 63 kW of solar PVPs.
- The Falcon 9 (F9) launch cost is $62 Million (published costs).
- The Falcon Heavy (FH) launch cost is $130 Million (published, Precept costs).
- The Plan_B Space Vehicle build-to cost is $5000 per pound (Precept).
- The Human Rated Space Vehicle build-to cost is $10,000 per pound (Precept). This includes the CCtCap vehicles and the CSCM.
- Other Mission Circle costs are $100 Million per assigned Mission.
- NASA Plan_A through 2024 costs are $25.9 Billion (Precept, published).
- FH Mission cycle minimum day count is 24 days (launches cannot occur faster than the minimum day count).
- FH Booster cycle minimum day count is 84 days (refurbished launch vehicles not ready for next launch faster than the minimum day count). Note SpaceX claims 60 days, so 84 days is a more conservative (easier to accomplish) refurbishment schedule.

- F9 Mission cycle minimum day count is 28 Days (launches cannot occur faster than the minimum day count).
- F9 Booster cycle minimum day count is 72 Days (new launch vehicles not ready for next launch faster than the minimum day count).
- Revision date of the Mission Matrix row of data. As of Plan_B V1.00 dates adjusted to a common benchmark date (2019-01-06). For future revisions, only if a row of data changes will that row's date update/change.

Chart 132: Plan B Defined Variables Used in Calculations								Plan B to the Moon Version 1.01 © Stephen W. Long

Item	Amount	Source	Design-to-Costs ($M)	Notes	Item	Amount	Source	Notes
C_CCtCap lbs	21,300	Confirmed	$213.00	Published	C_CCtCap Hab Vol ftcu	350	Confirmed	Published
C_PVM Shell lbs	10,000	Design (Calc)	$50.00	See Fuel Calcs Sheet	C_SPM Hab Vol ftcu	3,888	Design	Design (Calc)
C_PVM Per Linear Foot	455	Design (Calc)			C_CSCM Hab Vol ftcu	350	Design	Design (Calc)
C_PVM LLB LL Shell lbs	12,500	Design (Calc)	$62.50	See Fuel Calcs Sheet	C_CTUG Hab Vol ftcu	350	Design	Design (Calc)
C_CSCM lbs	10,650	Design (Calc)	$106.50	50% of CCtCap	C_Lunar EV Hab Vol ftcu	8,126	Design	Design (Calc)
C_Common Service Module lbs	30,086	Confirmed		Published less 4000	C_F9 Launch Cost $M	$62	Confirmed	Published
C_XDock lbs	10,650	Design (Calc)	$53.25	50% of CCtCap	C_FHR Launch Cost $M	$130	Precept	
C_CCE Dry lbs	29,700	Design (Calc)	$148.50		C_Space Veh Cost Per lb $M	$0.0050	Design (Calc)	Calc $5k
C_Fuel LEO to LLO lbs	200,400	Precept				$5,000	Design (Calc)	Precept $5k
C_CTugRV Dry lbs	43,950	Design (Calc)	$326.25	+CCE+CSM*3	C_Space Veh Cost 4 Per lb $M	$0.0040	Design (Calc)	Calc $4k
C_CTugRV Max Wet lbs	244,350	Design (Calc)		Fuel to LLO + Dry CTUG		$4,000	Design (Calc)	
C_SPM Hab lbs	43,700	Design (Calc)	$218.50	Orion Capsule	C_Space Veh Cost 7 Per lb $M	$0.0071	Design (Calc)	Calc $7k
C_OrionCap plus CSM lbs	56,985	Confirmed		Published		$7,063	Design (Calc)	
C_STM Dry lbs	3,750	Design (Calc)	$18.75		C_HR Space Veh Cost Per lb $M	$0.0100	Design (Calc)	Calc $10k
C_STM H20 Payload lbs	51,250	Design (Calc)				$10,000	Design (Calc)	
C_CSM Dry lbs	4,750	Design (Calc)	$23.75		C_Other Costs $M	$100	Design	
C_CSM Cryo Payload lbs	66,800	Design (Calc)		7 Tanks per CSM	C_Plan_A_Through_2024_$M	$25,900	Precept	Published
C_ASWTV Dry lbs	69,486	Design (Calc)	$347.43	+CSCM+CSSM+PVM+TSM*5	C_FH Mission Cycle Days	24	Design	
C_ASWTV Max lbs	325,736	Design (Calc)			C_FH Booster Cycle Days	84	Design	
C_Lunar Long Boat lbs	21,750	Design (Calc)	$108.75		C_F9 Mission Cycle Days	28	Design	
C_Lunar Lander lbs	21,750	Design (Calc)	$217.50		C_F9 Booster Cycle Days	72	Design	
C_PVP Power kW	21	Design (Calc)		See Power Sheet	C_Mission_Matrix_Date	2019-01-07		

10.3   Mission Matrix Interpretation: ID and Costs

Each Mission Circle Number (M Circle) is assigned within a Mission Phase (Phase). Circle Actions define activity and Vehicle/Milestone/Notes further define actions. Chart 133 depicts samples of Mission Matrix Circle Actions, Descriptions, and Cost data.

Phase	M Circle	Circle Action	Vehicles / Milestones / Notes	Cost FH Launch ($M)	Cost F9 Launch ($M)	Other Costs ($M)	HR Veh Costs ($M)	Veh Costs ($M)
		85.00%	$22,015	$3,962.0	$434.0	$5,212.2	$2,991.0	$9,415.8
1A	101	Launch FH, LEO Rendezvous	SPM-1+XDock-01+CSCM-01 [Phase 01A]	$130.0		$100.0	$106.5	$596.8
1A	102	Maneuver, Park LEO A	SPM-1+XDock-01+CSCM-01					
1A	103	Launch F9, LEO Rendezvous	CCTCap-01 (6 COB) [Phase 01F]		$62.0	$100.0	$213.0	$143.5
1A	104	Dock	CCTCap-01:XDock-01F					
1A	105	Launch FH, LEO Rendezvous	CTugRV-01(A) [Phase 01A]	$130.0		$100.0	$106.5	$219.8
1A	106	Dock	CTugRV-01:XDock-01D					
1A	107	LEOS-01 Assembly	LEOS-01 Assembly					
1A	127	Launch FH, LEO Rendezvous	SPM-02+XDock-02 [Phase 01A]	$130.0		$100.0	$0.0	$650.0
1A	129	Undock	CTugRV-01:XDock-01B					
1A	133	LEOS-01 Assembly	LEOS-01 Assembly					
1A	134	Launch FH, LEO Rendezvous	SPM-03+ 4STM (Two STM Full) [Phase 01A]	$130.0		$100.0	$0.0	$137.5
1A	135	Undock	CTugRV-01:XDock-02B					
1A	136	Maneuver	CTugRV-01					
1A	137	Dock	CTugRV-01:SPM-03+ 4STM					
1A	150	Partial H2O Load	LEOS-01 H2O Transfer to Shielding Tanks					
1A	151	LEOS-01 Ops Normal	LEOS-01 Ops Normal					
1B	152	Launch F9, LEO Rendezvous	CCTCap-02 (6 COB) [Phase 01B]	$62.0	$62.0	$100.0	$213.0	$143.5
1B	153	Dock	CCTCap-02:XDock-01D					
1B	154	LEOS-01 Ops Normal	LEOS-01 Ops Normal					
1B	155	Undock	CCTCap-01R:XDock-01F					
1B	156	Return to Earth	CCTCap-01R (6 COB)					
1B	157	LEOS-01 Ops Normal	LEOS-01 Ops Normal					
1B	158	Launch FH, LEO Rendezvous	ASWTV-01(A) [Phase 01B]	$130.0		$100.0	$0.0	$347.4
1B	159	SAR-H2O	ASWTV-01(A) / LEOS-01 / H2O for Shielding					
1B	160	H20 State Change	ASWTV-01(A)					
1B	161	H20 State Change	LEOS-01					
1B	162	Maneuver	ASWTV-01(A)					
1B	163	Dock	ASWTV-01(A):XDock-02F					
1B	164	Partial H2O Load	LEOS-01 H2O Transfer to Shielding Tanks					
1B	165	LEOS-01 Ops Normal	LEOS-01 Ops Normal					

Chart 133: Mission Matrix Interpretation: ID and Costs

Circle Actions (the primary action of the Mission Circle) include:
- Launch (purple background).
- Rendezvous / Maneuvers.
- Park (as in Move to a defined LEO orbit—see Chart 115).
- Dock (connect craft together via hatches designed for periodic / permanent joining). Note that during LEOS-01 and LLOG-01 assembly multiple segments sequentially dock, one docking after another, but each docking has its own number.
- Undock (opposite of Dock).
- Space Autonomous Refueling or SAR (Blue for Water*, Yellow for LCH4/LOX).
- State Change or Note (fluid exits one craft, enters other craft).
- Extract (as in Extract a craft or payload from a PVM shell).
- Attach (as in one vehicle attaches to another—typically CTugRV RRMA grasping another component to Extract the component).
- Milestone (Green).
- Reserved (dark gray, usually hidden for printing).

Vehicle/Milestone/Notes include vehicle designations and vehicle combinations:
- SPM-xx (where xx = 01, 02, 03) (see Chapter 4).
- CSCM-xx (where xx = 01, 02, 03).
- XDock-xx (where xx = 01, 02, 03).
- ASWTV-xx (where xx = 01, 02, 03).
- CTugRV-xx (where xx = 01, 02, 03).
- CCtCap-xx (where xx = 01, 02, 03).

- WTV-xx (where xx = 01, 02, 03) Simplest Water* delivery vehicle–Toroidal Storage Modules in a shell. Note that this is a perfect vehicle to launch on an end-of-life Falcon Heavy booster stack.
- CRV-xx (where xx = 01, 02, 03), Simplest vehicle–Cryogenic-Storage-Modules in a shell.

Plan_B Costs include five categories:
- Falcon Heavy (FH) launch costs (published, see Defined Variables).
- Falcon 9 (F9) launch costs (published, see Defined Variables).
- Other costs.
- Calculated Human Rated Vehicle costs (Calculated, based on Defined Variables Human Rated Spacecraft cost per pound).
- All other space Vehicle costs (Calculated, based on Defined Variables Other Spacecraft cost per pound–all vehicles or components not human rated, does NOT include fuel masses).

Each Phase tallies the sub-total costs for that Phase (Circle Action typically states Phase XX Complete). All the Phase sub-totals sum to a Total Plan_B Cost: value in Row 910.

10.4   Mission Matrix Interpretation: Schedule

Plan_B provides a detailed Waterfall Schedule–a day-by-day incremental count, beginning from Plan_B 2018-01-01 start to establishing the first colonists on the Lunar Surface (item 899). This calculates as of V1.00 to 2025-01-18.

Chart 134 provides samples of Circle Action and Schedule data:
- Schedule includes a Mission Circle Start Date, the Nominal (Nom Days) for the Circle Action, the Adjusted Days (Ajust Days) and the calculated End Date. The End Date is Start Date plus Nominal plus Adjusted Days.
- Nominal Days are the required day count for each Mission Circle item. Multiple Circle items can share a single day. Nominal days include physics-based day counts such as the number of days to travel from LEO to LLO.
- Adjustment Dates lengthen the Mission Circle day counts beyond the Nominal Day counts to provide sufficient time for the Mission Cycle intervals or the Launch Vehicle intervals. Plan_B V1.00 added only one adjustment interval to fit the F9 schedule. Adjustment Dates provide sufficient time for the FH Mission Cycle intervals.
- The Nominal Plan_B Authority-to-Proceed Date is 20191001 (start of US Government Fiscal Year 2020). Changing this date in the Defined Variables cell causes the MISSION MATRIX to recalculate all subsequent dates.

Chart 134: Mission Matrix Interpretation: Schedule

Plan B to the Moon
Version 1.01
© Stephen W. Long

Phase	M Circle	Circle Action	Start Date	Nom Days	Ajust Days	End Date	L #	L Veh	FH Miss Cycle	FH M Cyl Error	FH1 LV Avail	FH2 LV Avail	FH3 LV Avail	FH4 LV Avail	FH B Cycle Error	F9 Miss Cycle	FH L Cycle Error
		85.00%	Return to LS -->			2024-11-02				24					84		28
1A	101	Launch FH, LEO Rendezvous	2022-12-27	3	16	2023-01-15	3	FH1	2022-12-27	0	2022-12-27				0		
1A	102	Maneuver, Park LEO A	2023-01-15	1	0	2023-01-16											
1A	103	Launch F9, LEO Rendezvous	2023-01-16	3	0	2023-01-19	4	F91								2023-01-16	0
1A	104	Dock	2023-01-19	1	0	2023-01-20											
1A	105	Launch FH, LEO Rendezvous	2023-01-20	3	0	2023-01-23	5	FH2	2023-01-20	0		2023-01-20			0		
1A	106	Dock	2023-01-23	1	0	2023-01-24											
1A	107	LEOS-01 Assembly	2023-01-24	7	0	2023-01-31											
1A	127	Launch FH, LEO Rendezvous	2023-02-13	3	0	2023-02-16	6	FH3	2023-02-13	0			2023-02-13		0		
1A	129	Undock	2023-02-16	0	0	2023-02-16											
1A	133	LEOS-01 Assembly	2023-02-16	7	14	2023-03-09											
1A	134	Launch FH, LEO Rendezvous	2023-03-09	3	0	2023-03-12	7	FH4	2023-03-09	0				2023-03-09	0		
1A	135	Undock	2023-03-12	0	0	2023-03-12											
1A	136	Maneuver	2023-03-12	0	0	2023-03-12											
1A	137	Dock	2023-03-12	0	0	2023-03-12											
1A	150	Partial H2O Load	2023-03-28	0	0	2023-03-28											
1A	151	LEOS-01 Ops Normal	2023-03-28	1	0	2023-03-29											
1B	152	Launch F9, LEO Rendezvous	2023-03-29	3	0	2023-04-01	8	F92								2023-02-13	44
1B	153	Dock	2023-04-01	1	0	2023-04-02											
1B	154	LEOS-01 Ops Normal	2023-04-02	0	0	2023-04-02											
1B	155	Undock	2023-04-02	0	0	2023-04-02											
1B	156	Return to Earth	2023-04-02	0	0	2023-04-02											
1B	157	LEOS-01 Ops Normal	2023-04-02	0	0	2023-04-02											
1B	158	Launch FH, LEO Rendezvous	2023-04-02	3	0	2023-04-05	9	FH1	2023-04-02	0	2023-03-21				12		
1B	159	SAR-H2O	2023-04-05	0	0	2023-04-05											
1B	160	H20 State Change	2023-04-05	0	0	2023-04-05											
1B	161	H20 State Change	2023-04-05	0	0	2023-04-05											
1B	162	Maneuver	2023-04-05	0	0	2023-04-05											
1B	163	Dock	2023-04-05	0	0	2023-04-05											
1B	164	Partial H2O Load	2023-04-05	1	0	2023-04-06											
1B	165	LEOS-01 Ops Normal	2023-04-06	1	19	2023-04-26											

The Launch Number (L #) is the incremental count from Launch 1 (Mission Circle 019) to Launch 37 (Mission Circle 787). The Launch Vehicle (L Veh) is SpaceX Falcon Heavy booster (stack 1 or 2 or 3 or 4) or the F9 stack 1-6. Plan_B Version 1.00 only includes Falcon Heavy and Falcon 9 launch vehicles. Plan_B welcomes the eventual availability of other commercial launch vehicles—if such vehicles can meet the Plan_B cost per launch pound and related precepts.

The FH Mission Cycle Dates applies the defined FH M Cycle Defined Variable to the End Date. This calculated date is the earliest next launch date for the FH. As of Plan_B V1.00 this Cycle is 28 Days, meaning this is shortest interval (capacity) of the total mission cycle for FH Mission launches. If the entire process (delivery, payload integration, launch pad operations, etc.) is less than 28 Days, it shortens the end-to-end Plan_B schedule. If launch cycles need over 28 Days, such cycles lengthen the overall Plan_B schedule. This is a different day count from FH Booster reuse or recycling days.

The F9 Mission Cycle Date applies the defined F9 M Cycle Defined Variable to the End Date. This calculated date is the earliest next launch date for the F9. As of Plan_B V1.00 this Cycle is 44 days—the logic being, it will take longer to prepare for human crewed launches.

FH Launch Vehicle Availability calculates the interval for Falcon Heavy refurbishment and next launch. F9 Launch Vehicle Availability calculates (if F9s re-used), the interval for F9s refurbishment and next launch. The Vehicle Availability dates reflect a complex iterative math model to determine the minimum mission cycle times (how soon after a launch is the next launch). This is a Chapter 11.2 Defined Variable. Through iterative adjustments, the Plan_B Team believes four FH booster stacks, each re-used an average of 8 times will provide all the Plan_B Falcon Heavy launches. More (than 4) booster stacks will allow compression of the

Plan_B master schedule (fewer days from start to finish). As of Plan_B V1.00 an 84-day cycle is the minimum cycle interval. This is 24 days longer (more-time or easier to accomplish) than the SpaceX published goal of 60 days to recycle an FH launch system. Given that the F9s are all human launch missions, Plan_B assumes a new F9 (not previously flown) vehicle for every F9 launch. Plan_B V1.00 uses a 72-day cycle time (72-day F9 delivery cycle).

10.5 Mission Matrix Interpretation: Mass

Mission plans adjust the total launch mass of any Mission Circle Launches to fit within the FH 130,000-pound launch mass payload capabilities. Mission plans also adjust Masses for Mission Circle Launches using the F9 50,000-pound launch mass payload capability.

Chart 135 provides samples of Mission Matrix Mass data, including Vehicle Mass budgets for every system and included sub-element of the Mission Circle. A column total provides a cross check to verify that the total Mass fits within the Mass budget of the specified launch vehicle. For publication print reasons, the Falcon 9 (F9) mass budgets are at the far-right end of the Matrix (see Tutorial Chart 139).

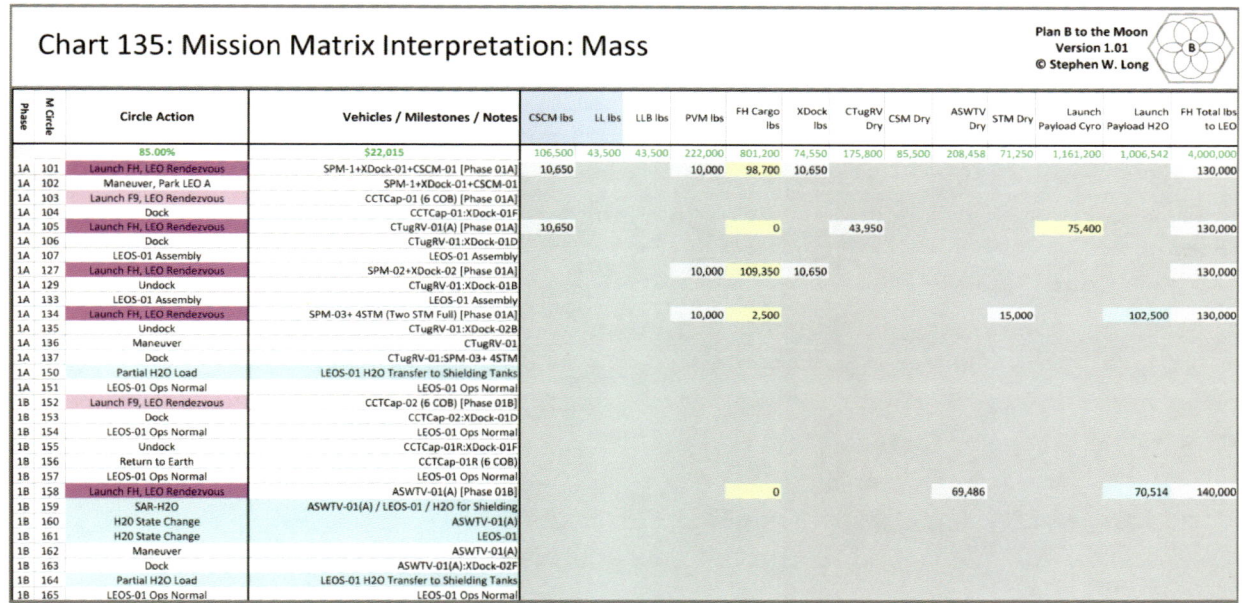

Chart 135: Mission Matrix Interpretation: Mass

Mission Matrix Mass entries are pointers back to the Chart 132 Defined Variables. Changing these values enables updates to Plan_B calculations to reflect the most current mass budgets over time for each Plan_B XPM/SPM and other component vehicle masses:
- CSCM mass, LL mass, LLB mass;
- FH Cargo Mass. The XPM parts catalog (such as FCCs, PVPs, QRMs…) define the assigned Cargo masses for each Plan_B FH mission;
- XDock has a design mass of 10,650 pounds (core vehicle mass);
- The CTugRV includes one CSCM (10,650 pounds, design cost $106.5 M), one CCE (29,700 pounds, design cost of $148.5 M), and five CSMs (dry weight 4,750 pounds each, $23.75 M).

For a maximum FH launch weight of 130,000 pounds the CTugRV can deliver a maximum of 75,400 pounds of cryogenic fuel into LEO. CTugRVs do NOT launch from Earth with a full fuel load to LEO. Upon reaching LEO, the CTugRV will have to be topped-off with fuel to enable ELTV missions. The total CTugRV dry mass is 64,100 pounds, with a design-to-cost of $379.5 Million per vehicle. This cost includes Human-Rated and Other Design cost masses.

- Mission Planners assign STMs and CSMs to Mission Circles as incremental launch vehicle payload capacities. Mass calculations include partial STM and CSM loads. Every available pound of launch capacity ships Water* and Cryogenic fuel to LEO when not needed for solid cargo. Attention-to-detail readers will note specific payloads are full 140,000-pound launches (instead of 130,000). The weight (mass) of such payloads already includes the full mass of the spacecraft components (such as the CCE and PVM shell). Therefore, these launches do not have to derate the launch mass ceiling to account for sending the extra payload faring mass into LEO.

- The Autonomous-Space Water-Transfer-Vehicle (ASWTV) has a design mass of 69,486 pounds (dry vehicle mass), and a design-to-cost of $347.43 Million per vehicle. The ASWTV includes one CSCM (10,650 pounds), one Common-Space-Service-Module (30,086 pounds), and five Storage-Toroidal-Modules (dry mass 3,750 pounds each). For a maximum FH launch mass of 140,000 pounds the ASWTV can carry a maximum of 70,514 pounds of Water* into LEO. ASWTVs do NOT launch from Earth with a full Water* load to LEO. After reaching LEO, the ASWTVs transfer Water* from other launches, moving the Water* to the SFT-01 or to LEO-Station Shielding-Storage-Tanks.

- A single Storage-Toroidal-Module (for Water*) has a dry mass of 3,750 pounds and can contain a maximum of 51,250 pounds of Water*. Five STMs will support a maximum of 256,250 pounds of Water*. This design enables one full ASWTV to fill the water shielding tanks of two SPMs. This design enables one full ASWTV to transport 121,785 pounds of CO_2 and 107,824 pounds of H_2O (total 239,609 pounds of Water*). This is the raw stock Water* mass for the Space-Fuel-Terminal to produce 200,400 pounds LCH4/LOX for an ELTV mission. Borrowing a concept from the ISS, the Plan_B Team believes future portions of the required CO_2 for fuel may eventually come from CO_2 scrubbing of SPM atmospheres.

- A single Cryogenic-Storage-Module (CSM) carries 7 cryogenic tanks, with a dry weight of 4750 pounds and can load a maximum of 66,800 pounds of LCH4/LOX fuel. Three CSMs will support a maximum of 200,400 pounds of LCH4/LOX fuel. This design enables the **Plan_B Precept** of 200,400 pounds of fuel to transport 100,000 pounds of lunar payloads.

10.6 Mission Matrix Interpretation: Cryogenic Fuel and Water*

Calculating fuel and how much Water* or LCH4/LOX fuel to deliver into LEO for fuel to return to the moon was the most difficult Plan_B sub-task. One Team member noted that in the ancient F-4 fighter bomber era, "fuel planning was everything." Bomb-delivery-mission planning had only minutes of timeline fuel margin. Fuel burns shaped every flight moment to carry six or more Mk-82s to their appointments with exothermic destiny.

As Plan_B writing and calculations neared completion, they got revised again (and again) based on new fuel production and launch payload data. There was a familiar sense of Deja Vue—where Moon fuel calculations drove every mission calculation. Defining and re-designing the fuel production requirements for the Space-Fuel-Terminal (see Chapter 7.2) was daunting. The Team realized that prognosticating fuel creation and consumption burns for the still-to-be-designed and built spacecraft was rather too fuzzy to satisfy our engineering meticulous souls. The agreed upon collective way forward was to re-work and adjust the entire Plan_B mission plan to incorporate significantly higher fuel margins. We added fuel for translunar orbit burns and other unknown-unknown fuel consumption. This caused a complete redesign of the CTug-RV fuel storage system (we enlarged the CTug-RV tanks to carry more fuel). This caused ripples of fuel delivery launches and revised fuel production and consumption in every mission circle. We had to re-write/re-calculate most of Plan_B. Oops—there went our late 2018 completion schedule deadline. This also led to a necessary revision of an early design decision. The revised ELTV translunar orbit missions will start with 200,400 pounds of LCH4/LOX fuel plus the CTugRV dry mass 54,600 pounds. The revised CTugRV wet mass is 255,000 pounds to deliver 100,000 pounds of lunar bound spacecraft mass to LLO.

Based on the hard-learning from the fuel calculations (and recalculations and recalculations) we adopted a strategy to start always-round-up cost, mass, and fuel calculations. We decided it was better to err on the side of producing or delivering too-much fuel rather than not enough fuel. We increased the fuel mass budget to 200,400 pounds which provides a significant margin (surplus) of 64,400 fuel pounds times four ELTV missions. This significantly larger fuel load provides fuel margin for the various unknown losses or production/delivery shortfalls of the entire Plan_B fuel life cycle. Also, we added one more CTugRV (dash 10) Mission Circle to deliver the fuel and significant fuel margins for LLO to Lunar Surface operations. If the Space-Fuel-Terminal produces its full (maximum) fuel deliveries, Plan_B V1.00 calculates a fuel-on-LEO surplus of up to 474,098 pounds of LCH4/LOX fuel. This is enough fuel for the next two ELTV delivery missions—beyond Plan_B 1.0 Mission Circle 899.

Based on what we learned as we recalculated and re-wrote the entire Mission plan, we made an editorial decision to publish / present ALL of our previously masked fuel calculations. We hope future mission planners will therefore more easily pick up where Plan_B V1.0 leaves off. They can see how we calculated fuel budgets and not have to re-invent their own fuel calculation matrix. Chart 136 provides the detailed calculation model for Plan_B mission fuel requirements.

Chart 136: Mission Matrix Interpretation: Cryogenic Fuel and Water* Calculations

Plan B to the Moon
Version 1.01
© Stephen W. Long

Molecular Table

kilo grams	Mole	Name
19,760.8	16	C_CH4_kg
14,820.6	12	C_C_kg
4,940.2	4	C_4H_kg
54,342.3	44	C_CO2_kg
14,820.6	12	C_Carbon_kg
39,521.7	32	C_Oxy2_kg
35,569.5	18	C_H2O_kg
3,952.2	2	C_H2_kg
31,617.3	16	C_O2_kg
8,892.4	18	
988.0	2	
7,904.3	16	

Plan B Fuel Calculations

Label	Value	Label (lb/lbs)	Value	Label (kg)	Value
C_Fuel_Required_Payload_Ratio	4.60	C_Total_Fuel_Required_lb	200,400	C_Total_Fuel_Required_kg	90,899.8
C_Fuel_Ratio_Conversion_Factor	19,760.83	C_Fuel_LCH4_Required_lbs	43,565	C_Fuel_LCH4_Required_kg	19,760.8
C_Ideal_Mass_Ratio_1kgCH4	1.00	C_Fuel_LOX_Required_lbs	156,835	C_Fuel_LOX_Required_kg	71,139.0
C_Ideal_Mass_Ratio_3.6kgO2	3.60	C_Fuel_CO2_Required_lbs	119,804	C_Fuel_CO2_Required_kg	54,342.3
C_Pounds_to_kg_Conversion	0.453592	C_Fuel_Total_H2O_Required_lbs	98,022	C_Carbon_from_CO2_Fuel_kg	14,820.6
		C_Water*_As_Fuel_Required_lbs	217,826	C_Oxygen_from_CO2_Fuel_kg	39,521.7
C_LC4H_Ratio	0.22	C_Fuel_110%_CO2_Required_lbs	131,785	C_Delta_Oxygen_Required_kg	31,617.3
C_LOX_Ratio	0.78	C_Fuel_110%_H2O_Required_lbs	107,824	C_H_from_H2O_Fuel_Required_kg	4,940.2
			239,609	C_Delta_Hydrogen_Required_kg	988.0
		C_Water*_As_Fuel_Plus_10%_lbs	239,609	C_Fuel_H_H2O_Required_kg	35,569.5
		C_Water*_As_Fuel_Ratio	0.83636	C_Fuel_O2_H2O_Required_kg	8,892.4
				C_Fuel_Total_H2O_Required_kg	44,461.9

SFT Electrolizer Reactors Specifications

Label	Value	Label (lbs/kW)	Value	Label (kg)	Value
Hydrogen_kg_Prod_per_50kWh	1.00	C_Surplus_O2_lbs	17,426	C_Surplus_O2_kg	7,904.3
Oxygen_kg_Prod_per_50kWh	8.00	Electrolizer_Module_Weight_lbs	110	Electrolizer_Module_Weight_kg	50.0
Hydrogen_kg_Prod_per_1kWh	0.0200	Electrolizer_Power_Requirements_kW	50		16329.3
Oxygen_kg_Prod_per_1kWh	0.1600	SPM_Electrolizer_Weight_Allowance	36,000		
SFT_Power_Prod_kW	460.00	Number_of_Electrolizers	327		
SFT_Power_24hr_Prod_kWh	11040	Pwr_Required_for_Required_Fuel_kWh	988,042		
		Calculated_SFT_Mission_Days	89.50		

SFT Sabatier Reactors Specifications

Label	Value	Label (lbs/kW/kg)	Value	Label (kg)	Value
C_Fuel_LCH4_Required_kg	19,760.8	Sabatier_Module_Weight_lbs	1,874	Sabatier_Module_Weight_kg	850.0
CH4_Prod_per_Module_per_24hrs_kg	21.0	Sabatier_Module_Power_Req_kW	10		
CH4_Prod_All_Modules_per_24hrs_kg	231.0	SPM_Sabatier_Weight_Allowance_lbs	20,000		
CH4_Mission_Cycle_Days	89.5	Number_of_Sabatier_Modules	11		
CH4_Prod_Per_Mission_Cycle_kg	20,673.7	Sabtier_All_Power_Req_kW	110		
SFT_Power_Allocation_kW	5.0	Calculated_Delta_Power_kWh	10.0	(Greater than zero is good)	
SFT_Power_24hr_Prod_kWh	120.0	Calculated_Delta_CH4_kg	912.9	(Greater than zero is good)	

SFT Cryogenic Liquification CH4 Specifications

Label	Value	Label (lbs/kW/kg)	Value	Label (kg)	Value
C_Fuel_LCH4_Required_kg	19,760.8	Cryo_CH4_Module_Weight_lbs	992	Cryo_CH4_Module_Weight_kg	450.0
CH4_Prod_per_Module_per_24hrs_kg	20.0	Cryo_CH4_Module_Power_Req_kW	10		
CH4_Prod_All_Modules_per_24hrs_kg	240.0	SPM_Cryo_CH4_Weight_Allowance_lbs	12,000		
CH4_Mission_Cycle_Days	89.5	Number_of_Cryo_CH4_Modules	12		
CH4_Prod_Per_Mission_Cycle_kg	21,479.2	Cryo_CH4_All_Power_Req_kW	120		
SFT_Power_Allocation_kW	10.0	Calculated_Delta_Power_kWh	120.0	(Greater than zero is good)	
SFT_Power_24hr_Production_kWh	240.0	Calculated_Delta_Cryo_CH4_kg	1,718.3	(Greater than zero is good)	

SFT Cryogenic Liquification O2 Specifications

Label	Value	Label (lbs/kW/kg)	Value	Label (kg)	Value
C_Fuel_LOX_Required_kg	71139.0	Cryo_O2_Module_Weight_lbs	1,213	Cryo_O2_Module_Weight_kg	550.0
O2_Prod_per_Module_per_24hrs_kg	20.0	Cryo_O2_Module_Power_Req_kW	10		
O2_Prod_All_Modules_per_24hrs_kg	860.0	SPM_Cryo_O2_Weight_Allowance_lbs	52,000		
O2_Mission_Cycle_Days	89.5	Number_of_Cryo_O2_Modules	43		
O2_Prod_Per_Mission_Cycle_kg	76,967.0	Cryo_O2_All_Power_Req_kW	430		
SFT_Power_Allocation_kW	29.0	Total_SFT_Power_Delta_kW	0.0		
SFT_Power_24hr_Production_kWh	696.0				

Calculated_Delta_Power_kWh	266.0	(Greater than zero is good)
Calculated_Delta_Cryo_O2_kg	5,828.0	(Greater than zero is good)
Sum_SFT_Factory_QRM_Weight_lbs	120,000	
Sum_SFT_Phase2_Weight_lbs	128,850	
	8,850	(Greater than zero is good)

The reported ideal mass ratio for LCH4 / LOX rocket engines is 1 kg of Methane for every 3.6 kg of Oxygen. The CTugRV max fuel payload of 200,400 pounds of LCH4 / LOX equates to 90,899.8 kg total mass, or 19,760.8 kg of LCH4 and 71,139.0 kg of LOX. Based on the masses of Hydrogen (H), Oxygen (O), Water (H_2O), Carbon Dioxide (CO_2), and Methane (CH_4); Chart 136 shows 54,342.3 kg of CO_2 plus 35,569 kg of H_2O will produce 19,760.8 kg of LCH4 and 71,139.0 kg of LOX.

Best commercial processes for Electrolysis converting liquid H_2O into gaseous H and O2 will produce 1 kg of Hydrogen and 8 kg of Oxygen for 50 kW of electric power. Commercial electrolizers (analogs for the eventual Plan_B Space-Fuel-Terminal factory modules / design) weigh 50 kg (110 pounds). For an SPM launch mass allowance of 36,000 pounds, 327 electrolizers would produce the required H and O2 gas in 89.4 Mission days. We allocate 460 kW of constant solar PV electric power for this production. The Space-Fuel-Terminal and its very large solar power array design meets these power requirements.

An SFT Sabatier single reactor produces 21 kg of CH_4 in 24 Hours. Commercial Sabatier reactor (analogs for the eventual Plan_B Space-Fuel-Terminal factory design) weigh 850 kg (1874 pounds). For an SPM launch mass allowance of 20,000 pounds, 11 reactors would produce the required CH_4 gas in 89.4 Mission days. We allocate only 10 kW of constant solar PV electric power because the design assumes the ~400 Celsius required for the Sabatier reactor would come from concentrated solar light (not electric heaters).

The estimated commercial processes for Plan_B SFT Factory Cryogenic CH_4 (using passive cooling to convert the gaseous CH_4 into cryogenic LCH4) will produce 20 kg of CH_4 in 24 Hours. A cryogenic cooler design (analogs for the eventual Plan_B Space-Fuel-Terminal design) have an allocated design mass of 450 kg (992 pounds). For an SPM launch mass allowance of 12,000 pounds, 12 coolers would produce the required LCH4 in 89.4 Mission days. 10 kW of constant solar PV electric power is allocated. This low power requirement assumes the bulk of the required LCH4 cooling would come from the native cold of shaded (shielded) space (see Chapter 4). Electric power allocations will support two-stage cooling if required. We note that LCH4 is warmer than LEO native space cold (this is a good thing).

Commercial processes for Plan_B SFT Factory Cryogenic O2 (primarily using passive cooling to convert the gaseous O2 into cryogenic LOX) will produce 20 kg of LOX in 24 Hours. An analog cryogenic cooler design has a design mass of 550 kg (1213 pounds). For an SPM launch mass allowance of 52,000 pounds, 43 coolers would produce the required LOX in 89.4 Mission days. We allocate 29 kW of constant solar PV electric power for cooling. This medium power requirement assumes the bulk of the required cooling would come from native cold of shaded space. The electric power allocations will support two-stage cooling if required.

Within Chart 136, the Fuel Production Ratio number translates all the complexity of the Water* as fuel creation calculations into a single summary number. This number (Plan_B V1.00 = 0.836364) defines how much LCH4/LOX you get for the amount of Water* you have. You need

119,804 pounds of CO_2 plus 98,022 pounds of H_2O for 239,609 pounds of Water* as fuel (which includes a 10% margin). 504 kW of power (for 90 days) will produce 200,400 pounds of LCH4/LOX. The Plan_B Team revised previous versions of our fuel calculations indicating more H_2O mass was needed to become LOX. We admit that our prior calculations did not account for the Sabatier reactor *waste* product known as H_2O. By harvesting (recycling) this wasted H_2O mass we could reduce the total mass of raw stock H_2O delivered on LEO.

Chart 137 provides samples of Cryogenic Fuel and Water* mass allocations. Examining the Mission Circle 105 launch, after on-orbit delivery, fuel is transferred (via Space Autonomous Refueling—Yellow for LCH4/LOX, Blue for Water*) to other Mission Circle spacecraft. For example, Mission Circle 105 launches CTugRV-01(A) into LEO, pre-loaded with 75,400 pounds of fuel. Mission Circle 188 launches CTugRV-02(A) into LEO, pre-loaded with 75,400 pounds of fuel. That fuel is SAR transferred to CTugRV-01, increasing the fuel load of CTugRV-01 from 75,400 pounds to 150,800 pounds. Fuel deliveries and SARs continue until a spacecraft has a full fuel load (maximum of 200,400 pounds).

There is a subtle simplification used in the Mission Matrix fuel SAR transfer calculations. For any surplus (an amount greater than 204,000 pounds) after SAR, the Matrix allocates that fuel to the NEXT mission (see Chart 137 Mission Circle 194). As of V1.00, these assignments are simple cumulative math calculations that may not always line up with a scheduled on-orbit future SAR Mission Circle defined event date. The Plan_B Team assumes future (after Plan_B V1.0) detailed mission fuel planning teams will more precisely allocate these fuel transfer load dates.

Using the same method used for cryogenic fuels, the Mission Matrix assigns Water* payloads. Water* assignments support SPM shielding (Plan_B V1.00 allocates 247,000 pounds of Water* for LEOS-01 shielding). Water* transferred to the Space-Fuel-Terminal becomes LCH4/LOX fuel (after 90 days).

Chart 138 introduces Mission Circle 363 where delivered raw stock Water* masses decrement to zero and the Water* to Fuel Conversion Ratio calculates the SFT-01 production fuel. For example, Mission Circle 363 converts 402,042 pounds of Water* raw stock into 336,253 pounds of LCH4/LOX. Some Mission Schedule dates use a combination of nominal and adjusted dates to allocate 90 days to produce 200,400 pounds of fuel.

Chart 137: Mission Matrix Interpretation: Cryogenic Fuel and Water*

Plan B to the Moon
Version 1.01
© Stephen W. Long

Phase	M Circle	Circle Action	Vehicles / Milestones / Notes	L Payload Cyro	Cumulative Launch CH4O2	CTug-01 Fuel	Trans ELTV-01 Fuel	Trans ELTV-02 Fuel	Trans ELTV-03 Fuel	Trans ELTV-04 Fuel	Trans CTug-10 Fuel	Cumu Allocated C4HH2O Fuel	C4HO2 Xcheck	L Payload H2O	Cumulative Launch H2O	H2O Allocated for Shielding	H2O Allocated for Fuel	Cumulative Allocated H2O	H2O Xcheck	Net Fuel Water
		$5.00% / $85.00%	$22,015	1,161,200	1,161,200	200,400	200,400	200,400	200,400	200,400	159,200	1,161,200	0	1,086,542	1,086,542	297,000	709,542	1,006,542	0	709,542
1A	101	Launch FH, LEO Rendezvous	SPM-1+XDock-01+CSCM-01 [Phase 01A]		0							0	0		0			0	0	0
1A	102	Maneuver, Park LEO A	SPM-1+XDock-01+CSCM-01		0							0	0		0			0	0	0
1A	103	Launch F9, LEO Rendezvous	CCTCap-01 (6 COb) [Phase 01A]		0							0	0		0			0	0	0
1A	104	Dock	CCTCap-01:XDock-01F		0							0	0		0			0	0	0
1A	105	Launch FH, LEO Rendezvous	CTugRV-01(A) [Phase 01A]	75,400	75,400	75,400						75,400	0		0			0	0	0
1A	134	Launch FH, LEO Rendezvous	SPM-03+ 4STM (Two STM Full) [Phase 01A]		75,400							75,400	0	102,500	102,500			0	102,500	0
1A	135	Undock	CTugRV-01:XDock-02B		75,400							75,400	0		102,500			0	102,500	0
1A	136	Maneuver	CTugRV-01		75,400							75,400	0		102,500			0	102,500	0
1A	137	Dock	CTugRV-01:SPM-03+ 4STM		75,400							75,400	0		102,500			0	102,500	0
1A	150	Partial H2O Load	LEOS-01 H2O Transfer to Shielding Tanks		75,400							75,400	0		102,500			102,500	0	0
1A	151	LEOS-01 Ops Normal	LEOS-01 Ops Normal		75,400							75,400	0		102,500	102,500		102,500	0	0
1B	158	Launch FH, LEO Rendezvous	ASWTV-01(A) [Phase 01B]		75,400							75,400	0	70,514	173,014			102,500	70,514	0
1B	159	SAR-H2O	ASWTV-01(A) / LEOS-01 / H2O for Shielding		75,400							75,400	0		173,014	70,514		173,014	0	0
1B	160	H2O State Change	ASWTV-01(A)		75,400							75,400	0		173,014			173,014	0	0
1B	161	H2O State Change	LEOS-01		75,400							75,400	0		173,014			173,014	0	0
1B	184	LEOS-01 Ops Normal	LEOS-01 Ops Normal		75,400							75,400	0		314,042			314,042	0	17,042
1B	185	Launch FH, LEO Rendezvous	CTugRV-02(A) [Phase 01B]	75,400	150,800	75,400						150,800	0		314,042			314,042	0	17,042
1B	186	Dock	CTugRV-02(A):XDock-02B		150,800							150,800	0		314,042			314,042	0	17,042
1B	187	Launch FH, LEO Rendezvous	CTugRV-03(A) [Phase 01B]	75,400	226,200		25,800					150,800	75,400		314,042			314,042	0	17,042
1B	188	SAR-CH4O2	CTugRV-03(A):CTugRV-01 / Top Off		226,200							150,800	75,400		314,042			314,042	0	17,042
1B	189	Fuel State Change	CTugRV-03(A) Decrease		226,200							150,800	75,400		314,042			314,042	0	17,042
1B	190	Fuel State Change	CTugRV-01 Fill		226,200							150,800	75,400		314,042			314,042	0	17,042
1B	191	Maneuver	CTugRV-03(A)		226,200							150,800	75,400		314,042			314,042	0	17,042
1B	192	SAR-CH4O2	CTugRV-03(A):CTugRV-02 / Partial Load		226,200							150,800	75,400		314,042			314,042	0	17,042
1B	193	Fuel State Change	CTugRV-02 Empty		226,200	49,600						226,200	0		314,042			314,042	0	17,042
1B	194	Fuel State Change	CTugRV-01 Fill		226,200							226,200	0		314,042			314,042	0	17,042
1B	195	Maneuver	CTugRV-03(A)		226,200							226,200	0		314,042			314,042	0	17,042
1B	196	Dock	CTugRV-03(A):XDock-02C		226,200							226,200	0		314,042			314,042	0	17,042
1B	198	LEOS-01 Ops Normal	LEOS-01 Ops Normal		226,200							226,200	0		314,042			314,042	0	17,042
1B	199	Phase-01 Complete	Reserved / $7,199		226,200							226,200	0		314,042			314,042	0	17,042

Chart 138: Mission Matrix Interpretation: SFT Water* to LCH4/LOX Fuel Transfers

Plan B to the Moon
Version 1.01
© Stephen W. Long

Phase	M Circle	Circle Action	Vehicles / Milestones / Notes	L Payload Cyro	C4HO2 Xcheck	L Payload H2O	Cumulative Launch H2O	H2O Allocated for Shielding	H2O Allocated for Fuel	Cumulative Allocated H2O	H2O Xcheck	Net Fuel Water	SFT CH4O2 Production	Cumu CH4 Production	Trans CTug-10 Fuel	Trans Future ELTV	Cumu Allocated CH4O2	Allocated CH4O2 Xcheck
		85.00%	$22,015	1,161,200	0	1006542	1006542	297000	709542	1006542		709542	593434	593434	41200	552234	593434	0
2B	277	Launch FH, LEO Rendezvous	Max Water Cargo WTV [Phase 02B]		0	128,750	596542			467792	128750	170792		0			0	0
2B	278	SAR-H2O	Water Cargo Mission		0		596542			467792	128750	170792		0			0	0
2B	279	H2O State Change	Water Cargo Mission		0		596542			467792	128750	170792		0			0	0
2B	280	H2O State Change	SFT-01		0		596542		128750	596542	0	299542		0			0	0
2B	281	Maneuver	Water Cargo Mission		0		596542			596542	0	299542		0			0	0
2B	282	Park	On-Orbit Spare		0		596542			596542	0	299542		0			0	0
2B	299	Phase-02 Complete	$2,423															
3A	301	Launch FH, LEO Rendezvous	SPM-08+STM+CSCM-03+XDock-04 ELTV-01		0	51,250	647792		51250	647792	0	350792		0			0	0
3A	355	Launch FH, LEO Rendezvous	Cryo Fuel Cargo CRV-05 [Phase 03B]	121,000	121,000		699042			699042	0	402042		0			0	0
3A	356	Undock	CTugRV-01:XDock-01D		121,000		699042			699042	0	402042		0			0	0
3A	357	Maneuver	CTugRV-01:RV-05(A)		121,000		699042			699042	0	402042		0			0	0
3A	363	SFT-01 Fuel Ready	SFT-01 Fuel Ready		0		699042			699042	0	402042	336253	336253			0	0
3A	364	Undock	CTugRV-03(A)		0		699042			699042	0	0		336253			0	0
3A	365	Maneuver	CTugRV-03(A)		0		699042			699042	0	0		336253			0	0
7B	765	LLOG-01 Ops Normal	LLOG-01 Ops Normal		0		1006542			1006542	0	102500		507707			0	0
7B	786	SFT-01 Fuel Ready	SFT-01 Fuel Ready		0		1006542			1006542	0	102500	85727	593434			0	0
7B	787	Launch FH, LEO Rendezvous	CTugRV-10(A) [Phase 06A]	75,400	0		1006542			1006542	0	0		593434			0	0
7B	788	SAR-CH4O2	CTugRV-03(A)		0		1006542			1006542	0	0		593434	41200	552234	593434	0
7B	789	Fuel State Change	CTugRV-03(A)		0		1006542			1006542	0	0		593434			593434	0
7B	790	Fuel State Change	SFT-01		0		1006542			1006542	0	0		593434			593434	0
7B	791	Maneuver / Depart	CTugRV-10(A) Depart for LLO		0		1006542			1006542	0	0		593434			593434	0
7B	792	LLO Arrive, Rendezvous	CTugRV-10(A)		0		1006542			1006542	0	0		593434			593434	0
7B	793	Maneuver	CTugRV-10(A)		0		1006542			1006542	0	0		593434			593434	0
7B	794	Dock	LLOG-01		0		1006542			1006542	0	0		593434			593434	0
7B	799	Phase-07 Complete	$756															

10.7   Mission Matrix Interpretation: F9 Launches and Crew Rotations

Chart 139 is a sample of Mission Matrix Circle Action, F9 Launch, Crew Rotations, and Summary Metric data.   CCtCap provides the dry mass (no humans, cargo, fuel, expendables) of the spacecraft, supporting calculations for the much higher cost-per-pound construction costs for Human Rated spacecraft.   For a defined CCtCap dry mass of 21,300 pounds, the build-to costs are $213 M.   For the remaining F9 maximum launch weight ceiling (defined as 50,000), the F9 Cargo mass is typically 50,000 minus 21,300 = 28,700 pounds.   The defined build-to cost is $143.5 Million for this other mass.   Plan_B separates these masses to account properly for the higher costs of human-rated craft masses.   We use the lower costs for other masses (such as cargo, fuel, and expendables).

Crew Counts calculate how many crew members and their locations during every Plan_B Mission Circle activity:

- To-From LEO, To-From LLO, To-From the Lunar Surface.
- Crew Counts drive the needed availability of lifeboat capacity at all times. No matter their location (LEO or LLO), every human on-orbit has a place to evacuate during a mission emergency. LLO crew members have a CTugRV vehicle to return to LEO. If there is an emergency in LEO, there are enough seats to return Astronauts to Earth's surface. Note that this necessary requirement causes Plan_B to have and keep more spacecraft on-orbit than needed to just complete the mission steps to return to the Moon.
- CCtCaps can carry up to 7 crew members (published max value). No Plan_B normal Launch or Return-to-Earth exceeds 6 crew members. Additional analysis will determine if entire crews should rotate in/out, or should rotations always leave one or more experienced crew members on-board. The stay-behind member would become the Mentor/Commander of the next crew.
- Following the common practice of aircraft crews, no mission occurs with less than two crew members in a space vehicle. Future semi-autonomous craft auto-pilots may enable routine single-crew-person (pilot) on-orbit operations. This is likely for piloting the CTugRV for LEO Station assembly tasks.

Chart 139: Mission Matrix Interpretation: F9 Launches and Crew Rotations

Plan B to the Moon
Version 1.01
© Stephen W. Long

Phase	M Circle	Circle Action	Vehicles / Milestones / Notes	CCtCap lbs	F9 Cargo lbs	F9 lbs to LEO	On Bd	Net Space	Crew Chge	Net Crew LEO	Crew Chge	Net Crew LLO	Crew Chge	Net Crew LS	Life boat S.	Net Seats	Habitable Vol (cuft)	PV-Pwr Prod (kW)	HR Veh No Fuel lbs	Veh No Fuel lbs	LLO/LS No Fuel lbs	Data Date
		85.00%	$22,015	149100	200900	350000											43780	798	299100	1883158	716700	
0	19	Launch FH, LEO Rendezvous	CSM+SPM-0+STM+XDock-00+CSCM-00 [Phase 000]				6	0		0		0		0		0						2019-01-07
0	20	Maneuver, Park LEO A	SPM-0+XDock-00+CSCM-00 [Phase 000]				-6	0		0		0		0		0	4238	21	10650	119350		2019-01-07
0	21	Launch F9, LEO Rendezvous	CCTCap-00 (6 COB) [Phase 000]	21300	28700	50000	6	6	6	6		0		0	7	7	350		21300	28700		2019-01-07
0	26	Return to Earth	CCTCap-00R (6 COB) [Phase 000]	21300			-6	0	-6	6		0		0	-7	0						2019-01-07
1A	103	Launch F9, LEO Rendezvous	CCTCap-01 (6 COB) [Phase 01A]	21300	28700	50000	6	6	6	6	0	0		0	7	7	350		21300	28700		2019-01-07
1A	105	Launch FH, LEO Rendezvous	CTugRV-01(A) [Phase 01A]				6	6	-6	12	0	0		0	7	7	350	21	10650	43950	54600	2019-01-07
1B	152	Launch F9, LEO Rendezvous	CCTCap-02 (6 COB) [Phase 01B]	21300	28700	50000	6	12	6	12	0	0		0	7	14	350		21300	28700		2019-01-07
1B	156	Return to Earth	CCTCap-01R (6 COB)				-6	6	-6	6	0	0		0	-7	7						2019-01-07
1B	182	Launch F9, LEO Rendezvous	CCTCap-03 (6 COB) [Phase 01B]	21300	28700	50000	6	12	6	12	0	0		0	7	14	350		21300	28700		2019-01-07
3B	394	Launch F9, LEO Rendezvous	CCTCap-04 (6 COB) [Phase 03B]	21300	28700	50000	6	18	6	12	6	6		0	7	21	350		21300	28700		2019-01-07
3B	398	Return to Earth	CCTCap-02R (6 COB)				-6	12	-6	6	6	6		0	-7	14						2019-01-07
4A	478	LEOS-01 Ops Normal	LEOS-01 Ops Normal					12		6		6		0		14						2019-01-07
4A	479	LLO Maneuver / Depart	ELTV-02 Depart for LLO					12	-4	2		6		0		14						2019-01-07
4A	480	Phase-04A Complete	Phase-04A Complete					12		2		6		0		14						2019-01-07
4B	481	LLO Arrive, Rendezvous	CTugRV-02: ELTV-02 / Phase 04B:					12	2	2	4	10		0		14						2019-01-07
4B	489	Maneuver, Return LEOS-01	CTugRV-02					12	2	2	-4	6		0		14						2019-01-07
4B	490	LLO Gateway-01 Ops Norm	CTugRV-02					12	2	2		6		0		14						2019-01-07
4B	491	LEO Arrive	CTugRV-02					12	4	6		6		0		14						2019-01-07
4B	492	Rendezvous	CTugRV-02					12		6		6		0		14						2019-01-07
4B	493	Dock	CTugRV-02:LEOS-01 XDock-01D					12		6		6		0		14						2019-01-07
4B	494	Launch F9, LEO Rendezvous	CCTCap-05 (6 COB) [Phase 04B]	21300	28700	50000	6	18	6	12		6		0	7	21			21300	28700		2019-01-07
4B	495	Dock	CCTCap-05:XDock-01F					18		12		6		0		21						2019-01-07
4B	496	LEO-01 Crew Swap	LEO-01 Ops Normal					18		12		6		0		21						2019-01-07
4B	497	Return to Earth	CCTCap-03R (4 COB)				-4	14	-4	8		6		0	-7	14						2019-01-07
4B	498	Undock	CCTCap-03R:LEOS-01 XDock-01F					14		8		6		0		14						2019-01-07
4B	499	Phase-04B Complete	Phase-04B Complete	$2,469				14		8		6		0		14						2019-01-07

Chapter 11: Plan_B Summary and Conclusions

11.1 Program Plan to Return to the Moon

In 2018, the Plan_B Team set out to develop a Program Plan to Return to the Moon that would:
- Meet a target expenditure of 85% of the costs of Plan_A.
- Enable Americans (as representatives of humanity) to land on the Moon by the end of 2024 (if funds available starting in FY 2020).
- Meet or exceed the top-level metric requirements of Plan_A—namely first return to Lunar Orbit, land on the surface and begin the longer-term colonization of the Moon.
- Develop the Plan_B documents using non-technical—laymen's language if possible. Returning to the Moon is for all of humanity, therefore all of humanity should be able to understand the Plan_B goals and execution elements.

11.2 Costs, Schedule and Performance

Throughout this document, Plan_B has attempted to define and document the Precepts (engineering and other assumptions) that informed the Plan_B designs, Mission Plans, and the resulting calculated costs. Over time, the Plan_B Team created and extensively updated the MISSION MATRIX. The Matrix eventually became a complex numerical model for all the total mission planning elements (Mission, Costs, Schedule, Mass, Water, Fuel, and Summary Totals) based on the given Defined Variables. If future planners want to change Defined Variables, such as the cost per pound to launch or build a craft, Defined Variable changes will automatically ripple through the Mission Matrix. The Matrix then automagically provides the new summary costs. If the Plan_B readers do not agree with Plan_B Precept values (say change the FH launch costs from $130 Million to $125 Million), changing that value does not invalidate the year of work that went into producing Plan_B. We believe we have captured every significant Plan_B Precept using defined financial variables that will enable straight forward recalculations of Plan_B costs based on future-defined cost assumptions.

Recall that Plan_B arose out of our early 2018 frustrations when we learned that America has spent / will spend over $86 Billion for Plan_A (through EM-10). For that large sum we do NOT land on the Moon. Chapter 2 estimates NASA spends $25.9 Billion through 2024.

Plan_B Version 1.01 update: Plan_B was published on March 14, 2019. On March, 26, 2019, Vice President Mike Pence ordered NASA to land on the Moon. "The National Space Council will send recommendations to the president that will launch a major course correction for NASA and reignite that spark of urgency that propelled America to the vanguard of space exploration 50 years ago," Pence said. "...If NASA is not currently capable of landing American astronauts on the moon in five years, we need to change the organization — not the mission." The space agency's initial plan was to send astronauts to the moon by 2028, 18 years after NASA began developing the Space Launch System (SLS) to make the journey. But SLS has suffered massive cost overruns and schedule delays that have put in question whether even 2028 was achievable.

"Ladies and gentlemen, that's just not good enough," Pence said. "...It took us eight years to get to the moon the first time 50 years ago when we had never done it before."

Plan_B answers the tough challenge to develop planning, design, and cost savings concepts. Plan_B defines how to land on the Moon for less than Plan_A's $25.9 B. This Plan_A amount became the Plan_B Maximum-Cost-Ceiling. The Plan_B working budget ceiling is 85% of Plan_A, or $22.015 B. In all candor we also report that our preliminary cost calculations were too low—so low we felt no one would believe our cost math. Therefore, we explicitly rounded-up all cost calculations to climb up to the imposed 85% of Plan_A cost objective. The largest example of this is where we added significant Other costs to the calculations. We made a conscious decision to in-plain-sight label these as Other Costs totaling $1.4 Billion for RDT&E (funds before Mission Circle 101). We also apportioned $100 Million per Plan_B Launch cycle (37 launches and for some Lunar Surface Mission Circles), totaling $5.212 B. As the old sad joke goes, even in Washington, DC this is a large number. Perhaps the best way to evaluate this number is to say this is the planning wedge or cost ceiling under which all other Plan_B activities fit under. Further reductions in this Other-Cost number (spend less money), will cause Plan_B to cost even less than stated in Version 1.0.

The Plan_B Team believes and notes here that even if Plan_B costs grow to the full amount of Plan_A through 2024 ($25.9 B), Plan_B is still a better value to the taxpayer than Plan_A because Plan_B lands on the Moon and establishes the first Lunar Colony. Chart 140 provides the Summary Cost, Schedule and Performance data for Plan_B.

Chart 140: Plan B Summary: Cost, Schedule and Performance

Plan B to the Moon
Version 1.01
© Stephen W. Long

Plan B Summary Costs:

	Plan B	Plan B $4k	Plan B $7k	No SFT
FH Payload Mass	4,000,000			
F9 Payload Mass	350,000	☑ Cost		
Human Veh No Fuel Mass	299,100			
Vehicle No Fuel Mass	1,883,158			
Human Veh Costs	$2,991	$2,991	$2,991	
Vehicle No Fuel Costs	$9,416	$7,533	$13,301	
FH Launch Costs	$3,962	$3,962	$3,962	
F9 Launch Costs	$434	$434	$434	
Plan B Other Costs	$5,212	$5,212	$5,212	-$2,423
Plan B Total Costs	$22,015	$20,132	$25,900	$19,592
Plan A Through 2024 $M	$25,900	$25,900	$25,900	$25,900
Plan B Savings $M	$3,885	$5,768	$0	$6,308
Plan B to A Ratio	85.00%	77.73%	100.00%	75.65%

Plan B Summary Schedule:

Authority to Proceed (FY2020):	2019-10-01
First Mission Payload Launch:	2022-12-27
Phase 01 - LEO Station-01 Complete:	2023-07-31
Phase 02 - SFT-01 Complete:	2023-11-28
Phase 03A - ELTV-01 Humans LLO:	2024-02-18
Phase 03B - LLO Gateway-01 IOC:	2024-03-03
Phase 04 - ELTV-02 Complete:	2024-06-07
Phase 05 - ELTV-03 Complete:	2024-09-11
Phase 06 - ELTV-04 Complete:	2024-10-20
Phase 07 - LLO Gateway-01 FOC:	2024-11-01
Phase 08A - Humans Land on Moon:	2024-11-02
Phase 08B - Lunar Base-01 IOC:	2025-01-18

☑ Schedule

Plan B Summary Metrics:

Launches (FHR and F9)	37		One (1) LEO Station-01V1, FOC
Cumulative Habitable Vol LEO/LLO/LS	43,780	cu ft	One (1) Space Fuel Terminal-01, FOC
Cumulative Mass Sent to LEO	4,350,000	lbs	One (1) LLO Gateway-01V1, FOC
Cumulative Mass No Fuel to LLO / LS	716,700	lbs	One (1) Lunar Surface Base-01V1, IOC
Water* (from Earth, Shielding/Stock)	297,000	lbs	Two (2) Lunar Landers (Human)
LCH4/LOX (from Earth)	1,161,200	lbs	Two (2) Lunar Long Boats (Cargo)
LCH4/LOX from SFT Production	593,434	lbs	Two (2) Lunar Habitat Vehicles
LCH4/LOX from Earth and SFT	1,754,634	lbs	
Power on Orbit (SFT Array is 504kW)	798	kW	☑ Performance

Plan B to the Moon: A How-to Guide to Return to and Colonize the Lunar Surface

Plan_B sends 4,000,000 payload mass pounds into LEO using FH rockets. Plan_B sends 350,000 pounds of payload mass into LEO using F9 rockets, which includes humans and human-rated vehicle masses at a total cost of $4.396 B. For cost calculations, together FH and F9 rockets deliver 299,100 pounds of human-rated (much more expensive to build) vehicle mass into LEO. Plan_B does not use Falcon Heavy for human launches, but Falcon Heavy delivers non-crewed CSCM and LL components to orbit.

Plan_B provides three cost estimates for the Plan_B Defined Variable Cost-Per-Pound-to-Build figures. The Plan_B nominal cost is $5000 per pound. Plan_B sends 1,883,158 pounds of non-human-rated vehicles into LEO. At $5000 per pound this cost is $9.4169 Billion (meeting the Plan_B 85% cost objective). If the Cost-Per-Pound-to-Build value is $4000, Plan_B costs are $7.533 Billion (Plan_B costs 77.73% of Plan_A). If the Cost-Per-Pound-to-Build value is $7000 per pound, total Plan_B costs increase and Plan_B becomes 100% (equivalent) with Plan_A costs. Therefore, the eventual real-world Plan_B costs will be highly sensitive to the cost-to-build-per-pound figure of merit and the specified Other Costs. Plan_B provides a fourth summary cost estimate where Plan_B skips (eliminates) building the Space-Fuel-Terminal (PHASE 2). Other Mission Circles already include and account for the costs to ship sufficient LCH4/LOX from Earth to LEO to return to the Moon. Eliminating the SFT saves $2.423 Billion and Plan_B costs become 75.64% of Plan_A.

During our extensive fuel cost calculations, the Plan_B Team noted it would be less expensive for Plan_B to just ship LCH4/LOX to LEO. It costs less to launch ready-fuel than it costs to launch Water* to LEO and convert it to fuel. Based on Chapter 3 ($1000 per pound) launch costs and Chapter 10 fuel production calculations, where the Water* to Fuel calculation ratio is 0.83636, shipping 239,609 pounds of Water* costs $240 M. Shipping 200,400 pounds of LCH4/LOX costs $200.4 M. Sending fuel instead of Water* saves $39 M. Plan_B Chapter 1 asserts that over the longer-term mining ice from the Moon or a comet will eliminate the need to ship Water* or LCH4/LOX up from Earth. But until we mine water from the Moon, we will have to ship water from the Earth to stock the Sabatier reaction engines and validate SFT technology. Making deep space fuel from ice remains a critical technology milestone in the next decade if humans are to leave Earth for other deep space destinations. Launching human-safety-shielding mass (Plan_B uses water) is important, no matter the raw source of the rocket fuel. Plan_B asserts that Now Is the Time to prove once and for all that Water* as fuel works. Once we prove we can produce fuel on-orbit, we can produce fuel on the Moon from Lunar ice.

Plan_B V1.00 presents an admittedly aggressive schedule for Mission Phases 1–8. The day-by-gritty-day incremental count starts from 2018-01-01 all the way to establishing the first colonists on the Lunar Surface (item 899), which calculates as 2025-01-18. The Plan_B team added 340 Adjustment-Days to the schedule to account for the Defined Variable Mission Cycle of 28-Days between mission launches. If (when) SpaceX Falcon Heavy launches become a routine production process, 28 Days could become 21 days. A significant amount of improved schedule margin is available for capture during the two years of Plan_B spacecraft development and testing before Mission Circle 101. Building many FH and F9 boosters up front will support rapid SPM launches. We also note we did NOT use explicit parallel mission scheduling. The V1.00

schedule is an old-fashioned Waterfall Schedule–everything happens one-after-the-other. As an example, if Phase 3 (first ELTV mission) began in parallel with PHASE 2 (SFT construction), up to 120 Days of master schedule improvements are possible.

Perhaps it is worthy to remind our readers that Plan_B team members came from impossible-mission backgrounds (admitted lifelong over-achievers). We offer our prior professional career accomplishments as existence-proof that the right team with the right leadership can accomplish difficult goals. We know it is possible to deliver results in significantly shorter time-lines than traditional PoR processes and teams. We also note that existing acquisition / procurement rules and regulations allow (and even encourage) rapid response approaches. As a further existence proof, consider the historical creation of the Secretary of Defense ISR Task Force whose heydays were from ~2008 to ~2010. The SECDEF created the ISR Task Force to rush advanced ISR capabilities to Afghanistan and Iraq to help save American soldier's lives. The traditional program offices of traditional PoRs hated how the Task Force was willing and able to get things done quickly. But the soldiers loved it, and many of those young men and women are alive today because the ISR Task Force would not accept the pace-of-PoRs. The Task Force operated at the Speed-of-Need. The Task Force spent authorized funds using the rapid-acquisition provisions in most all existing procurement laws. The Plan_B Team asserts the accelerated Plan_B Schedule can use existing published acquisition rules–it is a matter of will and prioritization. Perhaps we should also look back in history at the Apollo program. Consider only eight years passed from Kenney's, "We Choose to Go to the Moon" speech, to when we heard, "The Eagle has Landed." Plan_B shows us how to land on the Moon before the end of 2024–it is just a matter of CHOOSING to do so.

The nominal Plan_B Authority-to-Proceed Date is 20191001 (start of US Government Fiscal Year 2020). If future planners change this date (slip six months, etc.), all other Plan_B dates recalculate, based on the new Authority date. The Plan_B V1.00 schedule allows humans to Land on the Moon by the end of 2024. Plan_B posits that 2024-11-02 will become (is) a significant date in human history.

Chart 140 includes Summary Performance Metrics (measurable accomplishments) for the Plan_B mission designs. Plan_B's Metrics meet or exceed all published Plan_A metrics. Chart 141 is the Launch-Only Mission Circle Summary. Plan_B has 37 planned launches, with Launch 1 scheduled for 2021-07-21 and Launch 37 on 2024-10-29. One RDT&E FH launch (Launch 0) occurs on 2021-07-21 and one early F9 launch occurs on 2021-07-11.

Plan_B V1.00 delivers 43,780 cubic feet of human habitable volume on-orbit. For comparison, the 747-400 jumbo airliner has 31,285 of interior cubic feet and it has the largest passenger interior volume of any commercial airliner. A comfortable 2500 square foot home has 20,000 cubic feet. Said another way, Plan_B delivers breathing room for crew members equivalent to a spacious 5000+ square foot home.

Plan_B V1.00 delivers 4,350,000 pounds of mass to Low Earth Orbit. That is equivalent to 2,507 model 1969 VW Beetles (1735 pounds each). As depicted in Chapter 2, the Apollo 17 Saturn V mission launched 310,000 pounds into LEO. The Plan_B V1.00 total launch mass is equivalent to the mass of 14 Apollo 17 launches. Plan_B V1.00 delivers 2,074,858 pounds of non-fuel mass to LEO, LLO and the Lunar Surface. This is equivalent to 148 orbiting elephants (at 14,000 pounds each). Plan_B V1.00 delivers 607,500 pounds of non-fuel-mass to Low Lunar Orbit and the Lunar Surface. This is equivalent to ~17 Lunar Modules from the Apollo era (each 36,200 pounds). This is equivalent to 43 Lunar Surface elephants (at 14,000 pounds each). 43 elephants would require 22,500 pounds of food per day. If all the Plan_B fuel masses automagically transported and became elephant food on the Lunar Surface, we could feed the 43 lunar elephants for 108 days. Plan_B does not plan to send elephants to LEO, LLO or to the Lunar Surface—it would cost too much to feed them.

Plan_B V1.00 delivers 1,006,542 pounds of Water* to LEO. The LEOS-01 Shielding-Storage-Tanks use 297,000 pounds of this Water. 709,542 pounds of Water* Payload become LCH4 / LOX fuel in the SFT-01. Plan_B Version 3.0 develops the planning steps required to use Lunar Ice as shielding water for LLOG-01 and all Lunar surface craft.

As calculated in the Chapter 10 Mission Matrix, through Mission Circle 899 Plan_B launches 760,792 pounds of Water* as fuel. Plan_B builds the Space-Fuel-Terminal on LEO to convert that Water* into LCH4/LOX. The SFT will produce 593,434 pounds of LCH4/LOX for the provided Water* raw stock. Plan_B launches/delivers 1,161,200 pounds of LCH4/LOX fuel into LEO. Plan_B produces or launches a grand total of 1,754,634 pounds of LCH4/LOX. Plan_B delivers 1,202,400 pounds of LCH4/LOX across four ELTV Missions and two CTugRV-fuel-only delivery missions to deliver ~600,000 pounds of non-fuel mass to LLO.

As of Plan_B V1.00:
- The ELTV-01 fuel load is 200,400 pounds of LCH4 / LOX.
- The CTugRV-01 fuel load is 200,400 pounds of LCH4 / LOX to deliver itself and bring lunar operations fuel to LLO.
- The ELTV-02/03/04 fuel loads are 200,400 pounds of LCH4 / LOX.
- The CTugRV-10 fuel load is 200,400 pounds of LCH4 / LOX to deliver itself and bring lunar operations fuel to LLO.
- 474,098 pounds of LCH4 / LOX are spare on-orbit fuels for future ELTV missions.

Plan_B Version 3.0 develops the planning required to use Lunar Ice as raw stock to create LCH4 / LOX. Plan_B expects to move the on-orbit-spare cryogenic storage tanks from the CRV-04 through CRV-09 missions to LLO and LS via future ELTV missions.

Plan_B includes specific technical risk reduction strategies to build the Space-Fuel-Terminal to make fuel on-orbit and deliver cryogenic fuel from Earth to LEO. Once in LEO, Plan_B delivers to LLO all the cryogenic fuel needed for Lunar Missions. The Planning Team added Cryogenic Refueling Vehicle (CRV) fuel-tank (no other payload) missions to the Mission Circle timelines to address this need. Even with this added fuel redundancy, Plan_B still meets its cost ceiling objectives (85% of Plan_A).

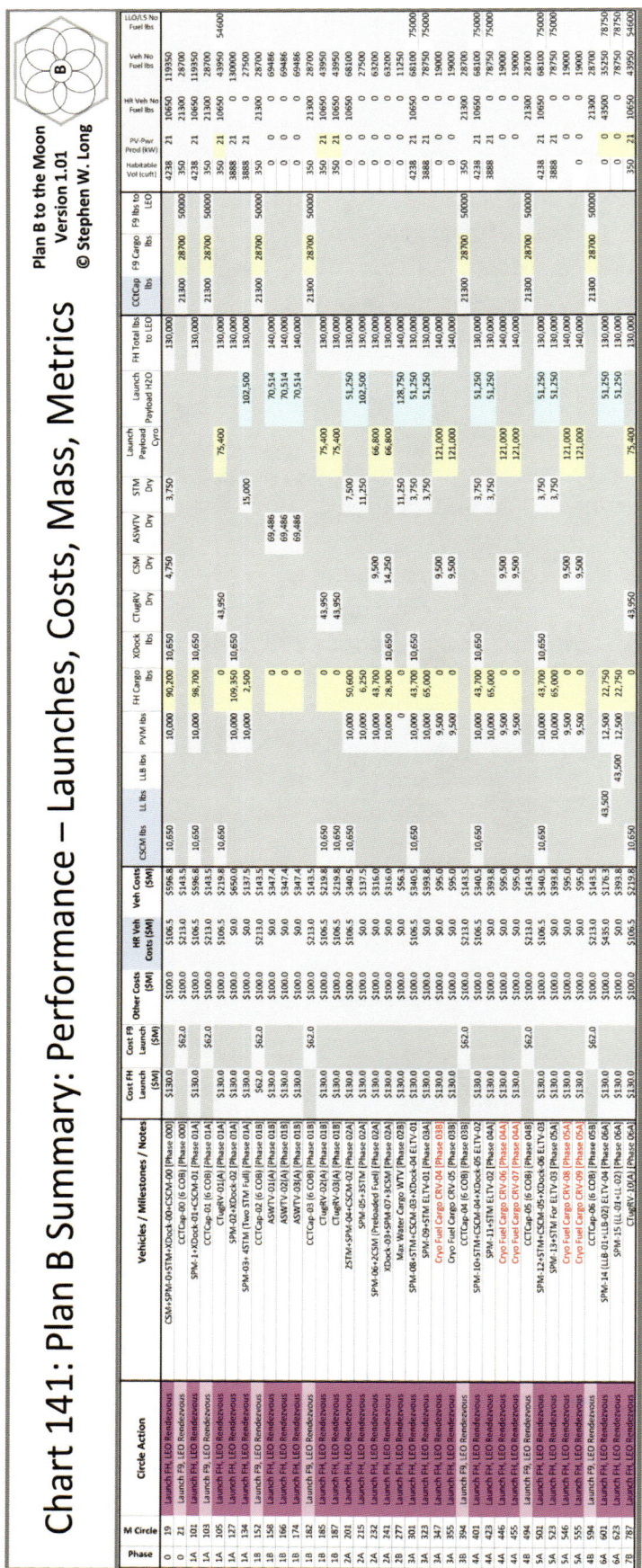

Chart 141: Plan B Summary: Performance – Launches, Costs, Mass, Metrics

Plan B to the Moon
Version 1.01
© Stephen W. Long

Plan B to the Moon: A How-to Guide to Return to and Colonize the Lunar Surface

Plan_B V1.00 delivers 798 kW of electric power on-orbit when the PVPs are solar illuminated. For comparison, the ISS produces 120 kW. The Plan_A Lunar Orbital Gateway will produce 40 kW. The summary power production number includes the Plan_B Version 1.00 (Mission PHASE 2) Space-Fuel-Terminal Array that makes 504 kW of electrical power on LEO.

How many crew members do you need to Return to the Moon and start the first Lunar colony? Plan_B V1.00 requires 42 crew members. Future planning may reduce this crew count number (fewer crew members) IF the crew members live in orbit for extended mission cycles (we don't recommend this). As a general rule, Plan_B cycles through a new crew after completion of a major milestone. Rotations occur after building LEOS-01, SFT-01, LLOG-01, Lunar Components, Land on Lunar Surface, and Start Colonization. There will never be less than 2 crew in LEO and 2 crew in LLO. After Lunar Base-01 raises its flag on the Moon, there will never be less than 4 crew on the Lunar surface.

We explicitly state here an additional **Plan_B Precept** that once we reach a shore, humans never leave that shore. Once we occupy LLO, it remains occupied forever. Once we occupy the Moon (LSHVs), it remains inhabited forever. Never again will people ask why we left the moon.

Summary Crew Rotations are in Chart 142. Crew Counts calculate how many and the locations of all Plan_B crews. During Phase 8B (Lunar Surface), there are Mission Circles that state Transition. Transitions define when crew members become the first Colonists. The Plan_B Team recommends that married (life committed couples) should be the designated Transition crew members. Perhaps this will be a firm Plan_B 2.0 or Plan 3.0 requirement. These couples should not leave previous children on Earth (don't leave children behind), but need to be young enough to be fertile. We gently remind the reader that the primary Plan_B tenant is that returning to the Moon is about colonization—not tourism. While Mark Watney (in the book and movie, "The Martian") claims he became the first colonist because he grows food on Mars, the Plan_B Team asserts you only become true Colonists when you conceive and raise children on the remote shore.

11.3   Summary and Conclusions

Coming full circle, on the cover page, the sub-title to Plan_B is, *"A How-to Guide to Return to and Colonize the Lunar Surface."* Recalling the profound childhood riddle, how do you eat an elephant, the Plan_B Team's answer about how-to-write a How-to Guide to Return to the Moon, was *one bite at a time*. We threw in a few Lunar Elephant mass calculations at the end of Chapter 10 just to see if you were paying attention.

Plan B to the Moon
Version 1.01
© Stephen W. Long

Chart 142: Plan B Summary: Performance - Crew Rotations

Phase	M Circle	Circle Action	Vehicles / Milestones / Notes	On Bd	Net Space	Crew Chge	Net Crew LEO	Crew Chge	Net Crew LLO	Crew Chge	Net Crew LS	Life boat S.	Net Seats
0	21	Launch F9, LEO Rendezvous	CCTCap-00 (6 COB) [Phase 000]	6	6	6	6		0		0	7	7
0	26	Return to Earth	CCTCap-00R (6 COB) [Phase 000]	-6	0	-6	0		0		0	-7	0
1A	103	Launch F9, LEO Rendezvous	CCTCap-01 (6 COB) [Phase 01A]	6	6	6	6		0		0	7	7
1B	152	Launch F9, LEO Rendezvous	CCTCap-02 (6 COB) [Phase 01B]	6	12	6	12		0		0	7	14
1B	156	Return to Earth	CCTCap-01R (6 COB)	-6	6	-6	6		0		0	-7	7
1B	182	Launch F9, LEO Rendezvous	CCTCap-03 (6 COB) [Phase 01B]	6	12	6	12		0		0	7	14
3A	378	LLO Maneuver / Depart	ELTV-01 Depart for LLO		12		6	6	6		0		14
3A	379	LLO Maneuver / Depart	CTugRV-01		12		4	-2	8		0		14
3B	381	LLO Arrive, Park	CTugRV-02: ELTV-01 / Humans Return to LLO	6	12		4		8		0		14
3B	382	LLO Arrive, Rendezvous	CTugRV-01		12		4		8		0		14
3B	390	LEO Arrive	CTugRV-02		12		6	2	8		0		14
3B	394	Launch F9, LEO Rendezvous	CCTCap-04 (6 COB) [Phase 03B]	6	18	6	12		8		0	7	21
3B	398	Return to Earth	CTugRV-02R (6 COB)	-6	12	-4	2		8		0	-7	14
4A	479	LLO Maneuver / Depart	ELTV-02 Depart for LLO		12		2	4	12		0		14
4B	481	LLO Arrive, Rendezvous	CTugRV-02: ELTV-02 / Phase 04B:	6	12		2	-4	8		0		14
4B	489	Maneuver, Return LEOS-01	CTugRV-02		12	4	6		8		0		14
4B	491	LEO Arrive	CTugRV-02		12		6		8		0		14
4B	494	Launch F9, LEO Rendezvous	CCTCap-05 (6 COB) [Phase 04B]	6	18	6	12		8		0	7	21
4B	497	Return to Earth	CCTCap-03R (4 COB)	-4	14	-4	8		8		0	-7	14
5A	579	LLO Maneuver / Depart	ELTV-03 Depart for LLO		14	-6	2	6	14		0		14
5B	581	LLO Arrive, Rendezvous	CTugRV-02: ELTV-03 / Phase 05B	6	14		2	-6	8		0		14
5B	589	Maneuver, Return LEOS-01	CTugRV-02		14	6	8		8		0		14
5B	591	LEO Arrive	CTugRV-02		14		8		8		0		14
4B	594	Launch F9, LEO Rendezvous	CCTCap-06 (6 COB) [Phase 05B]	6	20	6	14		8		0	7	21
5B	598	Return to Earth	CCTCap-04R (6 COB)	-6	14	-6	8		8		0	-7	14
6A	679	LLO Maneuver / Depart	ELTV-04 Depart for LLO		14		2	-6	14		0		14
6B	681	LLO Arrive, Rendezvous	CTug02:ELTV-04		14		2	6	14		0		14
8A	804	Humans Land on LS	LL-01		14	2	2	-4	10	4	4		14
8A	806	Liftoff, LS to LLO	LL-01		14	2	2	4	14	-4	0		14
8A	814	Humans Land on LS	LL-02		14	2	2	-4	10	4	4		14
8A	816	Liftoff, LS to LLO	LL-02		14	2	2	4	14	-4	0		14
8A	824	Humans Land on LS	LL-01		14	2	2	-4	10	4	4		14
8A	826	Liftoff, LS to LLO	LL-01		14	2	2	4	14	-4	0		14
8A	834	Humans Land on LS	LL-02		14	2	2	-4	10	4	4		14
8A	836	Liftoff, LS to LLO	LL-02		14	2	2	4	14	-4	0		14
8A	844	Humans Land on LS	LL-01		14	2	2	-4	10	4	4		14
8A	846	Liftoff, LS to LLO	LL-01		14	2	2	4	14	-4	0		14
8B	854	Land, Lunar Base-01	LL-01		14	2	2	-6	8	6	6		14
8B	867	LL Liftoff, LS to LLO	LL-01		14	2	2	2	10	-2	4		14
8B	874	Land, Lunar Base-01	LL-02		14	2	2	-6	4	6	10		14
8B	887	LL Liftoff, LS to LLO	LL-02		14	2	2	2	6	-2	8	8	14

Plan B to the Moon: A How-to Guide to Return to and Colonize the Lunar Surface

The aggregate body of work captured in the original Plan_B Program Plan and re-stated in this book is overwhelming—the Team felt overwhelmed throughout most of 2018. But we found a way to eat-the-elephant by dividing the return-to-the-moon problems into more bite-sized pieces, which became the Plan_B Six Pillars:

- Pillar 1: Mass into Orbit Affordably
- Pillar 2: Use X-Poly-Modules to Lower Spacecraft Costs
- Pillar 3: Leverage Existing Human Rated Spacecraft
- Pillar 4: Water* as Fuel—Common Conventional CH4/O2 Rocket and LARIT Engines
- Pillar 5: On-Orbit Refueling and Infrastructure
- Pillar 6: Catalog of Plan_B Spacecraft (Detailed Spacecraft designs, built from XPMs)

The Plan_B Team thanks our Readers for reading and working through all the text, Charts and math of Plan_B. If you remember nothing else, please take with you the two most important big ideas from our Plan_B spacecraft designs and mission planning:

- Apportion the mass required to return to the Moon into smaller, bite-sized payloads that fit within the maximum payload mass for available (already flying) commercial launch vehicles.
- Send compact building blocks into orbit and construct on-orbit much larger vehicles and habitats out of the building blocks.

We also hope that our thinking differently about how to solve problems affordably is the true magic found in Plan_B. Per our tasking, here at the end, in profession Program Plan fashion we report that Plan_B Accomplished its Cost, Schedule, and Performance Objectives. Only you the reader can judge if we met our fourth objective, to use non-technical—laymen's language. The Plan_B Team good-naturedly argued (and argued) about the need for more and more (and more) technical details, but the PD insisted on keeping descriptions as simple as practicable. We know the literal rocket scientist readers will say Plan_B does not have enough technical details. Casual readers may say there are too many details. We hope we found the right balance.

In conclusion, the motivation behind the 12-month development of Program Plan_B was to answer the hard question: What will we as a species do if NASA's Plan_A fails or continues to stumble? We did not write Plan_B to focus on negative outcomes. Instead, we wrote Plan_B as an insurance policy. If the NASA PoR stumbles, we as a species will have at least considered another way to return to the Moon and stay there permanently this time around.

Plan_B to the Moon is a love story about Returning to the Moon—and eventually striving for the Stars in this 21st century. Plan_B has attempted to put into words and straight-forward, accessible math the human spirit's desire and existential need to stand on another world. This book's Title Page image from Apollo 17 causes intense emotions in the Plan_B Team members every time we see it. For we all secretly wish we were young enough to be that future crew member in the spacesuit standing by the American Flag on that other world. Our hearts ache to see with our own eyes the blue marble that is humanities' birth place—the Earth—in the sky.

Plan B to the Moon: A How-to Guide to Return to and Colonize the Lunar Surface

The Earth may be our birthplace but the deeper exploration of the unknown is our life's destination.

We close our story by repeating a fundamental truth—we humans do our best work when we tackle the hardest problems. **Step-by-gritty-step, we will eventually stand on that far Lunar shore because it will be the hardest thing there is to do.**

Pages 201-224 provide summary literal spreadsheet data of all numerical details (except for elephants) for every Mission Circle. The entire numeric landscape (132 pages) is available for download at:

www.PlanBtotheMoon.space

We invite you to leave any comments or suggestions on our web page on how to improve Plan_B. Plan_B Version 2.0 is scheduled for publication in 2020.

Plan B to the Moon: A How-to Guide to Return to and Colonize the Lunar Surface

Phase	M Circle	Circle Action	Vehicles / Milestones / Notes	Cost FH Launch ($M)	Cost F9 Launch ($M)	Other Costs ($M)	HR Veh Costs ($M)	Veh Costs ($M)	Start Date	Nom Days	Ajust Days	End Date	L #	L Veh	FH Miss Cycle	F9 Miss Cycle	CCtCap lbs	F9 Cargo lbs	F9 lbs to LEO
		85.00%	**$22,015**	$3,962.0	$434.0	$5,212.2	$2,991.0	$9,415.8	Return to US ->	24	0	2024-11-02					149100	200900	350000
0	15	Phase 0: Approved Proceed	Plan B V1.x, Appr. to Proceed FY2020 [Phase 0]						2019-10-01	24	0	2019-10-25							
0	19	Launch FH, LEO Rendezvous	CSM+SPM-0+STM+XDock-00-CSCM-00 [Phase 0]	$130.0		$100.0			2021-07-21	3	0	2021-07-24	1	FH0	2021-07-21				
0	21	Launch F9, LEO Rendezvous	CCTCap-00 (6 COB) [Phase 0]		$62.0	$100.0	$213.0	$143.5	2021-07-25	3	0	2021-07-28	2	F90		2021-07-25	21300	28700	50000
0	27	Phase 0: Construct Comp.	Plan B, Exec. / Phase 0: Spacecraft Const Comp.			$100.0			2021-08-13	366	0	2022-08-14							
0	28	Phase 0: Testing Complete	Plan B, Exec. / Phase 0: Test & Launch Int. Comp.			$100.0			2022-08-14	109	0	2022-12-01							
0	35	PM Master Schedule Reserve	Plan B V1.x, Execution			$100.0			2022-12-01		26	2022-12-27							
0	99	Phase-00 Complete	Plan B V1.x, Execution **$2,652**						2022-12-27			2022-12-27							
1A	101	Launch FH, LEO Rendezvous	SPM-1+XDock-01+CSCM-01 [Phase 01A]	$130.0		$100.0	$106.5	$596.8	2022-12-27	3	16	2023-01-15	3	FH1	2022-12-27				
1A	102	Maneuver, Park LEO A	SPM-1+XDock-01+CSCM-01						2023-01-15	1	0	2023-01-16							
1A	103	Launch F9, LEO Rendezvous	CCTCap-01 (6 COB) [Phase 01A]		$62.0	$100.0	$213.0	$143.5	2023-01-16	3	0	2023-01-19	4	F91		2023-01-16	21300	28700	50000
1A	104	Dock	CCTCap-01:XDock-01F						2023-01-19	1	0	2023-01-20							
1A	105	Launch FH, LEO Rendezvous	CTugRV-01(A) [Phase 01A]	$130.0		$100.0	$106.5	$219.8	2023-01-20	3	0	2023-01-23	5	FH2	2023-01-20				
1A	106	Dock	CTugRV-01:XDock-01D						2023-01-23	1	0	2023-01-24							
1A	107	LEOS-01 Assembly	LEOS-01 Assembly						2023-01-24	7	0	2023-01-31							
1A	108	Undock	CTugRV-01:XDock-01D (2 COB)						2023-01-31	0	0	2023-01-31							
1A	109	Maneuver	CTugRV-01						2023-01-31	0	0	2023-01-31							
1A	110	Dock	CTugRV-01:XDock-01E						2023-01-31	0	0	2023-01-31							
1A	111	Undock	CTugRV-01:XDock-01E						2023-01-31	0	0	2023-01-31							
1A	112	Maneuver	CTugRV-01						2023-01-31	0	0	2023-01-31							
1A	113	Dock	CTugRV-01:XDock-01F						2023-01-31	0	0	2023-01-31							
1A	114	Undock	CTugRV-01:XDock-01F						2023-01-31	0	0	2023-01-31							
1A	115	Maneuver	CTugRV-01 / Test Docking Ports						2023-01-31	0	0	2023-01-31							
1A	116	Dock	CTugRV-01:XDock-01C						2023-01-31	0	0	2023-01-31							
1A	117	Undock	CTugRV-01:XDock-01C						2023-01-31	0	0	2023-01-31							
1A	118	Maneuver	CTugRV-01						2023-01-31	0	0	2023-01-31							
1A	119	Attach	CSCM-01:CSCM-01						2023-01-31	0	0	2023-01-31							
1A	120	Undock	CSCM-01:XDock-01B						2023-01-31	0	0	2023-01-31							
1A	121	Maneuver	CTugRV-01+CSCM-01						2023-01-31	0	0	2023-01-31							
1A	122	Dock	CSCM-01+CTugRV-01:XDock-01E						2023-01-31	0	0	2023-01-31							
1A	123	Unattach	CTugRV-01						2023-01-31	0	0	2023-01-31							
1A	124	Maneuver	CTugRV-01						2023-01-31	0	0	2023-01-31							
1A	125	Dock	CTugRV-01:XDock-01B						2023-01-31	0	0	2023-01-31							
1A	126	LEOS-01 Assembly	LEOS-01 Assembly						2023-01-31	7	6	2023-02-13							
1A	127	Launch FH, LEO Rendezvous	SPM-02+XDock-02 [Phase 01A]	$130.0		$100.0	$0.0	$650.0	2023-02-13	3	0	2023-02-16	6	FH3	2023-02-13				
1A	129	Undock	CTugRV-01:XDock-01B						2023-02-16	0	0	2023-02-16							
1A	130	Maneuver	CTugRV-01						2023-02-16	0	0	2023-02-16							
1A	131	Dock	CTugRV-01:XDock-02B						2023-02-16	0	0	2023-02-16							
1A	132	Dock	CTugRV-01+SPM-02:SPM-01						2023-02-16	0	0	2023-02-16							
1A	133	LEOS-01 Assembly	LEOS-01 Assembly						2023-02-16	7	14	2023-03-09							
1A	134	Launch FH, LEO Rendezvous	SPM-03+ 4STM (Two STM Full) [Phase 01A]	$130.0		$100.0	$0.0	$137.5	2023-03-09	3	0	2023-03-12	7	FH4	2023-03-09				
1A	135	Undock	CTugRV-01:XDock-02B						2023-03-12	0	0	2023-03-12							
1A	136	Maneuver	CTugRV-01						2023-03-12	0	0	2023-03-12							
1A	137	Dock	CTugRV-01:SPM-03+ 4STM						2023-03-12	0	0	2023-03-12							
1A	138	Dock	SPM-03+ 4STM:XDock-02B						2023-03-12	0	0	2023-03-12							
1A	139	LEOS-01 Assembly	LEOS-01 Assembly						2023-03-12	7	9	2023-03-28							
1A	140	Undock	CTugRV-01+SPM-03+4STM+XDock-02						2023-03-28	0	0	2023-03-28							
1A	141	Maneuver	CTugRV-01+SPM-03+4STM+XDock-02 / Pirouette						2023-03-28	0	0	2023-03-28							
1A	142	Undock	CTugRV-01:SPM-03+4STM+XDock-02						2023-03-28	0	0	2023-03-28							
1A	143	Maneuver	CTugRV-01						2023-03-28	0	0	2023-03-28							
1A	144	Dock	CTugRV-01:XDock-02B						2023-03-28	0	0	2023-03-28							

Plan B to the Moon: A How-to Guide to Return to and Colonize the Lunar Surface

Phase	M Circle	Circle Action	Vehicles / Milestones / Notes	Cost FH Launch ($M)	Cost F9 Launch ($M)	Other Costs ($M)	HR Veh Costs ($M)	Veh Costs ($M)	Start Date	Nom Days	Ajust Days	End Date	L #	L Veh	FH Miss Cycle	F9 Miss Cycle	CCICap lbs	F9 Cargo lbs	F9 lbs to LEO
1A	145	Maneuver	CTugRV-01+XDock-02+4STM+5PM3						2023-03-28	0		2023-03-28							
1A	146	Dock	CTugRV-01+XDock-02+4STM+5PM3:5PM-02						2023-03-28	0		2023-03-28							
1A	147	Undock	CTugRV-01:XDock-02						2023-03-28	0		2023-03-28							
1A	148	Maneuver	CTugRV-01						2023-03-28	0		2023-03-28							
1A	149	Dock	CTugRV-01:XDock-01						2023-03-28	0		2023-03-28							
1A	150	Partial H2O Load	LEOS-01 H2O Transfer to Shielding Tanks						2023-03-28	0		2023-03-28							
1A	151	LEOS-01 Ops Normal	LEOS-01 Ops Normal						2023-03-28	1	0	2023-03-29							
1B	152	Launch F9, LEO Rendezvous	CCTCap-02 (6 COB) [Phase 01B]		$62.0	$100.0	$213.0	$143.5	2023-03-29	3	0	2023-04-01	8	F92		2023-02-13	21300	28700	50000
1B	153	Dock	CCTCap-02:XDock-01D						2023-04-01	1	0	2023-04-02							
1B	154	LEOS-01 Ops Normal	LEOS-01 Ops Normal						2023-04-02	0	0	2023-04-02							
1B	155	Undock	CCTCap-01R:XDock-01F						2023-04-02	0	0	2023-04-02							
1B	156	Return to Earth	CCTCap-01R (6 COB)						2023-04-02	0	0	2023-04-02							
1B	157	LEOS-01 Ops Normal	LEOS-01 Ops Normal						2023-04-02	0	0	2023-04-02							
1B	158	Launch FH, LEO Rendezvous	ASWTV-01(A) [Phase 01B]	$130.0		$100.0	$0.0	$347.4	2023-04-02	3	0	2023-04-05	9	FH1	2023-04-02				
1B	159	SAR-H2O	ASWTV-01(A) / LEOS-01 / H2O for Shielding						2023-04-05	0		2023-04-05							
1B	160	H2O State Change	ASWTV-01(A)						2023-04-05	0		2023-04-05							
1B	161	H2O State Change	LEOS-01						2023-04-05	0		2023-04-05							
1B	162	Maneuver	ASWTV-01(A)						2023-04-05	0		2023-04-05							
1B	163	Dock	ASWTV-01(A):XDock-02F						2023-04-05	1		2023-04-06							
1B	164	Partial H2O Load	LEOS-01 H2O Transfer to Shielding Tanks						2023-04-06	0		2023-04-06							
1B	165	LEOS-01 Ops Normal	LEOS-01 Ops Normal						2023-04-06	1	19	2023-04-26							
1B	166	Launch FH, LEO Rendezvous	ASWTV-02(A) [Phase 01B]	$130.0		$100.0	$0.0	$347.4	2023-04-26	3	0	2023-04-29	10	FH2	2023-04-26				
1B	167	SAR-H2O	ASWTV-02(A) / LEOS-01 / H2O for Shielding						2023-04-29	0		2023-04-29							
1B	168	H2O State Change	ASWTV-02(A)						2023-04-29	0		2023-04-29							
1B	169	H2O State Change	LEOS-01						2023-04-29	0		2023-04-29							
1B	170	Maneuver	ASWTV-02(A)						2023-04-29	0		2023-04-29							
1B	171	Dock	ASWTV-02(A):XDock-02D						2023-04-29	0		2023-04-29							
1B	172	SAR-H2O	ASWTV-02(A) / LEOS-01 / H2O for Shielding						2023-04-29	0		2023-04-29							
1B	173	LEOS-01 Ops Normal	LEOS-01 Ops Normal						2023-04-29	1	20	2023-05-20							
1B	174	Launch FH, LEO Rendezvous	ASWTV-03(A) [Phase 01B]	$130.0		$100.0	$0.0	$347.4	2023-05-20	3	0	2023-05-23	11	FH3	2023-05-20				
1B	175	SAR-H2O	ASWTV-03(A) / LEOS-01 / H2O for Shielding						2023-05-23	0		2023-05-23							
1B	176	H2O State Change	ASWTV-03(A)						2023-05-23	0		2023-05-23							
1B	177	H2O State Change	LEOS-01						2023-05-23	0		2023-05-23							
1B	178	Maneuver	ASWTV-03(A)						2023-05-23	1		2023-05-24							
1B	179	Maneuver, Park LEO D	LEOS-01 H2O Shielding						2023-05-24	1		2023-05-25							
1B	180	LEOS-01 H2O Shielding	LEOS-01 H2O Shielding / Complete (297k lbs)						2023-05-25	0		2023-05-25							
1B	181	LEOS-01 Ops Normal	LEOS-01 Ops Normal						2023-05-25	1	14	2023-06-09							
1B	182	Launch F9, LEO Rendezvous	CCTCap-03 (6 COB) [Phase 01B]		$62.0	$100.0	$213.0	$143.5	2023-06-09	3	0	2023-06-12	12	F93		2023-03-13	21300	28700	50000
1B	183	Dock	CCTCap-03:XDock-01F						2023-06-12	0		2023-06-12							
1B	184	LEOS-01 Ops Normal	LEOS-01 Ops Normal						2023-06-12	1	0	2023-06-13							
1B	185	Launch FH, LEO Rendezvous	CTugRV-02(A) [Phase 01B]	$130.0		$100.0	$106.5	$219.8	2023-06-13	3	0	2023-06-16	13	FH4	2023-06-13				
1B	186	Dock	CTugRV-02(A):XDock-02B						2023-06-16	1	20	2023-07-07							
1B	187	Launch FH, LEO Rendezvous	CTugRV-03(A) [Phase 01B]	$130.0		$100.0	$106.5	$219.8	2023-07-07	3	0	2023-07-10	14	FH1	2023-07-07				
1B	188	SAR-CH4O2	CTugRV-03(A):CTugRV-01 / Top Off						2023-07-10	0		2023-07-10							
1B	189	Fuel State Change	CTugRV-03(A) Decrease						2023-07-10	0		2023-07-10							
1B	190	Fuel State Change	CTugRV-01 Fill						2023-07-10	0		2023-07-10							
1B	191	Maneuver	CTugRV-03(A)						2023-07-10	0		2023-07-10							
1B	192	SAR-CH4O2	CTugRV-03(A):CTugRV-02 / Partial Load						2023-07-10	0		2023-07-10							
1B	193	Fuel State Change	CTugRV-02 Empty						2023-07-10	0		2023-07-10							
1B	194	Fuel State Change	CTugRV-01 Fill						2023-07-10	0		2023-07-10							
1B	195	Maneuver	CTugRV-03(A)						2023-07-10	0		2023-07-10							
1B	196	Dock	CTugRV-03(A):XDock-02C						2023-07-10	0		2023-07-10							

Plan B to the Moon: A How-to Guide to Return to and Colonize the Lunar Surface

Mission Matrix Data Sheets

Phase	M Circle	Circle Action	Vehicles / Milestones / Notes	Cost FH Launch ($M)	Cost F9 Launch ($M)	Other Costs ($M)	HR Veh Costs ($M)	Veh Costs ($M)	Start Date	Nom Days	Ajust Days	End Date	L #	L Veh	FH Miss Cycle	F9 Miss Cycle	CDCargo lbs	F9 Cargo lbs	F9 lbs to LEO
1B	198	LEOS-01 Ops Normal	LEOS-01 Ops Normal						2023-07-10	1	20	2023-07-31							
1B	199	Phase-01 Complete	$7,199						2023-07-31	0	0	2023-07-31							
2A	201	Launch FH, LEO Rendezvous	2STM+SPM-04+CSCM-02 [Phase 02A]	$130.0		$100.0	$106.5	$340.5	2023-07-31	3	0	2023-08-03	15	FH2	2023-07-31				
2A	202	Undock	CTugRV-02:XDock-02B						2023-08-03	0	0	2023-08-03							
2A	204	Maneuver	CTugRV-02						2023-08-03	0	0	2023-08-03							
2A	205	Dock	CTugRV-02:2STM+SPM-04+CSCM-02						2023-08-03	0	0	2023-08-03							
2A	206	Maneuver	CTugRV-02+2STM+SPM-04+CSCM-02						2023-08-03	0	0	2023-08-03							
2A	207	Dock	CTugRV-02+2STM+SPM-04+CSCM-02:XDock-02E						2023-08-03	0	0	2023-08-03							
2A	208	Unattach	CTugRV-02+2STM+SPM-04:CSCM-02						2023-08-03	0	0	2023-08-03							
2A	209	Maneuver	CTugRV-02+2STM+SPM-04						2023-08-03	0	0	2023-08-03							
2A	210	Dock	CTugRV-02+2STM+SPM-04:XDock-02B						2023-08-03	0	0	2023-08-03							
2A	211	Undock	CTugRV-02:2STM+SPM-04						2023-08-03	0	0	2023-08-03							
2A	212	Maneuver	CTugRV-02						2023-08-03	0	0	2023-08-03							
2A	213	Dock	CTugRV-02:XDock-01D						2023-08-03	0	0	2023-08-03							
2A	214	SFT-01 Assembly	SFT-01 Assembly						2023-08-03	21	0	2023-08-24							
2A	215	Launch FH, LEO Rendezvous	SPM-05+3STM [Phase 02A]	$130.0		$100.0	$0.0	$137.5	2023-08-24	3	0	2023-08-27	16	FH3	2023-08-24				
2A	216	Undock	CTugRV-02:XDock-01D						2023-08-27	0	0	2023-08-27							
2A	218	Maneuver	CTugRV-02						2023-08-27	0	0	2023-08-27							
2A	219	Dock	CTugRV-02:SPM-05+3STM						2023-08-27	0	0	2023-08-27							
2A	220	Maneuver	CTugRV-02+SPM-05+3STM						2023-08-27	0	0	2023-08-27							
2A	221	Dock	CTugRV-02+SPM-05+3STM:STM						2023-08-27	0	0	2023-08-27							
2A	222	Undock	SPM-04:XDock-02B						2023-08-27	0	0	2023-08-27							
2A	223	Undock	SPM-05-STM						2023-08-27	0	0	2023-08-27							
2A	224	Maneuver	CTugRV-02+SPM-05						2023-08-27	0	0	2023-08-27							
2A	225	Dock	SPM-04:SPM-05						2023-08-27	0	0	2023-08-27							
2A	226	Maneuver	SPM-04+5STM / Pirouette Flip						2023-08-27	0	0	2023-08-27							
2A	227	Dock	SPM-05+SPM-04+5STM:XDock-02B						2023-08-27	0	0	2023-08-27							
2A	228	Undock	CTugRV-02:SPM-05						2023-08-27	0	0	2023-08-27							
2A	229	Maneuver	CTugRV-02						2023-08-27	0	0	2023-08-27							
2A	230	Dock	CTugRV-02:XDock-01D						2023-08-27	0	0	2023-08-27							
2A	231	SFT-01 Assembly	SFT-01 Assembly						2023-08-27	21	0	2023-09-17							
2A	232	Launch FH, LEO Rendezvous	SPM-06+2CSM (Preloaded Fuel) [Phase 02A]	$130.0		$100.0	$0.0	$316.0	2023-09-17	3	0	2023-09-20	17	FH4	2023-09-17				
2A	233	Undock	CTugRV-01:XDock-01B						2023-09-20	0	0	2023-09-20							
2A	235	Maneuver	CTugRV-01						2023-09-20	0	0	2023-09-20							
2A	236	Dock	CTugRV-01:CSM+CSM+SPM-06						2023-09-20	0	0	2023-09-20							
2A	237	Maneuver	2CSM+SPM-06:SPM-05						2023-09-20	0	0	2023-09-20							
2A	238	Undock	CTugRV-01						2023-09-20	0	0	2023-09-20							
2A	239	Maneuver	CTugRV-01						2023-09-20	0	0	2023-09-20							
2A	240	Dock	CTugRV-01						2023-09-20	1	20	2023-10-11							
2A	241	Launch FH, LEO Rendezvous	XDock-03+SPM-07+3CSM [Phase 02A]	$130.0		$100.0	$0.0	$316.0	2023-10-11	3	0	2023-10-14	18	FH1	2023-10-11				
2A	243	Undock	CTugRV-01:XDock-01B						2023-10-14	0	0	2023-10-14							
2A	245	Maneuver	CTugRV-01						2023-10-14	0	0	2023-10-14							
2A	246	Dock	CTugRV-01:XDock-03+SPM-07+3xCSM						2023-10-14	0	0	2023-10-14							
2A	247	Maneuver	CTugRV-01+XDock-03+SPM-07+3CSM						2023-10-14	0	0	2023-10-14							
2A	248	Dock	CTugRV-01+XDock-03+SPM-07+3CSM:CSM+...						2023-10-14	0	0	2023-10-14							
2A	249	Undock	CSM:SPM-06						2023-10-14	0	0	2023-10-14							
2A	250	Undock	XDock-03A:SPM-07						2023-10-14	0	0	2023-10-14							
2A	251	Maneuver	CTugRV-01+XDock-03 / Pirouette Flip						2023-10-14	0	0	2023-10-14							
2A	252	Dock	XDock-03B:SPM-07						2023-10-14	0	0	2023-10-14							
2A	253	Dock	SPM-07:SPM-06						2023-10-14	0	0	2023-10-14							
2A	254	Dock	CTugRV-01:XDock-03B						2023-10-14	0	0	2023-10-14							
2A	255	Maneuver	CTugRV-01						2023-10-14	0	0	2023-10-14							

Plan B to the Moon: A How-to Guide to Return to and Colonize the Lunar Surface

Phase	M Circle	Circle Action	Vehicles / Milestones / Notes	Cost FH Launch ($M)	Cost F9 Launch ($M)	Other Costs ($M)	HR Veh Costs ($M)	Veh Costs ($M)	Start Date	Nom Days	Ajust Days	End Date	L # Veh	FH Miss Cycle	F9 Miss Cycle
2A	256	Dock	CTugRV-01:XDock-02A						2023-10-14			2023-10-14			
2B	257	SFT-01 Assembly	SFT-01 / Phase 02B: SFT-01 Assembly Continue						2023-10-14	7	0	2023-10-21			
2B	258	SFT-01 Array Assemble	SFT-01						2023-10-21	7	0	2023-10-28			
2B	259	Undock	CTugRV-02						2023-10-28	0	0	2023-10-28			
2B	260	Maneuver	CTugRV-02						2023-10-28	0	0	2023-10-28			
2B	261	Dock	CTugRV-02:XDock-03B						2023-10-28	0	0	2023-10-28			
2B	262	Undock	XDock-02A / CTugRV-01 & CTugRV-02 Counter Pull						2023-10-28	0	0	2023-10-28			
2B	263	Undock	CTugRV-01						2023-10-28	0	0	2023-10-28			
2B	264	Maneuver	CTugRV-01						2023-10-28	0	0	2023-10-28			
2B	265	Dock	CTugRV-01:XDock-02A (SFT-01B)						2023-10-28	0	0	2023-10-28			
2B	266	Undock	SPM-04:STM (SFT-01B)						2023-10-28	0	0	2023-10-28			
2B	267	Maneuver	CTugRV-01+SFT-01A / Move SFT-01A to LEO Bravo						2023-10-28	0	0	2023-10-28			
2B	268	Maneuver	CTugRV-02+SFT-01B / Move SFT-01B to LEO Bravo						2023-10-28	0	0	2023-10-28			
2B	269	Dock	SFT-01A:SFT-01B						2023-10-28	0	0	2023-10-28			
2B	270	SFT-01 Checkout	SFT-01						2023-10-28	7	0	2023-11-04			
2B	271	Undock	CTugRV-01						2023-11-04	0	0	2023-11-04			
2B	272	Maneuver	CTugRV-01						2023-11-04	0	0	2023-11-04			
2B	273	Dock	CTugRV-01:XDock-01B						2023-11-04	0	0	2023-11-04			
2B	274	Undock	CTugRV-02						2023-11-04	0	0	2023-11-04			
2B	275	Maneuver	CTugRV-02						2023-11-04	0	0	2023-11-04			
2B	276	Dock	CTugRV-02:XDock-01D						2023-11-04	0	0	2023-11-04			
2B	277	Launch FH, LEO Rendezvous	Max Water Cargo WTV [Phase 02B]	$130.0		$100.0	$0.0	$56.3	2023-11-04	3	0	2023-11-07	19	FH2	2023-11-04
2B	278	SAR-H2O	Water Cargo Mission						2023-11-07	0	0	2023-11-07			
2B	279	H2O State Change	Water Cargo Mission						2023-11-07	0	0	2023-11-07			
2B	280	H2O State Change	SFT-01						2023-11-07	0	0	2023-11-07			
2B	281	Maneuver	Water Cargo Mission						2023-11-07	0	0	2023-11-07			
2B	282	Park	On-Orbit Spare						2023-11-07	1	20	2023-11-28			
2B	299	Phase-02 Complete	$2,423						2023-11-28			2023-11-28			
3A	301	Launch FH, LEO Rendezvous	SPM-08+STM+CSCM-03+XDock-04 ELTV-01	$130.0		$100.0	$106.5	$340.5	2023-11-28	3	0	2023-12-01	20	FH3	2023-11-28
3A	302	Undock	CTugRV-02						2023-12-01	0	0	2023-12-01			
3A	303	Maneuver	CTugRV-02						2023-12-01	0	0	2023-12-01			
3A	304	Maneuver	CTugRV-02						2023-12-01	0	0	2023-12-01			
3A	305	Maneuver	CTugRV-02+STM+SPM-08+CSCM-03+XDock-04						2023-12-01	0	0	2023-12-01			
3A	306	Dock	CTugRV-02+...+XDock-04:LEOS-01						2023-12-01	0	0	2023-12-01			
3A	307	Undock	CTugRV-02+STM+SPM-08+CSCM-03:XDock-04B						2023-12-01	0	0	2023-12-01			
3A	308	Maneuver	CTugRV-02+STM+SPM-08+CSCM-03						2023-12-01	0	0	2023-12-01			
3A	309	Dock	CTugRV-02+STM+SPM-08+CSCM-03:XDock-04E						2023-12-01	0	0	2023-12-01			
3A	310	Unattach	CTugRV-02+STM+SPM-08:CSCM-03						2023-12-01	0	0	2023-12-01			
3A	311	Maneuver	CTugRV-02+STM+SPM-08						2023-12-01	0	0	2023-12-01			
3A	312	Undock	CTugRV-02:STM+SPM-08						2023-12-01	0	0	2023-12-01			
3A	313	Maneuver	CTugRV-02						2023-12-01	0	0	2023-12-01			
3A	314	Dock	CTugRV-02:XDock-01D						2023-12-01	0	0	2023-12-01			
3A	315	Undock	ASWTV-02(A)						2023-12-01	0	0	2023-12-01			
3A	316	Maneuver	ASWTV-02(A)						2023-12-01	0	0	2023-12-01			
3A	317	SAR-H2O	ASWTV-02(A):STM+SPM-08						2023-12-01	0	0	2023-12-01			
3A	318	H2O State Change	ASWTV-02(A) Fill						2023-12-01	0	0	2023-12-01			
3A	319	H2O State Change	STM Empty						2023-12-01	0	0	2023-12-01			
3A	320	Maneuver, Park LEO A	ASWTV-02(A)						2023-12-01	0	0	2023-12-01			
3A	321	ELTV-01 Assembly	ELTV-01 Assembly						2023-12-01	21	0	2023-12-22			
3A	322	LEOS-01 Ops Normal	LEOS-01 Ops Normal						2023-12-22	0	0	2023-12-22			
3A	323	Launch FH, LEO Rendezvous	SPM-09+STM ELTV-01 [Phase 03A]	$130.0		$100.0	$0.0	$393.8	2023-12-22	3	0	2023-12-25	21	FH4	2023-12-22
3A	324	Undock	CTugRV-02:XDock-01D						2023-12-25	0	0	2023-12-25			

Plan B to the Moon: A How-to Guide to Return to and Colonize the Lunar Surface

Phase	M Circle	Circle Action	Vehicles / Milestones / Notes	Cost FH Launch ($M)	Cost F9 Launch ($M)	Other Costs ($M)	HR Veh Costs ($M)	Veh Costs ($M)	Start Date	Nom Days	Ajust Days	End Date	L #	L Veh	FH Miss Cycle	F9 Miss Cycle
3A	325	Maneuver	CTugRV-02						2023-12-25	0	0	2023-12-25				
3A	326	Dock	CTugRV-02:SPM-09+STM						2023-12-25	0	0	2023-12-25				
3A	327	Maneuver	CTugRV-02+SPM-09+STM						2023-12-25	0	0	2023-12-25				
3A	328	Maneuver	ASWTV-02(A)						2023-12-25	0	0	2023-12-25				
3A	329	SAR-H2O	ASWTV-02(A):STM+SPM-09						2023-12-25	0	0	2023-12-25				
3A	330	H20 State Change	ASWTV-02(A) Fill						2023-12-25	0	0	2023-12-25				
3A	331	H20 State Change	STM Empty						2023-12-25	0	0	2023-12-25				
3A	332	Maneuver	ASWTV-02(A)						2023-12-25	0	0	2023-12-25				
3A	333	SAR-H2O	ASWTV-02(A):SFT-01						2023-12-25	0	0	2023-12-25				
3A	334	H20 State Change	SFT-01 Fill						2023-12-25	0	0	2023-12-25				
3A	335	H20 State Change	ASWTV-02(A) Empty						2023-12-25	0	0	2023-12-25				
3A	336	Maneuver	ASWTV-02(A)						2023-12-25	0	0	2023-12-25				
3A	337	Dock	ASWTV-02(A):XDock-02D						2023-12-25	0	0	2023-12-25				
3A	338	Maneuver	CTugRV-02+SPM-09+STM						2023-12-25	0	0	2023-12-25				
3A	339	Dock	CTugRV-02+SPM-09+STM:STM+SPM-08+XDock-04						2023-12-25	0	0	2023-12-25				
3A	340	Undock	CTugRV-02+SPM-09+2STM+SPM-08:XDock-04B						2023-12-25	0	0	2023-12-25				
3A	341	Maneuver	CTugRV-02+SPM-09-2STM						2023-12-25	0	0	2023-12-25				
3A	342	Maneuver	CTugRV-02+SPM-09						2023-12-25	0	0	2023-12-25				
3A	343	Dock	CTugRV-02+SPM-09:SPM-08+2STM						2023-12-25	0	0	2023-12-25				
3A	344	Dock	CTugRV-02+SPM-09+SPM-08+2STM:XDock-04B						2023-12-25	0	0	2023-12-25				
3A	345	ELTV-01 Assembly	ELTV-01 Assembly						2023-12-25	14	7	2024-01-15				
3A	346	ELTV-01 Complete	ELTV-01 Complete						2024-01-15			2024-01-15				
3A	347	Launch FH, LEO Rendezvous	Cryo Fuel Cargo CRV-04 [Phase 03B]	$130.0		$100.0	$0.0	$95.0	2024-01-15	3	0	2024-01-18	22	FH1	2024-01-15	
3A	348	Undock	CTugRV-01:XDock-01D						2024-01-18	0	0	2024-01-18				
3A	349	Maneuver	CTugRV-01:CTugRV-04(A)						2024-01-18	0	0	2024-01-18				
3A	350	SAR-CH4O2	CRV-04(A):CTugRV-01						2024-01-18	0	0	2024-01-18				
3A	351	Fuel State Change	CTugRV-01 Fill						2024-01-18	0	0	2024-01-18				
3A	352	Fuel State Change	CTugRV-04(A) Empty						2024-01-18	0	0	2024-01-18				
3A	353	Maneuver	RV-04(A)						2024-01-18	0	0	2024-01-18				
3A	354	Dock	CRV-04(A):XDock-03E						2024-01-18	1	20	2024-02-08				
3A	355	Launch FH, LEO Rendezvous	Cryo Fuel Cargo CRV-05 [Phase 03B]	$130.0		$100.0	$0.0	$95.0	2024-02-08	3	0	2024-02-11	23	FH2	2024-02-08	
3A	356	Undock	CTugRV-01:XDock-01D						2024-02-11	0	0	2024-02-11				
3A	357	Maneuver	CTugRV-01:RV-05(A)						2024-02-11	0	0	2024-02-11				
3A	358	SAR-CH4O2	CRV-05(A):CTugRV-01						2024-02-11	0	0	2024-02-11				
3A	359	Fuel State Change	CTugRV-01 Fill						2024-02-11	0	0	2024-02-11				
3A	360	Fuel State Change	CTugRV-05(A) Empty						2024-02-11	0	0	2024-02-11				
3A	361	Maneuver	CRV-05(A)						2024-02-11	0	0	2024-02-11				
3A	362	Dock	CRV-05(A):XDock-03E						2024-02-11	0	0	2024-02-11				
3A	363	SFT-01 Fuel Ready	SFT-01 Fuel Ready						2024-02-11	0	0	2024-02-11				
3A	364	Undock	CTugRV-03(A)						2024-02-11	0	0	2024-02-11				
3A	365	Maneuver	CTugRV-03(A)						2024-02-11	0	0	2024-02-11				
3A	366	SAR-CH4O2	CTugRV-03(A):SFT-01						2024-02-11	0	0	2024-02-11				
3A	367	Fuel State Change	CTugRV-03(A) Fill						2024-02-11	0	0	2024-02-11				
3A	368	Fuel State Change	SFT-01 Empty						2024-02-11	0	0	2024-02-11				
3A	369	Maneuver	CTugRV-03(A)						2024-02-11	0	0	2024-02-11				
3A	370	Undock	CTugRV-02+...+XDock-04:LEOS-01						2024-02-11	0	0	2024-02-11				
3A	371	Maneuver	CTugRV-02+SPM-09+SPM-08+2STM+XDock-04						2024-02-11	0	0	2024-02-11				
3A	372	SAR-CH4O2	CTugRV-03(A):CTugRV-02						2024-02-11	0	0	2024-02-11				
3A	373	Fuel State Change	CTugRV-02 Fill						2024-02-11	0	0	2024-02-11				
3A	374	Fuel State Change	CTugRV-03(A) Empty						2024-02-11	0	0	2024-02-11				
3A	375	Maneuver	CTugRV-03(A)						2024-02-11	0	0	2024-02-11				
3A	376	Dock	CTugRV-03(A):XDock-02C						2024-02-11	1	0	2024-02-12				
3A	377	LEOS-01 Ops Normal	LEOS-01 Ops Normal													

Plan B to the Moon: A How-to Guide to Return to and Colonize the Lunar Surface

Phase	M Circle	Circle Action	Vehicles / Milestones / Notes	Cost FH Launch ($M)	Cost F9 Launch ($M)	Other Costs ($M)	HR Veh Costs ($M)	Veh Costs ($M)	Start Date	Nom Days	Ajust Days	End Date	L #	L Veh	FH Miss Cycle	F9 Miss Cycle	CCTCap lbs	F9 Cargo lbs	F9 lbs to LEO
3A	378	LLO Maneuver / Depart	ELTV-01 Depart for LLO						2024-02-12	3	0	2024-02-15							
3A	379	LLO Maneuver / Depart	CTugRV-01						2024-02-15	3	0	2024-02-18							
3A	380	Phase-03A Complete	Phase-03A Complete (SFT-01 FOC)						2024-02-18			2024-02-18							
3B	381	LLO Arrive, Park	CTugRV-02: ELTV-01 / Humans Return to LLO						2024-02-18	1	1	2024-02-19							
3B	382	LLO Arrive, Rendezvous	CTugRV-01						2024-02-19	3	0	2024-02-22							
3B	384	Maneuver	CTugRV-01						2024-02-22	0	0	2024-02-22							
3B	385	Dock	LLOG-01						2024-02-22	0	0	2024-02-22							
3B	386	LLOG-01A Assembly	LLOG-01A Assembly						2024-02-22	0	0	2024-02-22							
3B	387	Undock	CTugRV-02:LLOG-01						2024-02-22	0	0	2024-02-22							
3B	388	Maneuver, Return LEOS-01	CTugRV-02						2024-02-22	3	0	2024-02-25							
3B	389	LLOG-01 Ops Normal	LLOG-01 Ops Normal						2024-02-25	0	0	2024-02-25							
3B	390	LEO Arrive	CTugRV-02						2024-02-25	3	0	2024-02-28							
3B	391	Maneuver	CTugRV-02						2024-02-28	0	0	2024-02-28							
3B	392	Dock	CTugRV-02:LEOS-01 XDock-01D						2024-02-28	1	0	2024-02-29							
3B	393	LEOS-01 Ops Normal	LEOS-01 Ops Normal						2024-02-29	3	0	2024-03-03							
3B	394	Launch F9, LEO Rendezvous	CCTCap-04 (6 COB) [Phase 03B]		$62.0	$100.0	$213.0	$143.5	2024-02-29	3	0	2024-03-03	24	F94		2023-04-10	21300	28700	50000
3B	395	Dock	CCTCap-04:LEOS-01 XDock-01B						2024-03-03	0	0	2024-03-03							
3B	396	LLOG-01 Ops Normal	LLOG-01 Ops Normal						2024-03-03	0	0	2024-03-03							
3B	397	Undock	CCTCap-02R:LEOS-01 XDock-01B						2024-03-03	0	0	2024-03-03							
3B	398	Return to Earth	CCTCap-02R (6 COB)						2024-03-03	0	0	2024-03-03							
3B	399	Phase-03B Complete	Phase-03B Complete					$2,469	2024-03-03			2024-03-03							
4A	401	Launch FH, LEO Rendezvous	SPM-10+STM+CSCM-04+XDock-05 ELTV-02	$130.0		$100.0	$106.5	$340.5	2024-03-03	3	0	2024-03-06	25	FH3	2024-03-03				
4A	402	Undock	CTugRV-02:LEOS-01 XDock-01D						2024-03-06	0	0	2024-03-06							
4A	403	Maneuver	CTugRV-02						2024-03-06	0	0	2024-03-06							
4A	404	Dock	CTugRV-02:SPM-10+STM+CSCM-04+XDock-05						2024-03-06	0	0	2024-03-06							
4A	405	Maneuver	CTugRV-02+STM+SPM-10+CSCM-04+XDock-05						2024-03-06	0	0	2024-03-06							
4A	406	Dock	CTugRV-02+...+XDock-05:LEOS-01						2024-03-06	0	0	2024-03-06							
4A	407	Undock	CTugRV-02+STM+SPM-10+CSCM-04:XDock-05B						2024-03-06	0	0	2024-03-06							
4A	408	Maneuver	CTugRV-02+STM+SPM-10+CSCM-04						2024-03-06	0	0	2024-03-06							
4A	409	Dock	CTugRV-02+STM+SPM-10+CSCM-04:XDock-05E						2024-03-06	0	0	2024-03-06							
4A	410	Unattach	CTugRV-02+STM+SPM-10:CSCM-04						2024-03-06	0	0	2024-03-06							
4A	411	Maneuver	CTugRV-02+STM+SPM-10						2024-03-06	0	0	2024-03-06							
4A	412	Undock	CTugRV-02:STM+SPM-10						2024-03-06	0	0	2024-03-06							
4A	413	Maneuver	CTugRV-02						2024-03-06	0	0	2024-03-06							
4A	414	Dock	CTugRV-02:XDock-01D						2024-03-06	0	0	2024-03-06							
4A	415	Undock	ASWTV-02(A)						2024-03-06	0	0	2024-03-06							
4A	416	Maneuver	ASWTV-02(A)						2024-03-06	0	0	2024-03-06							
4A	417	SAR-H2O	ASWTV-02(A):STM+SPM-10						2024-03-06	0	0	2024-03-06							
4A	418	H2O State Change	ASWTV-02(A)						2024-03-06	0	0	2024-03-06							
4A	419	H2O State Change	STM+SPM-10						2024-03-06	0	0	2024-03-06							
4A	420	Maneuver, Park LEO A	ASWTV-02(A)						2024-03-06	0	0	2024-03-06							
4A	421	ELTV-02 Assembly	ELTV-02 Assembly						2024-03-06	14	6	2024-03-26							
4A	422	LEOS-01 Ops Normal	LEOS-01 Ops Normal						2024-03-26	1	0	2024-03-27							
4A	423	Launch FH, LEO Rendezvous	SPM-11+STM ELTV-02 [Phase 04A]	$130.0		$100.0	$0.0	$393.8	2024-03-27	3	0	2024-03-30	26	FH4	2024-03-27				
4A	424	Undock	CTugRV-02:XDock-01D						2024-03-30	0	0	2024-03-30							
4A	425	Maneuver	CTugRV-02						2024-03-30	0	0	2024-03-30							
4A	426	Dock	CTugRV-02:SPM-11+STM						2024-03-30	0	0	2024-03-30							
4A	427	Maneuver	CTugRV-02+SPM-11+STM						2024-03-30	0	0	2024-03-30							
4A	428	Maneuver	ASWTV-02(A)						2024-03-30	0	0	2024-03-30							
4A	429	SAR-H2O	ASWTV-02(A):STM+SPM-11						2024-03-30	0	0	2024-03-30							
4A	430	H2O State Change	ASWTV-02(A)						2024-03-30	0	0	2024-03-30							
4A	431	H2O State Change	STM						2024-03-30	0	0	2024-03-30							
4A	432	Maneuver	ASWTV-02(A)						2024-03-30	0	0	2024-03-30							

Plan B to the Moon: A How-to Guide to Return to and Colonize the Lunar Surface

Mission Matrix Data Sheets

Plan B Version 1.01 Page 207

Phase	M Circle	Circle Action	Vehicles / Milestones / Notes	Cost FH Launch ($M)	Other Costs ($M)	HR Veh Costs ($M)	Veh Costs ($M)	Start Date	Norm Days	Ajust Days	End Date	L# Veh	FH Miss Cycle	FH Miss Cycle
4A	433	SAR-H2O	ASWTV-02(A):SFT-01					2024-03-30	0	0	2024-03-30			
4A	434	H2O State Change	SFT-01					2024-03-30	0	0	2024-03-30			
4A	435	H2O State Change	ASWTV-02(A)					2024-03-30	0	0	2024-03-30			
4A	436	Maneuver	ASWTV-02(A)					2024-03-30	0	0	2024-03-30			
4A	437	Dock	ASWTV-02(A):XDock-02D					2024-03-30	0	0	2024-03-30			
4A	438	Maneuver	CTugRV-02+SPM-11+STM					2024-03-30	0	0	2024-03-30			
4A	439	Dock	CTugRV-02+SPM-11+STM:STM+SPM-10+XDock-05					2024-03-30	0	0	2024-03-30			
4A	440	Undock	CTugRV-02+SPM-11+2STM+SPM-10:XDock-05B					2024-03-30	0	0	2024-03-30			
4A	441	Undock	CTugRV-02+SPM-11:2STM					2024-03-30	0	0	2024-03-30			
4A	442	Maneuver	CTugRV-02+SPM-11					2024-03-30	0	0	2024-03-30			
4A	443	Dock	CTugRV-02+SPM-11:SPM-10+2STM					2024-03-30	0	0	2024-03-30			
4A	444	Dock	CTugRV-02+SPM-11+SPM-10+2STM:XDock-05B					2024-03-30	0	0	2024-03-30			
4A	445	ELTV-02 Assembly	ELTV-02 Assembly					2024-03-30	14	7	2024-04-20			
4A	446	Launch FH, LEO Rendezvous	Cryo Fuel Cargo CRV-06 [Phase 04A]	$130.0	$100.0	$0.0	$95.0	2024-04-20	3	0	2024-04-23	27	FH1	2024-04-20
4A	447	Undock	CTugRV-03:XDock-03D					2024-04-23	0	0	2024-04-23			
4A	448	Maneuver	CTugRV-03:RV-06					2024-04-23	0	0	2024-04-23			
4A	449	SAR-CH4O2	CTugRV-03(A):RV-06					2024-04-23	0	0	2024-04-23			
4A	450	Fuel State Change	CRV-06 Empty					2024-04-23	0	0	2024-04-23			
4A	451	Fuel State Change	CTugRV-03 Fill					2024-04-23	0	0	2024-04-23			
4A	452	Maneuver / Park	CRV-06(A)					2024-04-23	0	0	2024-04-23			
4A	453	Maneuver	CTugRV-03					2024-04-23	0	0	2024-04-23			
4A	454	Dock	CTugRV-03(A):XDock-03C					2024-04-23	0	21	2024-05-14			
4A	455	Launch FH, LEO Rendezvous	Cryo Fuel Cargo CRV-07 [Phase 04A]	$130.0	$100.0	$0.0	$95.0	2024-05-14	3	0	2024-05-17	28	FH2	2024-05-14
4A	456	Undock	CTugRV-03					2024-05-17	0	0	2024-05-17			
4A	457	Maneuver	CTugRV-03:RV-07(A)					2024-05-17	0	0	2024-05-17			
4A	458	SAR-CH4O2	CRV-07(A):CTugRV-03					2024-05-17	0	0	2024-05-17			
4A	459	Fuel State Change	CRV-07(A) Empty					2024-05-17	0	0	2024-05-17			
4A	460	Fuel State Change	CTugRV-03 Fill					2024-05-17	0	0	2024-05-17			
4A	461	Maneuver / Park	CRV-07(A)					2024-05-17	0	0	2024-05-17			
4A	462	Maneuver	CTugRV-03					2024-05-17	0	0	2024-05-17			
4A	463	Dock	CTugRV-03(A):XDock-03C					2024-05-17	0	0	2024-05-17			
4A	464	SFT-01 Fuel Ready	SFT-01 Fuel Ready					2024-05-17	0	0	2024-05-17			
4A	465	Undock	CTugRV-03(A)					2024-05-17	0	0	2024-05-17			
4A	466	Maneuver	CTugRV-03(A)					2024-05-17	0	0	2024-05-17			
4A	467	SAR-CH4O2	CTugRV-03(A):SFT-01					2024-05-17	0	0	2024-05-17			
4A	468	Fuel State Change	CTugRV-03(A)					2024-05-17	0	0	2024-05-17			
4A	469	Fuel State Change	CTugRV-03(A)					2024-05-17	0	0	2024-05-17			
4A	470	Maneuver	CTugRV-03(A)					2024-05-17	0	0	2024-05-17			
4A	471	Dock	CTugRV-02+...+XDock-05:LEOS-01					2024-05-17	0	0	2024-05-17			
4A	472	Maneuver	CTugRV-02+SPM-11+SPM-10+2STM+XDock-05					2024-05-17	0	0	2024-05-17			
4A	473	SAR-CH4O2	CTugRV-03(A):CTugRV-02					2024-05-17	0	0	2024-05-17			
4A	474	Fuel State Change	CTugRV-02					2024-05-17	0	0	2024-05-17			
4A	475	Fuel State Change	CTugRV-03(A)					2024-05-17	0	0	2024-05-17			
4A	476	Maneuver	CTugRV-03(A)					2024-05-17	0	0	2024-05-17			
4A	477	Dock	CTugRV-03(A):XDock-02C					2024-05-17	0	0	2024-05-17			
4A	478	LEOS-01 Ops Normal	LEOS-01 Ops Normal					2024-05-17	1	0	2024-05-18			
4A	479	LLO Maneuver / Depart	ELTV-02 Depart for LLO					2024-05-18	3	0	2024-05-21			
4A	480	Phase-04A Complete	Phase-04A Complete					2024-05-21	0	0	2024-05-21			
4B	481	LLO Arrive, Rendezvous	CTugRV-02: ELTV-02 / Phase 04B:					2024-05-21	3	0	2024-05-24			
4B	482	Undock	CTugRV-02+SPM-11					2024-05-24	0	0	2024-05-24			
4B	483	Maneuver	CTugRV-02+SPM-11					2024-05-24	0	0	2024-05-24			
4B	484	Dock	CTugRV-02+SPM-11:XDock-05A					2024-05-24	0	0	2024-05-24			
4B	485	Maneuver	CTugRV-02+SPM-11+XDock-05+2STM+SPM-10					2024-05-24	0	0	2024-05-24			

Plan B to the Moon: A How-to Guide to Return to and Colonize the Lunar Surface

Phase	M Circle	Circle Action	Vehicles / Milestones / Notes	Cost FH Launch ($M)	Cost F9 Launch ($M)	Other Costs ($M)	HR Veh Costs ($M)	Veh Costs ($M)	Start Date	Nom Days	Ajust Days	End Date	L # Veh	L Veh	FH Miss Cycle	F9 Miss Cycle	CCtCap lbs	F9 Cargo lbs	F9 lbs to LEO
4B	486	Dock	CTugRV-02+...+SPM-10:SPM-09...						2024-05-24	0	0	2024-05-24							
4B	487	LLOG-01B Assembly/Recon	LLOG-01B Assembly and LLO Polar Recon						2024-05-24	2	0	2024-05-26							
4B	488	Undock	CTugRV-02						2024-05-26	0	0	2024-05-26							
4B	489	Maneuver, Return LEOS-01	CTugRV-02						2024-05-26	3	0	2024-05-29							
4B	490	LLO Gateway-01 Ops Norm	CTugRV-02						2024-05-29	0	0	2024-05-29							
4B	491	LEO Arrive	CTugRV-02						2024-05-29	3	0	2024-06-01							
4B	492	Rendezvous	CTugRV-02						2024-06-01	0	0	2024-06-01							
4B	493	Dock	CTugRV-02::LEOS-01 XDock-01D						2024-06-01	0	0	2024-06-01							
4B	494	Launch F9, LEO Rendezvous	CCTCap-05 (6 COB) [Phase 04B]		$62.0	$100.0	$213.0	$143.5	2024-06-01	3	0	2024-06-04	29	F95		2023-05-08	21300	28700	50000
4B	495	Dock	CCTCap-05:XDock-01F						2024-06-04	0	0	2024-06-04							
4B	496	LEO-01 Crew Swap	LEO-01 Ops Normal						2024-06-04	0	3	2024-06-07							
4B	497	Return to Earth	CCTCap-03R (4 COB)						2024-06-07	0	0	2024-06-07							
4B	498	Undock	CCTCap-03R:LEOS-01 XDock-01F						2024-06-07	0	0	2024-06-07							
4B	499	Phase-04B Complete						$2,469											
5A	501	Launch FH, LEO Rendezvous	SPM-12+STM+CSCM-05+XDock-06 ELTV-03	$130.0		$100.0	$106.5	$340.5	2024-06-07	3	0	2024-06-10	30	FH3	2024-06-07				
5A	502	Undock	CTugRV-02						2024-06-10	0	0	2024-06-10							
5A	503	Maneuver	CTugRV-02						2024-06-10	0	0	2024-06-10							
5A	504	Dock	CTugRV-02:SPM-12+STM+CSCM-05+XDock-06						2024-06-10	0	0	2024-06-10							
5A	505	Maneuver	CTugRV-02+STM+SPM-12+CSCM-05+XDock-06						2024-06-10	0	0	2024-06-10							
5A	506	Dock	CTugRV-02+...+XDock-06:LEOS-01						2024-06-10	0	0	2024-06-10							
5A	507	Undock	CTugRV-02+STM+SPM-12+CSCM-05:XDock-06B						2024-06-10	0	0	2024-06-10							
5A	508	Maneuver	CTugRV-02+STM+SPM-12+CSCM-05						2024-06-10	0	0	2024-06-10							
5A	509	Dock	CTugRV-02+STM+SPM-12+CSCM-05:XDock-05C						2024-06-10	0	0	2024-06-10							
5A	510	Unattach	CTugRV-02+STM+SPM-12:CSCM-05						2024-06-10	0	0	2024-06-10							
5A	511	Maneuver	CTugRV-02+STM+SPM-12						2024-06-10	0	0	2024-06-10							
5A	512	Undock	CTugRV-02:STM+SPM-12						2024-06-10	0	0	2024-06-10							
5A	513	Maneuver	CTugRV-02						2024-06-10	0	0	2024-06-10							
5A	514	Dock	CTugRV-02:XDock-01D						2024-06-10	0	0	2024-06-10							
5A	515	Undock	ASWTV-02(A)						2024-06-10	0	0	2024-06-10							
5A	516	Maneuver	ASWTV-02(A)						2024-06-10	0	0	2024-06-10							
5A	517	SAR-H2O	ASWTV-02(A):STM+SPM-12						2024-06-10	0	0	2024-06-10							
5A	518	H2O State Change	ASWTV-02(A)						2024-06-10	0	0	2024-06-10							
5A	519	H2O State Change	STM						2024-06-10	0	0	2024-06-10							
5A	520	Maneuver, Park LEO A	ASWTV-02(A)						2024-06-10	14	0	2024-06-24							
5A	521	LLO Polar Recon	LLO Polar Recon						2024-06-24	1	6	2024-07-01							
5A	522	LEOS-01 Ops Normal	LEOS-01 Ops Normal						2024-07-01	3	0	2024-07-04							
5A	523	Launch FH, LEO Rendezvous	SPM-13+STM For ELTV-03 [Phase 05A]	$130.0		$100.0	$0.0	$393.8	2024-07-01	3	0	2024-07-04	31	FH4	2024-07-01				
5A	524	Undock	CTugRV-02:XDock-01D						2024-07-04	0	0	2024-07-04							
5A	525	Maneuver	CTugRV-02						2024-07-04	0	0	2024-07-04							
5A	526	Dock	CTugRV-02:SPM-13+STM						2024-07-04	0	0	2024-07-04							
5A	527	Maneuver	CTugRV-02+SPM-13+STM						2024-07-04	0	0	2024-07-04							
5A	528	Maneuver	ASWTV-02(A)						2024-07-04	0	0	2024-07-04							
5A	529	SAR-H2O	ASWTV-02(A):STM+SPM-13						2024-07-04	0	0	2024-07-04							
5A	530	H2O State Change	ASWTV-02(A)						2024-07-04	0	0	2024-07-04							
5A	531	H2O State Change	STM						2024-07-04	0	0	2024-07-04							
5A	532	Maneuver	ASWTV-02(A)						2024-07-04	0	0	2024-07-04							
5A	533	SAR-H2O	ASWTV-02(A):SFT-01						2024-07-04	0	0	2024-07-04							
5A	534	H2O State Change	SFT-01						2024-07-04	0	0	2024-07-04							
5A	535	H2O State Change	ASWTV-02(A)						2024-07-04	0	0	2024-07-04							
5A	536	Maneuver	ASWTV-02(A)						2024-07-04	0	0	2024-07-04							
5A	537	Dock	ASWTV-02(A):XDock-02D						2024-07-04	0	0	2024-07-04							
5A	538	Maneuver	CTugRV-02+SPM-13+STM						2024-07-04	0	0	2024-07-04							
5A	539	Dock	CTugRV-02+SPM-13+STM+SPM-12+XDock-06						2024-07-04	0	0	2024-07-04							

Plan B to the Moon: A How-to Guide to Return to and Colonize the Lunar Surface

Mission Matrix Data Sheets

Phase	M Circle	Circle Action	Vehicles / Milestones / Notes	Cost FH Launch ($M)	Cost F9 Launch ($M)	Other Costs ($M)	HR Veh Costs ($M)	Veh Costs ($M)	Start Date	Nom Days	Ajust Days	End Date	L#	L Veh	FH Miss Cycle	F9 Miss Cycle
5A	540	Undock	CTugRV-02+SPM-13+2STM+SPM-12:XDock-06B							0	0	2024-07-04				
5A	541	Undock	CTugRV-02+SPM-13:2STM							0	0	2024-07-04				
5A	542	Maneuver	CTugRV-02+SPM-13							0	0	2024-07-04				
5A	543	Dock	CTugRV-02+SPM-13:SPM-12+2STM							0	0	2024-07-04				
5A	544	Dock	CTugRV-02+SPM-13+SPM-12+2STM:XDock-06B							0	0	2024-07-04				
5A	545	ELTV-03 Assembly	ELTV-03 Assembly						2024-07-04	14	7	2024-07-25				
5A	546	Launch FH, LEO Rendezvous	Cryo Fuel Cargo CRV-08 [Phase 05A]	$130.0		$100.0	$0.0	$95.0	2024-07-25	3	21	2024-08-18	32	FH1	2024-07-25	
5A	547	Undock	CTugRV-03:XDock-01D						2024-08-18	0	0	2024-08-18				
5A	548	Maneuver	CTugRV-03(A):RV-08(A)						2024-08-18	0	0	2024-08-18				
5A	549	SAR-CH4O2	CTugRV-03(A)						2024-08-18	0	0	2024-08-18				
5A	550	Fuel State Change	CTugRV-03(A) Fill						2024-08-18	0	0	2024-08-18				
5A	551	Fuel State Change	CTugRV-03(A) Empty						2024-08-18	0	0	2024-08-18				
5A	552	Maneuver / Park	CRV-08(A)						2024-08-18	0	0	2024-08-18				
4A	553	Maneuver	CTugRV-03						2024-08-18	0	0	2024-08-18				
4A	554	Dock	CTugRV-03(A):XDock-03C						2024-08-18	0	0	2024-08-18				
5A	555	Launch FH, LEO Rendezvous	Cryo Fuel Cargo CRV-09 [Phase 05A]	$130.0		$100.0	$0.0	$95.0	2024-08-18	3	0	2024-08-21	33	FH2	2024-08-18	
5A	556	Undock	CTugRV-03:XDock-01D						2024-08-21	0	0	2024-08-21				
5A	557	Maneuver	CTugRV-03:RV-09(A)						2024-08-21	0	0	2024-08-21				
5A	558	SAR-CH4O2	CRV-09(A):CTugRV-03						2024-08-21	0	0	2024-08-21				
5A	559	Fuel State Change	CTugRV-03 Fill						2024-08-21	0	0	2024-08-21				
5A	560	Fuel State Change	CTugRV-09(A) Empty						2024-08-21	0	0	2024-08-21				
4A	561	Maneuver / Park	CRV-09(A)						2024-08-21	0	0	2024-08-21				
5A	562	Maneuver	CTugRV-03						2024-08-21	0	0	2024-08-21				
5A	563	Dock	CTugRV-03(A):XDock-03C						2024-08-21	0	0	2024-08-21				
5A	564	SFT-01 Fuel Ready	SFT-01 Fuel Ready						2024-08-21	0	0	2024-08-21				
5A	565	Undock	CTugRV-03(A)						2024-08-21	0	0	2024-08-21				
5A	566	Maneuver	CTugRV-03(A)						2024-08-21	0	0	2024-08-21				
5A	567	SAR-CH4O2	CTugRV-03(A):SFT-01						2024-08-21	0	0	2024-08-21				
5A	568	Fuel State Change	CTugRV-03(A)						2024-08-21	0	0	2024-08-21				
5A	569	Fuel State Change	SFT-01						2024-08-21	0	0	2024-08-21				
5A	570	Maneuver	CTugRV-03(A)						2024-08-21	0	0	2024-08-21				
5A	571	Dock	CTugRV-03(A):XDock-03C						2024-08-21	0	0	2024-08-21				
5A	572	Maneuver	CTugRV-02+...xXDock-06:LEOS-01						2024-08-21	1	0	2024-08-21				
5A	573	SAR-CH4O2	CTugRV-02+SPM-13+SPM-12+2STM+XDock-06						2024-08-21	3	0	2024-08-21				
5A	574	Fuel State Change	CTugRV-03(A):CTugRV-02						2024-08-21	0	0	2024-08-21				
5A	575	Fuel State Change	CTugRV-02						2024-08-21	0	0	2024-08-21				
5A	576	Maneuver	CTugRV-03(A)						2024-08-21	0	0	2024-08-21				
5A	577	Dock	CTugRV-03(A):XDock-02C						2024-08-21	0	0	2024-08-21				
5A	578	LEOS-01 Ops Normal	LEOS-01 Ops Normal						2024-08-21	0	0	2024-08-21				
5A	579	LLO Maneuver / Depart	ELTV-03 Depart for LLO						2024-08-21	1	0	2024-08-22				
5A	580	Phase-05A Complete	Phase-05A Complete						2024-08-22	3	0	2024-08-25				
5B	581	LLO Arrive, Rendezvous	CTugRV-02: ELTV-03 / Phase 05B						2024-08-25	3	0	2024-08-28				
5B	582	Undock	CTugRV-02+SPM-13						2024-08-28	0	0	2024-08-28				
5B	583	Maneuver	CTugRV-02+SPM-13						2024-08-28	0	0	2024-08-28				
5B	584	Dock	CTugRV-02+SPM-13:XDock-06+...+SPM-12						2024-08-28	0	0	2024-08-28				
5B	585	Maneuver	CTugRV-02+...+SPM-12						2024-08-28	0	0	2024-08-28				
5B	586	Dock	CTugRV-02+...+SPM-12:SPM-11... LLOG-01						2024-08-28	0	0	2024-08-28				
5B	587	LLOG-01C Assembly	LLOG-01C Assembly						2024-08-28	0	1	2024-08-29				
5B	588	Undock	CTugRV-02						2024-08-28	0	0	2024-08-29				
5B	589	Maneuver, Return LEOS-01	CTugRV-02						2024-08-29	3	0	2024-09-01				
5B	590	LLOG-01 Ops Normal	LLOG-01 Ops Normal						2024-09-01	0	0	2024-09-01				
5B	591	LEO Arrive	CTugRV-02						2024-09-01	3	0	2024-09-04				
5B	592	Maneuver	CTugRV-02						2024-09-04	0	0	2024-09-04				

Plan B to the Moon: A How-to Guide to Return to and Colonize the Lunar Surface

Mission Matrix Data Sheets

Phase	M Circle	Circle Action	Vehicles / Milestones / Notes	Cost FH Launch ($M)	Cost F9 Launch ($M)	Other Costs ($M)	HR Veh Costs ($M)	Veh Costs ($M)	Start Date	Nom Days	Ajust Days	End Date	L # Veh	L Veh	FH Miss Cycle	F9 Miss Cycle	CCBCap lbs	F9 Cargo lbs	F9 lbs to LEO
5B	593	Dock	CTugRV-02:LEOS-01 XDock-01D									2024-09-04							
4B	594	Launch F9, LEO Rendezvous	CCTCap-06 (6 COB) [Phase 05B]		$62.0	$100.0	$213.0	$143.5	2024-09-04	3	0	2024-09-07	34	F96		2023-06-05	21300	28700	50000
5B	595	Dock	CCTCap-05:XDock-01F									2024-09-08							
4B	596	LEO-01 Crew Swap	LEO-01 Ops Normal						2024-09-07	1	0	2024-09-08							
5B	597	Undock	CCTCap-04R:LEOS-01 XDock-01B						2024-09-08	2	0	2024-09-10							
5B	598	Return to Earth	CCTCap-04R (6 COB)						2024-09-10	1	0	2024-09-11							
5B	599	Phase-05B Complete	$2,469									2024-09-11							
6A	601	Launch FH, LEO Rendezvous	SPM-14 (LLB-01+LLB-02) ELTV-04 [Phase 06A]	$130.0		$100.0	$435.0	$176.3	2024-09-11	3	0	2024-09-14	35	FH3	2024-09-11				
6A	622	LEOS-01 Ops Normal	LEOS-01 Ops Normal						2024-09-14	14	7	2024-10-05							
6A	623	Launch FH, LEO Rendezvous	SPM-15 (LL-01+LL-02) [Phase 06A]	$130.0		$100.0	$0.0	$393.8	2024-10-05	3	0	2024-10-08	36	FH4	2024-10-05				
6A	624	Undock	CTugRV-02						2024-10-08	0	0	2024-10-08							
6A	625	Maneuver	CTugRV-02						2024-10-08	0	0	2024-10-08							
6A	626	Dock	CTugRV-02:SPM-14						2024-10-08	0	0	2024-10-08							
6A	627	Maneuver	CTugRV-02+SPM-14						2024-10-08	0	0	2024-10-08							
6A	628	Dock	CTugRV-02+SPM-14:SPM-15						2024-10-08	0	0	2024-10-08							
6A	629	Maneuver	CTugRV-02+SPM-14+SPM-15						2024-10-08	0	0	2024-10-08							
6A	630	Dock	CTugRV-02+SPM-14+SPM-15:LEOS-01						2024-10-08	0	0	2024-10-08							
6A	631	LEOS-01 Ops Normal	LEOS-01 Ops Normal						2024-10-08	1	0	2024-10-09							
6A	664	SFT-01 Fuel Ready	SFT-01 Fuel Ready						2024-10-09	0	0	2024-10-09							
6A	665	Undock	CTugRV-03(A):XDock-02D						2024-10-09	0	0	2024-10-09							
6A	666	Maneuver	CTugRV-03(A)						2024-10-09	0	0	2024-10-09							
6A	667	SAR-CH4O2	CTugRV-03(A)						2024-10-09	0	0	2024-10-09							
6A	668	Fuel State Change	CTugRV-03(A)						2024-10-09	0	0	2024-10-09							
6A	669	Fuel State Change	SFT-01						2024-10-09	0	0	2024-10-09							
6A	670	Maneuver	CTugRV-03(A)						2024-10-09	0	0	2024-10-09							
6A	671	Undock	CTugRV-02+SPM-14+SPM-15						2024-10-09	0	0	2024-10-09							
6A	672	Maneuver	CTugRV-02+SPM-14+SPM-15						2024-10-09	0	0	2024-10-09							
6A	673	SAR-CH4O2	CTugRV-03(A)						2024-10-09	0	0	2024-10-09							
6A	674	Fuel State Change	CTugRV-03(A)						2024-10-09	0	0	2024-10-09							
6A	675	Fuel State Change	SFT-01						2024-10-09	0	0	2024-10-09							
6A	676	Maneuver	CTugRV-03(A)						2024-10-09	0	0	2024-10-09							
6A	677	Dock	CTugRV-03(A):XDock-02D						2024-10-09	0	0	2024-10-09							
6A	678	LEOS-01 Ops Normal	LEOS-01 Ops Normal						2024-10-09	1	0	2024-10-10							
6A	679	LLO Maneuver / Depart	ELTV-04 Depart for LLO						2024-10-10	3	0	2024-10-13							
6A	680	Phase-06A Complete	Phase-06A Complete									2024-10-13							
6B	681	LLO Arrive, Rendezvous	CTug02:ELTV-04						2024-10-13	0	0	2024-10-13							
6B	682	Dock	CTugRV-02+SPM-14+SPM-15:XDock-04D						2024-10-13	0	0	2024-10-13							
6B	683	Undock	CTugRV-02+SPM-14:SPM-15						2024-10-13	0	0	2024-10-13							
6B	686	Maneuver	CTugRV-02+SPM-14						2024-10-13	0	0	2024-10-13							
6B	687	Dock	CTugRV-02+SPM-14:XDock-04F						2024-10-13	0	0	2024-10-13							
6B	688	Undock	CTugRV-02						2024-10-13	0	0	2024-10-13							
6B	689	Maneuver	CTugRV-02						2024-10-13	0	0	2024-10-13							
6B	690	Dock	CTugRV-02:XDock-05D						2024-10-13	0	0	2024-10-13							
6B	691	LLOG-01 Ops Normal	LLOG-01 Ops Normal ELTV-04 Assem. Cont						2024-10-13	7	0	2024-10-20							
6B	699	Phase-06 Complete	$1,465									2024-10-20							
7B	701	Undock	CTugRV-02:XDock-05D / Phase 07B: LB-01 Est.						2024-10-20	0	0	2024-10-20							
7B	702	Maneuver	CTugRV-02						2024-10-20	0	0	2024-10-20							
7B	703	Dock	CTugRV-02:SPM-13:SPM-13						2024-10-20	0	0	2024-10-20							
7A	704	Undock	...SPM-10:SPM-09...						2024-10-20	0	0	2024-10-20							
7A	705	Undock	XDock-05A...						2024-10-20	0	0	2024-10-20							
7A	706	Undock	XDock-06+STM+STM:SPM-12...						2024-10-20	0	0	2024-10-20							
7A	707	Undock	CTugRV-02+SPM13:XDock-06+STM+STM						2024-10-20	0	0	2024-10-20							
7A	708	Maneuver	CTugRV-02+SPM13						2024-10-20	0	0	2024-10-20							

Plan B to the Moon: A How-to Guide to Return to and Colonize the Lunar Surface

Mission Matrix Data Sheets

Phase	M Circle	Circle Action	Vehicles / Milestones / Notes	Cost FH Launch ($M)	Cost F9 Launch ($M)	Other Costs ($M)	HR Veh Costs ($M)	Veh Costs ($M)	Start Date	Nom Days	Ajust Days	End Date	L #	L Veh	FH Miss Cycle	F9 Miss Cycle	CCtCap lbs	F9 Cargo lbs	F9 lbs to LEO
7A	709	Dock	CTugRV-02+SPM13:CSCM-05						2024-10-20	0	0	2024-10-20							
7A	710	Undock	CTugRV-02+SPM13+CSCM-05:XDock-06C						2024-10-20	0	0	2024-10-20							
7A	711	Maneuver	CTugRV-02+SPM13+CSCM-05						2024-10-20	0	0	2024-10-20							
7A	712	Dock	CTugRV-02+SPM13+CSCM-05:XDock-05C						2024-10-20	0	0	2024-10-20							
7A	713	Undock	CTugRV-02:SPM13+CSCM-05						2024-10-20	0	0	2024-10-20							
7A	714	Maneuver	CTugRV-02						2024-10-20	0	0	2024-10-20							
7A	715	Dock	CTugRV-02:SPM-12+SPM-11						2024-10-20	0	0	2024-10-20							
7A	716	Maneuver	CTugRV-02:SPM-12+SPM-11						2024-10-20	0	0	2024-10-20							
7A	717	Dock	CTugRV-02+SPM-12+SPM-11:CSCM-04						2024-10-20	0	0	2024-10-20							
7A	718	Undock	CTugRV-02:SPM-12						2024-10-20	0	0	2024-10-20							
7A	719	Maneuver	CTugRV-02						2024-10-20	0	0	2024-10-20							
7A	720	Dock	CTugRV-02:SPM-10						2024-10-20	0	0	2024-10-20							
7A	721	Maneuver	CTugRV-02+SPM-10...SPM-09...						2024-10-20	0	0	2024-10-20							
7A	722	Dock	CTugRV-02+SPM-10:2STM+XDock-05B						2024-10-20	0	0	2024-10-20							
7A	723	Undock	CTugRV-02+SPM-10						2024-10-20	0	0	2024-10-20							
7A	724	Maneuver	CTugRV-02+SPM-10:SPM-13...						2024-10-20	0	0	2024-10-20							
7A	725	Dock	CTugRV-02:SPM-10...						2024-10-20	0	0	2024-10-20							
7A	726	Undock	CTugRV-02:SPM-10...						2024-10-20	0	0	2024-10-20							
7A	727	Maneuver	CTugRV-02						2024-10-20	0	0	2024-10-20							
7B	728	Dock	CTugRV-02:XDock-06A						2024-10-20	0	0	2024-10-20							
7B	729	Maneuver	CTugRV-02...						2024-10-20	0	0	2024-10-20							
7B	730	Dock	CTugRV-02+XDock-06+2STM:2STM+XDock-05B...						2024-10-20	0	0	2024-10-20							
7B	731	Undock	CTugRV-02:XDock-06A						2024-10-20	0	0	2024-10-20							
7B	732	Maneuver	CTugRV-02						2024-10-20	0	0	2024-10-20							
7B	733	Attach	CTugRV-02:SPM-14:LL-02						2024-10-20	0	0	2024-10-20							
7B	734	Extract	CTugRV-02:LL-02						2024-10-20	0	0	2024-10-20							
7B	735	Maneuver	CTugRV-02:LL-02						2024-10-20	0	0	2024-10-20							
7B	736	Dock	CTugRV-02:LL-02:XDock-06C						2024-10-20	0	0	2024-10-20							
7B	737	Detach	CTugRV-02:LL-02						2024-10-20	0	0	2024-10-20							
7B	738	Maneuver	CTugRV-02						2024-10-20	0	0	2024-10-20							
7B	739	Attach	CTugRV-02:SPM-14:LL-01						2024-10-20	0	0	2024-10-20							
7B	740	Extract	CTugRV-02:LL-01						2024-10-20	0	0	2024-10-20							
7B	741	Detach	CTugRV-02:LL-01						2024-10-20	0	0	2024-10-20							
7B	742	Maneuver / Pivot	CTugRV-02						2024-10-20	0	0	2024-10-20							
7B	743	Attach	CTugRV-02:LL-01						2024-10-20	0	0	2024-10-20							
7B	744	Maneuver	CTugRV-02:LL-01						2024-10-20	0	0	2024-10-20							
7B	745	Dock	CTugRV-02:LL-01:XDock-06E						2024-10-20	0	0	2024-10-20							
7B	746	Detach	CTugRV-02:LL-01						2024-10-20	0	0	2024-10-20							
7B	747	Maneuver	CTugRV-02						2024-10-20	0	0	2024-10-20							
7B	748	Dock	CTugRV-02:XDock-06A						2024-10-20	0	0	2024-10-20							
7B	749	LLOG-01 Ops Normal	LLOG-01 Ops Normal			$100.0			2024-10-20	2	0	2024-10-22							
7B	750	Undock	CTugRV-02:XDock-06A						2024-10-22	0	0	2024-10-22							
7B	751	Maneuver	CTugRV-02						2024-10-22	0	0	2024-10-22							
7B	752	Attach	CTugRV-02:SPM-15:LLB-02						2024-10-22	0	0	2024-10-22							
7B	753	Extract	CTugRV-02:LLB-02						2024-10-22	0	0	2024-10-22							
7B	754	Maneuver	CTugRV-02:LLB-02						2024-10-22	0	0	2024-10-22							
7B	755	Attach	CTugRV-02:LLB-02:SPM-10+SPM-13						2024-10-22	0	0	2024-10-22							
7B	756	Detach	CTugRV-02:LLB-02						2024-10-22	0	0	2024-10-22							
7B	757	Maneuver	CTugRV-02						2024-10-22	0	0	2024-10-22							
7B	758	Attach	CTugRV-02:SPM-15:LLB-01						2024-10-22	0	0	2024-10-22							
7B	759	Extract	CTugRV-02:LLB-01						2024-10-22	0	0	2024-10-22							
7B	760	Maneuver	CTugRV-02:LLB-01						2024-10-22	0	0	2024-10-22							

Plan B to the Moon: A How-to Guide to Return to and Colonize the Lunar Surface

Phase	M Circle	Circle Action	Vehicles / Milestones / Notes	Cost FH Launch ($M)	Cost F9 Launch ($M)	Other Costs ($M)	HR Veh Costs ($M)	Veh Costs ($M)	Start Date	Nom Days	Ajust Days	End Date	L #	Veh	FH Miss Cycle	F9 Miss Cycle
7B	761	Attach	CTugRV-02:LLB-01:SPM-11+SPM-12						2024-10-22			2024-10-22				
7B	762	Detach	CTugRV-02:LLB-01						2024-10-22			2024-10-22				
7B	763	Maneuver	CTugRV-02						2024-10-22			2024-10-22				
7B	764	Dock	CTugRV-02:XDock-06A						2024-10-22			2024-10-22				
7B	765	LLOG-01 Ops Normal	LLOG-01 Ops Normal			$100.0			2024-10-22	7		2024-10-29				
7B	786	SFT-01 Fuel Ready	SFT-01 Fuel Ready						2024-10-29			2024-10-29				
7B	787	Launch FH, LEO Rendezvous	CTugRV-10(A) [Phase 06A]	$130.0		$100.0	$106.5	$219.8	2024-10-29	3	0	2024-11-01	37	FH1	2024-10-29	
7B	788	SAR-CH4O2	CTugRV-03(A)						2024-11-01			2024-11-01				
7B	789	Fuel State Change	CTugRV-03(A)						2024-11-01			2024-11-01				
7B	790	Fuel State Change	SFT-01						2024-11-01			2024-11-01				
7B	791	Maneuver / Depart	CTugRV-10(A) Depart for LLO						2024-11-01			2024-11-01				
7B	792	LLO Arrive, Rendezvous	CTugRV-10(A)						2024-11-01			2024-11-01				
7B	793	Maneuver	CTugRV-10(A)						2024-11-01			2024-11-01				
7B	794	Dock	LLOG-01						2024-11-01			2024-11-01				
7B	799	Phase-07 Complete	$756						2024-11-01			2024-11-01				
8A	801	Humans Ready to Return LS	Humans Ready to Return to Lunar Surface						2024-11-01			2024-11-01				
8A	802	Undock	LL-01:XDock-06E						2024-11-01	1		2024-11-01				
8A	803	Descend to LS	LL-01						2024-11-01	1		2024-11-02				
8A	804	Humans Land on LS	LL-01			$112.2			2024-11-02	1	0	2024-11-03				
8A	805	Prospecting Mission-01	Prospecting Mission-01						2024-11-03	7		2024-11-10				
8A	806	Liftoff, LS to LLO	LL-01						2024-11-10	1		2024-11-11				
8A	807	LLO Rendezvous	LL-01						2024-11-11	1		2024-11-12				
8A	808	Dock	LL-01:XDock-06E						2024-11-12			2024-11-12				
8A	809	LL-01 LLOG-01 Refueling	LL-01 LLOG-01 Refueling						2024-11-12	1		2024-11-13				
8A	810	LLOG-01 Ops Normal	LLOG-01 Ops Normal						2024-11-13	7		2024-11-20				
8A	812	Undock	LL-02:XDock-06C						2024-11-20	1		2024-11-21				
8A	813	Descend to LS	LL-02						2024-11-21	1		2024-11-22				
8A	814	Humans Land on LS	LL-02						2024-11-22	7		2024-11-29				
8A	815	Prospecting Mission-02	Prospecting Mission-02						2024-11-29	1		2024-11-30				
8A	816	Liftoff, LS to LLO	LL-02						2024-11-30	1		2024-12-01				
8A	817	LLO Rendezvous	LL-02						2024-12-01			2024-12-01				
8A	818	Dock	LL-02:XDock-06C						2024-12-01	1		2024-12-02				
8A	819	LL-02 LLOG-01 Refueling	LL-02 LLOG-01 Refueling						2024-12-02	7		2024-12-09				
8A	820	LLOG-01 Ops Normal	LLOG-01 Ops Normal						2024-12-09	1		2024-12-09				
8A	822	Undock	LL-01:XDock-06E						2024-12-09	1		2024-12-10				
8A	823	Descend to LS	LL-01						2024-12-10	1		2024-12-11				
8A	824	Humans Land on LS	LL-01						2024-12-11	7		2024-12-11				
8A	825	Prospecting Mission-03	Prospecting Mission-03						2024-12-11	1		2024-12-18				
8A	826	Liftoff, LS to LLO	LL-01						2024-12-18	1		2024-12-19				
8A	827	LLO Rendezvous	LL-01						2024-12-19	1		2024-12-20				
8A	828	Dock	LL-01:XDock-06E						2024-12-20	1		2024-12-20				
8A	829	LL-01 LLOG-01 Refueling	LL-01 LLOG-01 Refueling						2024-12-20	7		2024-12-21				
8A	830	LLOG-01 Ops Normal	LLOG-01 Ops Normal						2024-12-21			2024-12-28				
8A	832	Undock	LL-02:XDock-06C						2024-12-28	1		2024-12-28				
8A	833	Descend to LS	LL-02						2024-12-28			2024-12-28				
8A	834	Humans Land on LS	LL-02						2024-12-28			2024-12-28				
8A	835	Prospecting Mission-04	Prospecting Mission-04						2024-12-28			2024-12-28				
8A	836	Liftoff, LS to LLO	LL-02						2024-12-28			2024-12-28				
8A	837	LLO Rendezvous	LL-02						2024-12-28			2024-12-28				
8A	838	Dock	LL-02:XDock-06C						2024-12-28			2024-12-28				
8A	839	LL-02 LLOG-01 Refueling	LL-02 LLOG-01 Refueling						2024-12-28			2024-12-28				
8A	840	LLOG-01 Ops Normal	LLOG-01 Ops Normal						2024-12-28			2024-12-28				

Plan B to the Moon: A How-to Guide to Return to and Colonize the Lunar Surface

Phase	M Circle	Circle Action	Vehicles / Milestones / Notes	Cost FH Launch ($M)	Cost F9 Launch ($M)	Other Costs ($M)	HR Veh Costs ($M)	Veh Costs ($M)	Start Date	Nom Days	Ajust Days	End Date	L # Veh	CCICap lbs	F9 Cargo lbs	F9 lbs to LEO
8A	842	Undock	LL-01:XDock-06E						2024-12-28	1	0	2024-12-29				
8A	843	Descend to LS	LL-01						2024-12-29	1	0	2024-12-30				
8A	844	Humans Land on LS	LL-01						2024-12-30	7	0	2025-01-06				
8A	845	Prospecting Mission-05	Prospecting Mission-05						2025-01-06	1	0	2025-01-07				
8A	846	Liftoff, LS to LLO	LL-01						2025-01-07	1	0	2025-01-08				
8A	847	LLO Rendezvous	LL-01						2025-01-08	0	0	2025-01-08				
8A	848	Dock	LL-01:XDock-06E						2025-01-08	1	0	2025-01-09				
8A	849	LL-01 LLOG-01 Refueling	LL-01 LLOG-01 Refueling						2025-01-09	7	0	2025-01-16				
8A	850	LLOG-01 Ops Normal	LLOG-01 Ops Normal						2025-01-16	1	0	2025-01-17				
8A	851	Lunar Base-01 Selected	Lunar Base-01 Selected						2025-01-17	1	0	2025-01-18				
8B	852	Undock	LL-01:XDock-06C						2025-01-17	1	0	2025-01-18				
8B	853	Descend to LS	LL-01						2025-01-18	0	0	2025-01-18				
8B	854	Land, Lunar Base-01	LL-01						2025-01-18	0	0	2025-01-18				
8B	855	Undock	LHV-01						2025-01-18	0	0	2025-01-18				
8B	856	Descend to LS	LHV-01						2025-01-18	0	0	2025-01-18				
8B	857	Land, Lunar Base-01	LHV-01						2025-01-18	0	0	2025-01-18				
8B	858	LHV-01 Deploys	LHV-01 Deploys						2025-01-18	0	0	2025-01-18				
8B	859	LHV-01 Mission A1	LHV-01 Mission						2025-01-18	0	0	2025-01-18				
8B	860	LHV-01 Mission A2	LHV-01 Mission						2025-01-18	0	0	2025-01-18				
8B	861	LHV-01 Transition / LB Ops	Transition to Lunar Habitation						2025-01-18	0	0	2025-01-18				
8B	862	Liftoff, LS to LLO	LLB-01						2025-01-18	0	0	2025-01-18				
8B	863	LLO Rendezvous	LLB-01						2025-01-18	0	0	2025-01-18				
8B	864	Maneuver / Mate	LLB-01						2025-01-18	0	0	2025-01-18				
8B	865	LLB-01 LLOG-01 Refueling	LLB-01 LLOG-01 Refueling						2025-01-18	0	0	2025-01-18				
8B	866	LLOG-01 Ops Normal	LLOG-01 Ops Normal						2025-01-18	0	0	2025-01-18				
8B	867	LL Liftoff, LS to LLO	LL-01						2025-01-18	0	0	2025-01-18				
8B	868	LLO Rendezvous	LL-01						2025-01-18	0	0	2025-01-18				
8B	869	Dock	LL-01:XDock-06C						2025-01-18	0	0	2025-01-18				
8B	870	LL-01 LLOG-01 Refueling	LL-01 LLOG-01 Refueling						2025-01-18	0	0	2025-01-18				
8B	871	LLOG-01 Ops Normal	LLOG-01 Ops Normal						2025-01-18	0	0	2025-01-18				
8B	872	Undock	LL-02:XDock-06E						2025-01-18	0	0	2025-01-18				
8B	873	Descend to LS	LL-02						2025-01-18	0	0	2025-01-18				
8B	874	Land, Lunar Base-01	LL-02						2025-01-18	0	0	2025-01-18				
8B	875	Undock	LHV-02						2025-01-18	0	0	2025-01-18				
8B	876	Descend to LS	LHV-02						2025-01-18	0	0	2025-01-18				
8B	877	Land, Lunar Base-01	LHV-02						2025-01-18	0	0	2025-01-18				
8B	878	LHV-02 Deploys	LHV-02 Deploys						2025-01-18	0	0	2025-01-18				
8B	879	LHV-02 Mission B1	LHV-02 Mission						2025-01-18	0	0	2025-01-18				
8B	880	LHV-02 Mission B2	LHV-02 Mission						2025-01-18	0	0	2025-01-18				
8B	881	LHV-02 Transition / LB Ops	Transition to Lunar Habitation						2025-01-18	0	0	2025-01-18				
8B	882	Liftoff, LS to LLO	LLB-02						2025-01-18	0	0	2025-01-18				
8B	883	LLO Rendezvous	LLB-02						2025-01-18	0	0	2025-01-18				
8B	884	Maneuver / Mate	LLB-02						2025-01-18	0	0	2025-01-18				
8B	885	LLB-01 LLOG-01 Refueling	LLB-02 LLOG-01 Refueling						2025-01-18	0	0	2025-01-18				
8B	886	LLOG-01 Ops Normal	LLOG-01 Ops Normal						2025-01-18	0	0	2025-01-18				
8B	887	LL Liftoff, LS to LLO	LL-02						2025-01-18	0	0	2025-01-18				
8B	888	LLO Rendezvous	LL-02						2025-01-18	0	0	2025-01-18				
8B	889	Dock	LL-02:XDock-06E						2025-01-18	0	0	2025-01-18				
8B	890	LL-02 LLOG-01 Refueling	LL-02 LLOG-01 Refueling						2025-01-18	0	0	2025-01-18				
8B	891	LLOG-01 Ops Normal	LLOG-01 Ops Normal						2025-01-18	0	0	2025-01-18				
8B	899	Phase-08 Complete (LB-IOC)	$112						2025-01-18	0	0	2025-01-18	37			
99	910	**Total Plan B Cost:**	**$22,015**	$3,962.0	$434.0	$5,212.2	$2,991.0	$9,415.8						149100	209900	350000

Plan B to the Moon: A How-to Guide to Return to and Colonize the Lunar Surface

Phase	M Circle	#	Circle Action	Vehicles / Milestones / Notes	CSCM lbs	LL lbs	LLB lbs	PVM lbs	FH Cargo lbs	XDock lbs	CTugRV Dry	CSM Dry	ASWTV Dry	STM Dry	Launch Payload Cyro	Launch Payload H2O	FH Total lbs to LEO	Habitable Volume (cuft)	PV-Power Pred (kW)	Vehicle No-Fuel lbs	LLO / 15 New Fuel lbs
			85.00%	$22,015	106,500	43,500	43,500	222,000	801,200	74,550	175,800	85,500	208,458	71,250	1,161,200	1,006,542	4,000,000	43780	798	1883158	716700
0	15		Phase 0: Approved Proceed	Plan B V1.x, Appr. to Proceed FY2020 [Phase 0]																	
0	19		Launch FH, LEO Rendezvous	CSM+SPM-0+STM+XDock-00+CSCM-00 [Phase 0]	10,650			10,000	90,200	10,650		4,750		3,750			130,000	4238	21	119350	
0	21		Launch F9, LEO Rendezvous	CCTCap-00 (6 COB) [Phase 0]														350		28700	
0	27		Phase 0: Construct Comp.	Plan B, Exec. / Phase 0: Spacecraft Const Comp.																	
0	28		Phase 0: Testing Complete	Plan B, Exec. / Phase 0: Test & Launch Int. Comp.																	
0	35		PM Master Schedule Reserve	Plan B V1.x, Execution																	
0	99		Phase-00 Complete	$2,652																	
1A	101		Launch FH, LEO Rendezvous	SPM-1+XDock-01+CSCM-01 [Phase 01A]	10,650			10,000	98,700	10,650							130,000	4238	21	119350	
1A	102		Maneuver, Park LEO A	SPM-1+XDock-01+CSCM-01																	
1A	103		Launch F9, LEO Rendezvous	CCTCap-01 (6 COB) [Phase 01A]														350		28700	
1A	104		Dock	CCTCap-01:XDock-01F																	
1A	105		Launch FH, LEO Rendezvous	CTugRV-01(A) [Phase 01A]	10,650				0		43,950				75,400		130,000	350	21	43950	54600
1A	106		Dock	CTugRV-01:XDock-01D																	
1A	107		LEOS-01 Assembly	LEOS-01 Assembly																	
1A	108		Undock	CTugRV-01:XDock-01D (2 COB)																	
1A	109		Maneuver	CTugRV-01																	
1A	110		Dock	CTugRV-01:XDock-01E																	
1A	111		Undock	CTugRV-01:XDock-01E																	
1A	112		Maneuver	CTugRV-01																	
1A	113		Dock	CTugRV-01:XDock-01F																	
1A	114		Undock	CTugRV-01:XDock-01F																	
1A	115		Maneuver	CTugRV-01 / Test Docking Ports																	
1A	116		Dock	CTugRV-01:XDock-01C																	
1A	117		Undock	CTugRV-01:XDock-01C																	
1A	118		Maneuver	CTugRV-01																	
1A	119		Attach	CTugRV-01:CSCM-01																	
1A	120		Undock	CSCM-01:XDock-01B																	
1A	121		Maneuver	CTugRV-01+CSCM-01																	
1A	122		Dock	CSCM-01+CTugRV-01:XDock-01E																	
1A	123		Unattach	CTugRV-01																	
1A	124		Maneuver	CTugRV-01																	
1A	125		Dock	CTugRV-01:XDock-01B																	
1A	126		LEOS-01 Assembly	LEOS-01 Assembly																	
1A	127		Launch FH, LEO Rendezvous	SPM-02+XDock-02 [Phase 01A]				10,000	109,350	10,650							130,000	3888	21	130000	
1A	129		Undock	CTugRV-01:XDock-01B																	
1A	130		Maneuver	CTugRV-01																	
1A	131		Dock	CTugRV-01:XDock-02B																	
1A	132		Dock	CTugRV-01+SPM-02:SPM-01																	
1A	133		LEOS-01 Assembly	LEOS-01 Assembly																	
1A	134		Launch FH, LEO Rendezvous	SPM-03+ 4STM (Two STM Full) [Phase 01A]				10,000	2,500					15,000		102,500	130,000	3888	21	27500	
1A	135		Undock	CTugRV-01:XDock-02B																	
1A	136		Maneuver	CTugRV-01																	
1A	137		Dock	CTugRV-01:SPM-03+ 4STM																	
1A	138		Dock	SPM-03+ 4STM:XDock-02B																	
1A	139		LEOS-01 Assembly	LEOS-01 Assembly																	
1A	140		Undock	CTugRV-01+SPM-03-4STM+XDock-02																	
1A	141		Maneuver	CTugRV-01+SPM-03-4STM+XDock-02 / Pirouette																	
1A	142		Undock	CTugRV-01:SPM-03+4STM+XDock-02																	
1A	143		Maneuver	CTugRV-01																	
1A	144		Dock	CTugRV-01:XDock-02B																	

Plan B to the Moon: A How-to Guide to Return to and Colonize the Lunar Surface

Phase	M Circle	Circle Action	Vehicles / Milestones / Notes	CSCM lbs	LL lbs	LLB lbs	PVM lbs	FH Cargo lbs	XDock lbs	CTugRV Dry	CSM Dry	ASWTV Dry	STM Dry	Launch Payload Cyro	Launch Payload H2O	FH Total lbs to LEO	Habitable Volume (cuft)	PV-Power Pind (kW)	Vehicle No-Fuel lbs	LLO/LS No Fuel lbs
1A	145	Maneuver	CTugRV-01+XDock-02+4STM+SPM3																	
1A	146	Dock	CTugRV-01+XDock-02+4STM+SPM3:SPM-02																	
1A	147	Undock	CTugRV-01:XDock-02																	
1A	148	Maneuver	CTugRV-01																	
1A	149	Dock	CTugRV-01:XDock-01														350		28700	
1A	150	Partial H2O Load	LEOS-01 H2O Transfer to Shielding Tanks																	
1A	151	LEOS-01 Ops Normal	LEOS-01 Ops Normal																	
1B	152	Launch F9, LEO Rendezvous	CCTCap-02 (6 COB) [Phase 01B]																	
1B	153	Dock	CCTCap-02:XDock-01D																	
1B	154	LEOS-01 Ops Normal	LEOS-01 Ops Normal																	
1B	155	Undock	CCTCap-01R:XDock-01F																	
1B	156	Return to Earth	CCTCap-01R (6 COB)																	
1B	157	LEOS-01 Ops Normal	LEOS-01 Ops Normal																	
1B	158	Launch FH, LEO Rendezvous	ASWTV-01(A) [Phase 01B]					0				69,486			70,514	140,000	0	0	69486	
1B	159	SAR-H2O	ASWTV-01(A) / LEOS-01 / H2O for Shielding																	
1B	160	H2O State Change	ASWTV-01(A)																	
1B	161	H2O State Change	LEOS-01																	
1B	162	Maneuver	ASWTV-01(A)																	
1B	163	Dock	ASWTV-01(A):XDock-02F																	
1B	164	Partial H2O Load	LEOS-01 H2O Transfer to Shielding Tanks																	
1B	165	LEOS-01 Ops Normal	LEOS-01 Ops Normal																	
1B	166	Launch FH, LEO Rendezvous	ASWTV-02(A) [Phase 01B]					0				69,486			70,514	140,000	0	0	69486	
1B	167	SAR-H2O	ASWTV-02(A) / LEOS-01 / H2O for Shielding																	
1B	168	H2O State Change	ASWTV-02(A)																	
1B	169	H2O State Change	LEOS-01																	
1B	170	Maneuver	ASWTV-02(A)																	
1B	171	Dock	ASWTV-02(A):XDock-02D																	
1B	172	SAR-H2O	ASWTV-02(A) / LEOS-01 / H2O for Shielding																	
1B	173	LEOS-01 Ops Normal	LEOS-01 Ops Normal																	
1B	174	Launch FH, LEO Rendezvous	ASWTV-03(A) [Phase 01B]					0				69,486			70,514	140,000	0	0	69486	
1B	175	SAR-H2O	ASWTV-03(A) / LEOS-01 / H2O for Shielding																	
1B	176	H2O State Change	ASWTV-03(A)																	
1B	177	H2O State Change	LEOS-01																	
1B	178	Maneuver, Park LEO D	ASWTV-03(A)																	
1B	179	LEOS-01 H2O Shielding	LEOS-01 H2O Shielding Tanks																	
1B	180	LEOS-01 H2O Shielding	LEOS-01 H2O Shielding / Complete (297k lbs)																	
1B	181	LEOS-01 Ops Normal	LEOS-01 Ops Normal																	
1B	182	Launch F9, LEO Rendezvous	CCTCap-03 (6 COB) [Phase 01B]																	
1B	183	Dock	CCTCap-03:XDock-01F														350		28700	
1B	184	LEOS-01 Ops Normal	LEOS-01 Ops Normal																	
1B	185	Launch FH, LEO Rendezvous	CTugRV-02(A) [Phase 01B]	10,650				0		43,950				75,400		130,000	350	21	43950	
1B	186	Dock	CTugRV-02(A):XDock-02B																	
1B	187	Launch FH, LEO Rendezvous	CTugRV-03(A) [Phase 01B]	10,650				0		43,950				75,400		130,000	350	21	43950	
1B	188	SAR-CH4O2	CTugRV-03(A):CTugRV-01 / Top Off																	
1B	189	Fuel State Change	CTugRV-03(A) Decrease																	
1B	190	Fuel State Change	CTugRV-01 Fill																	
1B	191	Maneuver	CTugRV-03(A)																	
1B	192	SAR-CH4O2	CTugRV-03(A):CTugRV-02 / Partial Load																	
1B	193	Fuel State Change	CTugRV-02 Empty																	
1B	194	Fuel State Change	CTugRV-01 Fill																	
1B	195	Maneuver	CTugRV-03(A)																	
1B	196	Dock	CTugRV-03(A):XDock-02C																	

Plan B to the Moon: A How-to Guide to Return to and Colonize the Lunar Surface

Mission Matrix Data Sheets

Phase	M Circle	Circle Action	Vehicles / Milestones / Notes	CSCM lbs	LL lbs	LLB lbs	PVM lbs	FH Cargo lbs	XDock lbs	CTugRV Dry	CSM Dry	ASWTV Dry	STM Dry	Launch Payload Cyro	Launch Payload H2O	FH Total lbs to LEO	Habitable Volume (cuft)	PV-Power Prod (kW)	Vehicle No Fuel lbs	LUO / LS No Fuel lbs
1B	198		LEOS-01 Ops Normal																	
1B	199	Phase-01 Complete	$7,199																	
2A	201	Launch FH, LEO Rendezvous	2STM+SPM-04+CSCM-02 [Phase 02A]	10,650			10,000	50,600					7,500		51,250	130,000	0	0	68100	
2A	202	Undock	CTugRV-02:XDock-02B																	
2A	204	Maneuver	CTugRV-02																	
2A	205	Dock	CTugRV-02:2STM+SPM-04+CSCM-02																	
2A	206	Maneuver	CTugRV-02+2STM+SPM-04+CSCM-02																	
2A	207	Dock	CTugRV-02+2STM+SPM-04+CSCM-02:XDock-02E																	
2A	208	Unattach	CTugRV-02+2STM+SPM-04:CSCM-02																	
2A	209	Maneuver	CTugRV-02+2STM+SPM-04																	
2A	210	Dock	CTugRV-02+2STM+SPM-04:XDock-02B																	
2A	211	Undock	CTugRV-02+2STM+SPM-04																	
2A	212	Maneuver	CTugRV-02																	
2A	213	Undock	CTugRV-02:XDock-01D																	
2A	214	SFT-01 Assembly	SFT-01 Assembly																	
2A	215	Launch FH, LEO Rendezvous	SPM-05+3STM [Phase 02A]				10,000	6,250					11,250		102,500	130,000	0	0	27500	
2A	216	Undock	CTugRV-02:XDock-01D																	
2A	218	Maneuver	CTugRV-02																	
2A	219	Dock	CTugRV-02:SPM-05+3STM																	
2A	220	Maneuver	CTugRV-02+SPM-05+3STM																	
2A	221	Dock	CTugRV-02+SPM-05+3STM:STM																	
2A	222	Undock	SPM-04:XDock-02B																	
2A	223	Dock	SPM-05:STM																	
2A	224	Maneuver	CTugRV-02+SPM-05																	
2A	225	Dock	SPM-04:SPM-05																	
2A	226	Maneuver	SPM-04+5STM / Pirouette Flip																	
2A	227	Dock	SPM-05+SPM-04+5STM:XDock-02B																	
2A	228	Undock	CTugRV-02:SPM-05																	
2A	229	Maneuver	CTugRV-02																	
2A	230	Dock	CTugRV-02:XDock-01D																	
2A	231	SFT-01 Assembly	SFT-01 Assembly																	
2A	232	Launch FH, LEO Rendezvous	SPM-06+2CSM (Preloaded Fuel) [Phase 02A]				10,000	43,700			9,500			66,800		130,000	0	0	63200	
2A	233	Undock	CTugRV-01:XDock-01B																	
2A	235	Maneuver	CTugRV-01																	
2A	236	Dock	CTugRV-01:CSM+CSM+SPM-06																	
2A	237	Dock	2CSM+SPM-06:SPM-05																	
2A	238	Undock	CSM:SPM-06																	
2A	239	Maneuver	CTugRV-01																	
2A	240	Dock	CTugRV-01																	
2A	241	Launch FH, LEO Rendezvous	XDock-03+SPM-07+3CSM [Phase 02A]				10,000	28,300	10,650		14,250			66,800		130,000	0	0	63200	
2A	243	Undock	CTugRV-01:XDock-01B																	
2A	245	Undock	CTugRV-01																	
2A	246	Maneuver	CTugRV-01:XDock-03+SPM-07+3xCSM																	
2A	247	Dock	CTugRV-01+XDock-03+SPM-07+3CSM																	
2A	248	Dock	CTugRV-01+XDock-03+SPM-07+3CSM:CSM+...																	
2A	249	Undock	CSM:SPM-06																	
2A	250	Maneuver	XDock-03A:SPM-07																	
2A	251	Undock	CTugRV-01+XDock-03 / Pirouette Flip																	
2A	252	Dock	XDock-03B:SPM-07																	
2A	253	Dock	SPM-07:SPM-06																	
2A	254	Undock	CTugRV-01:XDock-03B																	
2A	255	Maneuver	CTugRV-01																	

Plan B to the Moon: A How-to Guide to Return to and Colonize the Lunar Surface

Phase	M Circle	Circle Action	Vehicles / Milestones / Notes	CSCM lbs	LL lbs	LLB lbs	PVM lbs	FH Cargo lbs	XDock lbs	CTugRV Dry	CSM Dry	ASWTV Dry	STM Dry	Launch Payload Cyro	Launch Payload H2O	FH Total lbs to LEO	Habitable Volume (cuft)	PV: Power Prod (kW)	Vehicle No Fuel lbs	LLO / LS No Fuel lbs
2A	256	Dock	CTugRV-01:XDock-02A																	
2B	257	SFT-01 Assembly	SFT-01 / Phase 02B: SFT-01 Assembly Continue															504		
2B	258	SFT-01 Array Assemble	SFT-01																	
2B	259	Undock	CTugRV-02																	
2B	260	Maneuver	CTugRV-02																	
2B	261	Dock	CTugRV-02:XDock-03B																	
2B	262	Undock	XDock-02A / CTugRV-01 & CTugRV-02 Counter Pull																	
2B	263	Undock	CTugRV-01																	
2B	264	Maneuver	CTugRV-01																	
2B	265	Dock	CTugRV-01:XDock-02A (SFT-01B)																	
2B	266	Undock	SPM-04:STM (SFT-01B)																	
2B	267	Maneuver	CTugRV-01+SFT-01A / Move SFT-01A to LEO Bravo																	
2B	268	Maneuver	CTugRV-02+SFT-01B / Move SFT-01B to LEO Bravo																	
2B	269	Dock	SFT-01A:SFT-01B																	
2B	270	SFT-01 Checkout	SFT-01																	
2B	271	Undock	CTugRV-01																	
2B	272	Maneuver	CTugRV-01																	
2B	273	Dock	CTugRV-01:XDock-01B																	
2B	274	Undock	CTugRV-02																	
2B	275	Maneuver	CTugRV-02																	
2B	276	Dock	CTugRV-02:XDock-01D																	
2B	277	Launch FH, LEO Rendezvous	Max Water Cargo WTV [Phase 02B]				0	0	0				11,250		128,750	140,000	0	0	11250	
2B	278	SAR-H2O	Water Cargo Mission																	
2B	279	H20 State Change	Water Cargo Mission																	
2B	280	H20 State Change	SFT-01																	
2B	281	Maneuver	Water Cargo Mission																	
2B	282	Park	On-Orbit Spare																	
2B	299	Phase-02 Complete	$2,423																	
3A	301	Launch FH, LEO Rendezvous	SPM-08+STM+CSCM-03+XDock-04 ELTV-01	10,650			10,000	43,700	10,650				3,750		51,250	130,000	4238	21	68100	75000
3A	302	Undock	CTugRV-02																	
3A	303	Maneuver	CTugRV-02																	
3A	304	Dock	CTugRV-02																	
3A	305	Maneuver	CTugRV-02+STM+SPM-08+CSCM-03+XDock-04																	
3A	306	Dock	CTugRV-02+...+XDock-04:LEOS-01																	
3A	307	Undock	CTugRV-02+STM+SPM-08+CSCM-03:XDock-04B																	
3A	308	Maneuver	CTugRV-02+STM+SPM-08+CSCM-03																	
3A	309	Dock	CTugRV-02+STM+SPM-08+CSCM-03:XDock-04E																	
3A	310	Unattach	CTugRV-02+STM+SPM-08:CSCM-03																	
3A	311	Maneuver	CTugRV-02+STM+SPM-08																	
3A	312	Undock	CTugRV-02+STM+SPM-08																	
3A	313	Maneuver	CTugRV-02																	
3A	314	Dock	CTugRV-02:XDock-01D																	
3A	315	Undock	CTugRV-02:XDock-01D																	
3A	316	Maneuver	ASWTV-02(A)																	
3A	317	SAR-H2O	ASWTV-02(A):STM+SPM-08																	
3A	318	H20 State Change	ASWTV-02(A) Fill																	
3A	319	H20 State Change	STM Empty																	
3A	320	Maneuver, Park LEO A	ASWTV-02(A)																	
3A	321	ELTV-01 Assembly	ELTV-01 Assembly																	

Plan B to the Moon: A How-to Guide to Return to and Colonize the Lunar Surface

Mission Matrix Data Sheets

Phase	M Circle	Circle Action	Vehicles / Milestones / Notes	CSCM lbs	LL lbs	LLB lbs	PVM lbs	FH Cargo lbs	XDock lbs	CTugRV Dry	CSM Dry	ASWTV Dry	STM Dry	Launch Payload Cyro	Launch Payload H2O	FH Total lbs to LEO	H2O Usable Volume (cuft)	PV-Power Prod (kW)	Vehicle No Fuel lbs	LLO / LS No Fuel lbs
3A	322	LEOS-01 Ops Normal	LEOS-01 Ops Normal																	
3A	323	Launch FH, LEO Rendezvous	SPM-09+STM ELTV-01 [Phase 03A]				10,000	65,000					3,750		51,250	130,000	3888	21	78750	75000
3A	324	Undock	CTugRV-02:XDock-01D																	
3A	325	Maneuver	CTugRV-02																	
3A	326	Dock	CTugRV-02:SPM-09+STM																	
3A	327	Maneuver	CTugRV-02+SPM-09+STM																	
3A	328	Maneuver	ASWTV-02(A)																	
3A	329	SAR-H2O	ASWTV-02(A):STM+SPM-09																	
3A	330	H2O State Change	ASWTV-02(A) Fill																	
3A	331	H2O State Change	STM Empty																	
3A	332	Maneuver	ASWTV-02(A)																	
3A	333	SAR-H2O	ASWTV-02(A):SFT-01																	
3A	334	H2O State Change	SFT-01 Fill																	
3A	335	H2O State Change	ASWTV-02(A) Empty																	
3A	336	Maneuver	ASWTV-02(A)																	
3A	337	Dock	ASWTV-02(A):XDock-02D																	
3A	338	Maneuver	CTugRV-02+SPM-09+STM																	
3A	339	Dock	CTugRV-02+SPM-09+STM:STM+SPM-08+XDock-04																	
3A	340	Undock	CTugRV-02+SPM-09+2STM+SPM-08:XDock-04B																	
3A	341	Undock	CTugRV-02+SPM-09:2STM																	
3A	342	Maneuver	CTugRV-02+SPM-09																	
3A	343	Dock	CTugRV-02+SPM-09:SPM-08+2STM																	
3A	344	Dock	CTugRV-02+SPM-09+SPM-08+2STM:XDock-04B																	
3A	345	ELTV-01 Assembly	ELTV-01 Assembly																	
3A	346	ELTV-01 Complete	ELTV-01 Complete																	
3A	347	Launch FH, LEO Rendezvous	Cryo Fuel Cargo CRV-04 [Phase 03B]				9,500	0			9,500			121,000		140,000	0	0	19000	
3A	348	Undock	CTugRV-01:XDock-01D																	
3A	349	Maneuver	CTugRV-01:CTugRV-04(A)																	
3A	350	SAR-CH4O2	CRV-04(A):CTugRV-01																	
3A	351	Fuel State Change	CTugRV-01 Fill																	
3A	352	Fuel State Change	CTugRV-04(A) Empty																	
3A	353	Maneuver	RV-04(A)																	
3A	354	Dock	CRV-04(A):XDock-03E																	
3A	355	Launch FH, LEO Rendezvous	Cryo Fuel Cargo CRV-05 [Phase 03B]				9,500	0			9,500			121,000		140,000	0	0	19000	
3A	356	Undock	CTugRV-01:XDock-01D																	
3A	357	Maneuver	CTugRV-01:RV-05(A)																	
3A	358	SAR-CH4O2	CRV-05(A):CTugRV-01																	
3A	359	Fuel State Change	CTugRV-01 Fill																	
3A	360	Fuel State Change	CTugRV-05(A) Empty																	
3A	361	Maneuver	CRV-05(A)																	
3A	362	Dock	CRV-05(A):XDock-03E																	
3A	363	SFT-01 Fuel Ready	SFT-01 Fuel Ready																	
3A	364	Undock	CTugRV-03(A)																	
3A	365	Maneuver	CTugRV-03(A)																	
3A	366	SAR-CH4O2	CTugRV-03(A):SFT-01																	
3A	367	Fuel State Change	CTugRV-03(A) Fill																	
3A	368	Fuel State Change	SFT-01 Empty																	
3A	369	Maneuver	CTugRV-03(A)																	
3A	370	Undock	CTugRV-02+...+XDock-04:LEOS-01																	
3A	371	Maneuver	CTugRV-02+SPM-09+SPM-08+2STM+XDock-04																	
3A	372	SAR-CH4O2	CTugRV-03(A):CTugRV-02																	
3A	373	Fuel State Change	CTugRV-02 Fill																	

Plan B to the Moon: A How-to Guide to Return to and Colonize the Lunar Surface

Phase	M Circle	Circle Action	Vehicles / Milestones / Notes	CSCM lbs	LL lbs	LLB lbs	PVM lbs	FH Cargo lbs	XDock lbs	CTugRV Dry	CSM Dry	ASWTV Dry	STM Dry	Launch Payload Cyro	Launch Payload H2O	FH Total lbs to LEO	Habitation Volume (cuft)	PV-Power Prod (kW)	Vehicle No Fuel lbs	LLO / LS No Fuel lbs
3A	374	Fuel State Change	CTugRV-03(A) Empty																	
3A	375	Maneuver	CTugRV-03(A)																	
3A	376	Dock	CTugRV-03(A):XDock-02C																	
3A	377	LEOS-01 Ops Normal	LEOS-01 Ops Normal																	
3A	378	LLO Maneuver / Depart	ELTV-01 Depart for LLO																	
3A	379	LLO Maneuver / Depart	CTugRV-01																	
3A	380	Phase-03A Complete	Phase-03A Complete (SFT-01 FOC)																	
3B	381	LLO Arrive, Park	CTugRV-02: ELTV-01 / Humans Return to LLO																	
3B	382	LLO Arrive, Rendezvous	CTugRV-01																	
3B	384	Maneuver	CTugRV-01																	
3B	385	Maneuver	LLOG-01																	
3B	386	LLOG-01A Assembly	LLOG-01A Assembly																	
3B	387	Undock	CTugRV-02:LLOG-01																	
3B	388	Maneuver, Return LEOS-01	CTugRV-02																	
3B	389	LLOG-01 Ops Normal	LLOG-01 Ops Normal																	
3B	390	LEO Arrive	CTugRV-02																	
3B	391	Maneuver	CTugRV-02																	
3B	392	Dock	CTugRV-02:LEOS-01 XDock-01D																	
3B	393	LEOS-01 Ops Normal	LEOS-01 Ops Normal																	
3B	394	Launch F9, LEO Rendezvous	CCTCap-04 (6 COB) [Phase 03B]														350		28700	
3B	395	Dock	CCTCap-04:LEOS-01 XDock-01B																	
3B	396	LLOG-01 Ops Normal	LLOG-01 Ops Normal																	
3B	397	Undock	CCTCap-02R:LEOS-01 XDock-01B																	
3B	398	Return to Earth	CCTCap-02R (6 COB)																	
3B	399	Phase-03B Complete	$2,469																	
4A	401	Launch FH, LEO Rendezvous	SPM-10+STM+CSCM-04+XDock-05 ELTV-02	10,650			10,000	43,700	10,650				3,750		51,250	130,000	4238	21	68100	75000
4A	402	Undock	CTugRV-02:LEOS-01 XDock-01D																	
4A	403	Maneuver	CTugRV-02																	
4A	404	Dock	CTugRV-02:SPM-10+STM+CSCM-04+XDock-05																	
4A	405	Maneuver	CTugRV-02+STM+SPM-10+CSCM-04+XDock-05																	
4A	406	Dock	CTugRV-02+...+XDock-05:LEOS-01																	
4A	407	Undock	CTugRV-02+STM+SPM-10+CSCM-04:XDock-05B																	
4A	408	Maneuver	CTugRV-02+STM+SPM-10+CSCM-04																	
4A	409	Dock	CTugRV-02+STM+SPM-10+CSCM-04:XDock-05E																	
4A	410	Unattach	CTugRV-02+STM+SPM-10:CSCM-04																	
4A	411	Maneuver	CTugRV-02+STM+SPM-10:CSCM-04																	
4A	412	Undock	CTugRV-02+STM+SPM-10																	
4A	413	Maneuver	CTugRV-02																	
4A	414	Dock	CTugRV-02:XDock-01D																	
4A	415	Undock	ASWTV-02(A)																	
4A	416	Maneuver	ASWTV-02(A)																	
4A	417	SAR-H2O	ASWTV-02(A):STM+SPM-10																	
4A	418	H2O State Change	ASWTV-02(A)																	
4A	419	H2O State Change	STM+SPM-10																	
4A	420	Maneuver, Park LEO A	ASWTV-02(A)																	
4A	421	ELTV-02 Assembly	ELTV-02 Assembly																	
4A	422	LEOS-01 Ops Normal	LEOS-01 Ops Normal																	
4A	423	Launch FH, LEO Rendezvous	SPM-11+STM ELTV-02 [Phase 04A]				10,000	65,000					3,750		51,250	130,000	3888	21	78750	75000
4A	424	Undock	CTugRV-02:XDock-01D																	
4A	425	Maneuver	CTugRV-02																	
4A	426	Dock	CTugRV-02:SPM-11+STM																	
4A	427	Maneuver	CTugRV-02+SPM-11+STM																	

Plan B to the Moon: A How-to Guide to Return to and Colonize the Lunar Surface

Phase	M Circle	Circle Action	Vehicles / Milestones / Notes	CSCM lbs	LL lbs	LLB lbs	PVM lbs	FH Cargo lbs	XDock lbs	CTugRV Dry	CSM Dry	ASWTV Dry	STM Dry	Launch Payload Cyro	Launch Payload H2O	FH Total lbs to LEO	Hab/Habitat Volume (cuft)	PV-Power Prod (kW)	Vehicle No Fuel lbs	LLO / LS Return Fuel lbs
4A	428	Maneuver	ASWTV-02(A)																	
4A	429	SAR-H2O	ASWTV-02(A):STM+SPM-11																	
4A	430	H20 State Change	ASWTV-02(A)																	
4A	431	H20 State Change	STM																	
4A	432	Maneuver	ASWTV-02(A)																	
4A	433	SAR-H2O	ASWTV-02(A):SFT-01																	
4A	434	H20 State Change	SFT-01																	
4A	435	H20 State Change	ASWTV-02(A)																	
4A	436	Maneuver	ASWTV-02(A)																	
4A	437	Dock	ASWTV-02(A):XDock-02D																	
4A	438	Maneuver	CTugRV-02+SPM-11+STM																	
4A	439	Dock	CTugRV-02+SPM-11+STM:STM+SPM-10+XDock-05																	
4A	440	Undock	CTugRV-02+SPM-11+2STM+SPM-10:XDock-05B																	
4A	441	Undock	CTugRV-02+SPM-11:2STM																	
4A	442	Maneuver	CTugRV-02+SPM-11																	
4A	443	Dock	CTugRV-02+SPM-11:SPM-10-2STM																	
4A	444	Dock	CTugRV-02+SPM-11+SPM-10-2STM:XDock-05B																	
4A	445	ELTV-02 Assembly	ELTV-02 Assembly																	
4A	446	Launch FH, LEO Rendezvous	Cryo Fuel Cargo CRV-06 [Phase 04A]				9,500	0		9,500				121,000		140,000		0	19000	
4A	447	Undock	CTugRV-03:XDock-03D																	
4A	448	Maneuver	CTugRV-03:RV-06																	
4A	449	SAR-CH4O2	CTugRV-03(A):RV-06																	
4A	450	Fuel State Change	CRV-06 Empty																	
4A	451	Fuel State Change	CTugRV-03 Fill																	
4A	452	Maneuver / Park	CRV-06(A)																	
4A	453	Maneuver	CTugRV-03																	
4A	454	Dock	CTugRV-03(A):XDock-03C																	
4A	455	Launch FH, LEO Rendezvous	Cryo Fuel Cargo CRV-07 [Phase 04A]				9,500	0		9,500				121,000		140,000		0	19000	
4A	456	Undock	CTugRV-03																	
4A	457	Maneuver	CTugRV-03:RV-07(A)																	
4A	458	SAR-CH4O2	CRV-07(A):CTugRV-03																	
4A	459	Fuel State Change	CTugRV-03 Fill																	
4A	460	Fuel State Change	CRV-07(A) Empty																	
4A	461	Maneuver / Park	CRV-07(A)																	
4A	462	Maneuver	CTugRV-03																	
4A	463	Dock	CTugRV-03(A):XDock-03C																	
4A	464	SFT-01 Fuel Ready	SFT-01 Fuel Ready																	
4A	465	Undock	CTugRV-03(A)																	
4A	466	Maneuver	CTugRV-03(A)																	
4A	467	SAR-CH4O2	CTugRV-03(A):SFT-01																	
4A	468	Fuel State Change	CTugRV-03(A)																	
4A	469	Fuel State Change	SFT-01																	
4A	470	Maneuver	CTugRV-03(A)																	
4A	471	Undock	CTugRV-03(A):XDock-02C																	
4A	472	Maneuver	CTugRV-02+...+XDock-05+SPM-10+2STM+XDock-05																	
4A	473	SAR-CH4O2	CTugRV-03(A):CTugRV-02																	
4A	474	Fuel State Change	CTugRV-02																	
4A	475	Fuel State Change	CTugRV-03(A)																	
4A	476	Maneuver	CTugRV-03(A)																	
4A	477	Dock	CTugRV-03(A):XDock-02C																	
4A	478	LEOS-01 Ops Normal	LEOS-01 Ops Normal																	
4A	479	LLO Maneuver / Depart	ELTV-02 Depart for LLO																	
4A	480	Phase-04A Complete	Phase-04A Complete																	

Plan B to the Moon: A How-to Guide to Return to and Colonize the Lunar Surface

Phase	M Circle	Circle Action	Vehicles / Milestones / Notes	CSCM lbs	LL lbs	LLB lbs	PVM lbs	FH Cargo lbs	XDock lbs	CTugRV Dry	CSM Dry	ASWTV Dry	STM Dry	Launch Payload Cyro	Launch Payload H2O	FH Total lbs to LEO	PV-Habitable Volume (cuft)	PV-Power Prod (kW)	Vehicle No Fuel lbs	LLO / LS Max Fuel lbs
4B	481	LLO Arrive, Rendezvous	CTugRV-02: ELTV-02 / Phase 04B-																	
4B	482	Undock	CTugRV-02+SPM-11																	
4B	483	Maneuver	CTugRV-02+SPM-11																	
4B	484	Dock	CTugRV-02+SPM-11:XDock-05A																	
4B	485	Maneuver	CTugRV-02+SPM-11+XDock-05+2STM+SPM-10																	
4B	486	Dock	CTugRV-02+...+SPM-10+SPM-09...																	
4B	487	LLOG-01B Assembly/Recon	LLOG-01B Assembly and LLO Polar Recon																	
4B	488	Undock	CTugRV-02																	
4B	489	Maneuver, Return LEOS-01	CTugRV-02																	
4B	490	LLO Gateway-01 Ops Norm	CTugRV-02																	
4B	491	LEO Arrive	CTugRV-02																	
4B	492	Rendezvous	CTugRV-02																	
4B	493	Dock	CTugRV-02::LEOS-01 XDock-01D																	
4B	494	Launch F9, LEO Rendezvous	CCTCap-05 (6 COB) [Phase 04B]																28700	
4B	495	Dock	CCTCap-05:XDock-01F																	
4B	496	LEO-01 Crew Swap	LEO-01 Ops Normal																	
4B	497	Return to Earth	CCTCap-03R (4 COB)																	
4B	498	Undock	CCTCap-03R:LEOS-01 XDock-01F																	
4B	499	Phase-04B Complete	$2,469																	
5A	501	Launch FH, LEO Rendezvous	SPM-12+STM+CSCM-05+XDock-06 ELTV-03	10,650			10,000	43,700	10,650				3,750		51,250	130,000	4238	21	68100	75000
5A	502	Undock	CTugRV-02																	
5A	503	Maneuver	CTugRV-02																	
5A	504	Dock	CTugRV-02+SPM-12+STM+CSCM-05+XDock-06																	
5A	505	Maneuver	CTugRV-02+STM+SPM-12+CSCM-05+XDock-06																	
5A	506	Dock	CTugRV-02+...+XDock-06:LEOS-01																	
5A	507	Undock	CTugRV-02+STM+SPM-12+CSCM-05:XDock-06B																	
5A	508	Maneuver	CTugRV-02+STM+SPM-12+CSCM-05																	
5A	509	Dock	CTugRV-02+STM+SPM-12+CSCM-05:XDock-05C																	
5A	510	Unattach	CTugRV-02+STM+SPM-12:CSCM-05																	
5A	511	Maneuver	CTugRV-02+STM+SPM-12																	
5A	512	Undock	CTugRV-02+STM+SPM-12																	
5A	513	Maneuver	CTugRV-02																	
5A	514	Undock	CTugRV-02:XDock-01D																	
5A	515	Maneuver	ASWTV-02(A)																	
5A	516	Maneuver	ASWTV-02(A)																	
5A	517	SAR-H2O	ASWTV-02(A):STM+SPM-12																	
5A	518	H20 State Change	ASWTV-02(A)																	
5A	519	H20 State Change	STM																	
5A	520	Maneuver, Park LEO A	ASWTV-02(A)																	
5A	521	LLO Polar Recon	LLO Polar Recon																	
5A	522	LEOS-01 Ops Normal	LEOS-01 Ops Normal																	
5A	523	Launch FH, LEO Rendezvous	SPM-13+STM For ELTV-03 [Phase 05A]				10,000	65,000					3,750		51,250	130,000	3888	21	78750	75000
5A	524	Undock	CTugRV-02:XDock-01D																	
5A	525	Maneuver	CTugRV-02																	
5A	526	Dock	CTugRV-02+SPM-13+STM																	
5A	527	Maneuver	CTugRV-02+SPM-13+STM																	
5A	528	Maneuver	ASWTV-02(A)																	
5A	529	SAR-H2O	ASWTV-02(A):STM+SPM-13																	
5A	530	H20 State Change	ASWTV-02(A)																	
5A	531	H20 State Change	STM																	
5A	532	Maneuver	ASWTV-02(A)																	
5A	533	SAR-H2O	ASWTV-02(A):SFT-01																	
5A	534	H20 State Change	SFT-01																	

Plan B to the Moon: A How-to Guide to Return to and Colonize the Lunar Surface

Phase	M Circle	Circle Action	Vehicles / Milestones / Notes	CSCM lbs	LL lbs	LLB lbs	PVM lbs	FH Cargo lbs	XDock lbs	CTugRV Dry	CSM Dry	ASWTV Dry	STM Dry	Launch Payload Cyro	Launch Payload H2O	FH Total lbs to LEO	PV: Power Prod (kW)	Habitable Volume (cu.ft)	Vehicle No Fuel lbs
SA	535	H20 State Change	ASWTV-02(A)																
SA	536	Maneuver	ASWTV-02(A)																
SA	537	Dock	ASWTV-02(A):XDock-02D																
SA	538	Maneuver	CTugRV-02+SPM-13+STM																
SA	539	Dock	CTugRV-02+SPM-13+STM:STM+SPM-12+XDock-06																
SA	540	Undock	CTugRV-02+SPM-13+2STM+SPM-12:XDock-06B																
SA	541	Undock	CTugRV-02+SPM-13:2STM																
SA	542	Maneuver	CTugRV-02+SPM-13																
SA	543	Dock	CTugRV-02+SPM-13+SPM-12+2STM																
SA	544	Dock	CTugRV-02+SPM-13+SPM-12+2STM:XDock-06B																
SA	545	ELTV-03 Assembly	ELTV-03 Assembly																
SA	546	Launch FH, LEO Rendezvous	Cryo Fuel Cargo CRV-08 [Phase 05A]				9,500	0			9,500			121,000		140,000	0		19000
SA	547	Undock	CTugRV-03:XDock-01D																
SA	548	Maneuver	CTugRV-03(A):RV-08(A)																
SA	549	SAR-CH4O2	CTugRV-03(A)																
SA	550	Fuel State Change	CTugRV-03(A) Fill																
SA	551	Fuel State Change	CTugRV-03(A) Empty																
SA	552	Maneuver / Park	CRV-08(A)																
4A	553	Maneuver	CTugRV-03																
4A	554	Dock	CTugRV-03(A):XDock-03C																
SA	555	Launch FH, LEO Rendezvous	Cryo Fuel Cargo CRV-09 [Phase 05A]				9,500	0			9,500			121,000		140,000	0	0	19000
SA	556	Undock	CTugRV-03:XDock-01D																
SA	557	Maneuver	CTugRV-03:RV-09(A)																
SA	558	SAR-CH4O2	CRV-09(A):CTugRV-03																
SA	559	Fuel State Change	CTugRV-03 Fill																
SA	560	Fuel State Change	CTugRV-09(A) Empty																
SA	561	Maneuver / Park	CRV-09(A)																
SA	562	Maneuver	CTugRV-03																
SA	563	Dock	CTugRV-03(A):XDock-03C																
SA	564	SFT-01 Fuel Ready	SFT-01 Fuel Ready																
SA	565	Undock	CTugRV-03(A)																
SA	566	Maneuver	CTugRV-03(A)																
SA	567	SAR-CH4O2	CTugRV-03(A):SFT-01																
SA	568	Fuel State Change	CTugRV-03(A)																
SA	569	Fuel State Change	SFT-01																
SA	570	Maneuver	CTugRV-03(A)																
SA	571	Undock	CTugRV-03(A):XDock-02C																
SA	572	Maneuver	CTugRV-02+...+XDock-06:LEOS-01																
SA	573	Maneuver	CTugRV-02+SPM-13+SPM-12+2STM+XDock-06																
SA	574	SAR-CH4O2	CTugRV-03(A):CTugRV-02																
SA	575	Fuel State Change	CTugRV-02																
SA	576	Fuel State Change	CTugRV-03(A)																
SA	577	Maneuver	CTugRV-03(A)																
SA	578	Dock	CTugRV-03(A):XDock-02C																
SA	579	LEOS-01 Ops Normal	LEOS-01 Ops Normal																
SA	580	LLO Maneuver / Depart	ELTV-03 Depart for LLO																
SA		Phase-05A Complete	Phase-05A Complete																
SB	581	LLO Arrive, Rendezvous	CTugRV-02: ELTV-03 / Phase 05B																
SB	582	Undock	CTugRV-02+SPM-13																
SB	583	Maneuver	CTugRV-02+SPM-13																
SB	584	Dock	CTugRV-02+SPM-13:XDock-06+...+SPM-12																
SB	585	Maneuver	CTugRV-02+...+SPM-12																
SB	586	Dock	CTugRV-02+...+SPM-12:SPM-11... LLOG-01																

Plan B to the Moon: A How-to Guide to Return to and Colonize the Lunar Surface

Phase	M Circle	Circle Action	Vehicles / Milestones / Notes	CSCM lbs	LL lbs	LLB lbs	PVM lbs	FH Cargo lbs	XDock lbs	CTugRV Dry	CSM Dry	ASWTV Dry	STM Dry	Launch Payload Cyro	Launch Payload H2O	FH Total lbs to LEO	PV Habitable Volume (cuft)	Power Prod (kW)	Vehicle No Fuel lbs	LLO / LS New Fuel lbs
5B	587	LLOG-01C Assembly	LLOG-01C Assembly																	
5B	588	Undock	CTugRV-02																	
5B	589	Maneuver, Return LEOS-01	CTugRV-02																	
5B	590	LLOG-01 Ops Normal	LLOG-01 Ops Normal																	
5B	591	LEO Arrive	CTugRV-02																	
5B	592	Maneuver	CTugRV-02																	
5B	593	Dock	CTugRV-02:LEOS-01 xDock-01D																	
4B	594	Launch F9, LEO Rendezvous	CCTCap-06 (6 COB) [Phase 05B]													28700			
5B	595	Dock	CCTCap-05:XDock-01F																	
4B	596	LEO-01 Crew Swap	LEO-01 Ops Normal																	
5B	597	Undock	CCTCap-04R:LEOS-01 XDock-01B																	
5B	598	Return to Earth	CCTCap-04R (6 COB)																	
5B	599	Phase-05B Complete	$2,469																	
6A	601	Launch FH, LEO Rendezvous	SPM-14 (LLB-01+LLB-02) ELTV-04 [Phase 06A]		43,500		12,500	22,750							51,250	130,000	0	0	35,250	78750
6A	622	LEOS-01 Ops Normal	LEOS-01 Ops Normal																	
6A	623	Launch FH, LEO Rendezvous	SPM-15 (LL-01+LL-02) [Phase 06A]			43,500	12,500	22,750							51,250	130,000	0	0	78750	78750
6A	624	Undock	CTugRV-02																	
6A	625	Maneuver	CTugRV-02																	
6A	626	Dock	CTugRV-02:SPM-14																	
6A	627	Maneuver	CTugRV-02+SPM-14																	
6A	628	Dock	CTugRV-02+SPM-14:SPM-15																	
6A	629	Maneuver	CTugRV-02+SPM-14+SPM-15																	
6A	630	Dock	CTugRV-02+SPM-14+SPM-15:LEOS-01																	
6A	631	LEOS-01 Ops Normal	LEOS-01 Ops Normal																	
6A	664	SFT-01 Fuel Ready	SFT-01 Fuel Ready																	
6A	665	Undock	CTugRV-03(A):XDock-02D																	
6A	666	Maneuver	CTugRV-03(A)																	
6A	667	SAR-CH4O2	CTugRV-03(A)																	
6A	668	Fuel State Change	CTugRV-03(A)																	
6A	669	Fuel State Change	SFT-01																	
6A	670	Maneuver	CTugRV-03(A)																	
6A	671	Undock	CTugRV-02+SPM-14+SPM-15																	
6A	672	Maneuver	CTugRV-02+SPM-14+SPM-15																	
6A	673	SAR-CH4O2	CTugRV-03(A)																	
6A	674	Fuel State Change	CTugRV-03(A)																	
6A	675	Fuel State Change	SFT-01																	
6A	676	Maneuver	CTugRV-03(A)																	
6A	677	Dock	CTugRV-03(A):XDock-02D																	
6A	678	LEOS-01 Ops Normal	LEOS-01 Ops Normal																	
6A	679	LLO Maneuver / Depart	ELTV-04 Depart for LLO																	
6A	680	Phase-06A Complete	Phase-06A Complete																	
6B	681	LLO Arrive, Rendezvous	CTug02:ELTV-04																	
6B	682	Dock	CTugRV-02+SPM-14+SPM-15:XDock-04D																	
6B	683	Undock	CTugRV-02+SPM-14:SPM-15																	
6B	686	Maneuver	CTugRV-02+SPM-14																	
6B	687	Dock	CTugRV-02+SPM-14:XDock-04F																	
6B	688	Undock	CTugRV-02																	
6B	689	Maneuver	CTugRV-02																	
6B	690	Dock	CTugRV-02:XDock-05D																	
6B	691	LLOG-01 Ops Normal	LLOG-01 Ops Normal ELTV-04 Assem. Cont																	
6B	699	Phase-06 Complete	$1,465																	
7A	701	Undock	CTugRV-02:XDock-05D / Phase 07B: LB-01 Est.																	
7A	702	Maneuver	CTugRV-02																	

Plan B to the Moon Copyright 2019 Stephen W. Long

Mission Matrix Data Sheets

Table Page 224 Deleted, No Useful Data

Phase	M Circle	Circle Action	Vehicles / Milestones / Notes	CSCM lbs	LL lbs	LLB lbs	PVM lbs	FH Cargo lbs	XDock lbs	CTugRV Dry	CSM Dry	ASWTV Dry	STM Dry	Launch Payload Cyro	Launch Payload H2O	FH Total lbs to LEO	Habitat Volume (cu ft)	PV-Power Prod (kW)	Vehicle No Fuel lbs	LLO / LS No Fuel lbs
7B	755	Attach	CTugRV-02:LLB-02:SPM-10+SPM-13																	
7B	756	Detach	CTugRV-02:LLB-02																	
7B	757	Maneuver	CTugRV-02																	
7B	758	Attach	CTugRV-02:SPM-15:LLB-01																	
7B	759	Extract	CTugRV-02:LLB-01																	
7B	760	Maneuver	CTugRV-02:LLB-01																	
7B	761	Attach	CTugRV-02:LLB-01:SPM-11+SPM-12																	
7B	762	Detach	CTugRV-02:LLB-01																	
7B	763	Maneuver	CTugRV-02																	
7B	764	Dock	CTugRV-02:XDock-06A																	
7B	765	LLOG-01 Ops Normal	LLOG-01 Ops Normal																	
7B	786	SFT-01 Fuel Ready	SFT-01 Fuel Ready																	
7B	787	Launch FH, LEO Rendezvous	CTugRV-10(A) [Phase 06A]	10,650				0		43,950				75,400		130,000	350	21	43950	54600
7B	788	SAR-CH4O2	CTugRV-03(A)																	
7B	789	Fuel State Change	CTugRV-03(A)																	
7B	790	Fuel State Change	SFT-01																	
7B	791	Maneuver / Depart	CTugRV-10(A) Depart for LLO																	
7B	792	LLO Arrive, Rendezvous	CTugRV-10(A)																	
7B	793	Maneuver	CTugRV-10(A)																	
7B	794	Dock	LLOG-01																	
7B	799	Phase-07 Complete	$756																	
8A	801	Humans Ready to Return LS	Humans Ready to Return to Lunar Surface																	
8A	802	Undock	LL-01:XDock-06E																	
8A	803	Descend to LS	LL-01																	
8A	804	Humans Land on LS	LL-01																	
8A	805	Prospecting Mission-01	Prospecting Mission-01																	
8A	806	Liftoff, LS to LLO	LL-01																	
8A	807	LLO Rendezvous	LL-01																	
8A	808	Dock	LL-01:XDock-06E																	
8A	809	LL-01 LLOG-01 Refueling	LL-01 LLOG-01 Refueling																	
8A	810	LLOG-01 Ops Normal	LLOG-01 Ops Normal																	
8A	812	Undock	LL-02:XDock-06C																	
8A	813	Descend to LS	LL-02																	
8A	814	Humans Land on LS	LL-01																	
8A	815	Prospecting Mission-02	Prospecting Mission-02																	
8A	816	Liftoff, LS to LLO	LL-02																	
8A	817	LLO Rendezvous	LL-02																	
8A	818	Dock	LL-02:XDock-06C																	
8A	819	LL-02 LLOG-01 Refueling	LL-02 LLOG-01 Refueling																	
8A	820	LLOG-01 Ops Normal	LLOG-01 Ops Normal																	
8A	822	Undock	LL-01:XDock-06E																	
8A	823	Descend to LS	LL-01																	
8A	824	Humans Land on LS	LL-01																	
8A	825	Prospecting Mission-03	Prospecting Mission-03																	
8A	826	Liftoff, LS to LLO	LL-01																	
8A	827	LLO Rendezvous	LL-01																	
8A	828	Dock	LL-01:XDock-06E																	
8A	829	LL-01 LLOG-01 Refueling	LL-01 LLOG-01 Refueling																	
8A	830	LLOG-01 Ops Normal	LLOG-01 Ops Normal																	
8A	832	Undock	LL-02:XDock-06C																	
8A	833	Descend to LS	LL-02																	
8A	834	Humans Land on LS	LL-02																	